FRAMING DISEASE

**Health and Medicine
in American Society**
series

Editors
Judith Walzer Leavitt
Morris Vogel

FRAMING DISEASE

Studies in
Cultural History

edited by
CHARLES E. ROSENBERG
and JANET GOLDEN

RUTGERS UNIVERSITY PRESS
New Brunswick, New Jersey

Second paperback printing, 1997

Library of Congress Cataloging-in-Publication Data

Framing disease : studies in cultural history / edited by Charles E.
Rosenberg and Janet Golden.
p.   cm. — (Health and medicine in American society)
Consists partly of revised versions of papers originally presented
at a conference held in 1988 at the College of Physicians of
Philadelphia, organized by the Francis C. Wood Institute for the
History of Medicine.
Includes bibliographical references and index.
ISBN 0-8135-1756-7 (cloth) — ISBN 0-8135-1757-5 (pbk.)
1. Social medicine.   2. Medicine—History.   I. Rosenberg, Charles
E.   II. Golden, Janet Lynne, 1951– .   III. Francis Clark Wood
Institute for the History of Medicine.   IV. Series.
[DNLM: 1. Culture—history.   2. Culture—history—congresses.
3. Disease—history.   4. Disease—history—congresses.   5. History
of Medicine.   6. History of Medicine—congresses.   QZ 11.1 F813]
RA418.F665   1991
610'.9—dc20
DNLM/DLC
for Library of Congress   91-19164
                                                                            CIP

British Cataloging-in-Publication information available

For

Owsei Temkin,

and in memory of Erwin H. Ackerknecht,

students of the history of disease

CONTENTS

Acknowledgments **xi**

Introduction   Framing Disease: Illness, Society, and History **xiii**

Charles E. Rosenberg

PART 1

FRAMING DISEASE

1   From Bright's Disease to End-Stage Renal Disease   **3**

Steven J. Peitzman

2   Emergence of Rheumatic Fever in the Nineteenth Century   **20**

Peter C. English

3   Parasites and the Germ Theory of Disease   **33**

John Farley

4   "Definite and Material": Coronary Thrombosis and Cardiologists in the 1920s   **50**
Christopher Lawrence

PART 2

DISEASE AS FRAME

5   The Medicalization of Suicide in England: Laymen, Physicians, and Cultural Change, 1500–1870   **85**
Michael MacDonald

6   American Physicians' "Discovery" of Homosexuals, 1880–1900: A New Diagnosis in a Changing Society   **104**
Bert Hansen

7   From Psychiatric Syndrome to "Communicable" Disease: The Case of Anorexia Nervosa   **134**
Joan Jacobs Brumberg

8   From Myalgic Encephalitis to Yuppie Flu: A History of Chronic Fatigue Syndromes   **155**
Robert A. Aronowitz

PART 3

NEGOTIATING DISEASE: THE PUBLIC ARENA

9   The Illusion of Medical Certainty: Silicosis and the Politics of Industrial Disability, 1930–1960   **185**
Gerald Markowitz and David Rosner

10 The Legal Art of Psychiatric Diagnosis: Searching for
Reliability   **206**

Janet A. Tighe

PART 4

MANAGING DISEASE: INSTITUTIONS AS MEDIATORS

---

11 Quid pro Quo in Chronic Illness: Tuberculosis in Pennsylvania,
1876–1926   **229**

Barbara Bates

12 Stories of Epilepsy, 1880–1930   **248**

Ellen Dwyer

PART 5

DISEASE AS SOCIAL DIAGNOSIS

---

13 The Sick Poor and the State: Arthur Newsholme on Poverty,
Disease, and Responsibility   **275**

John M. Eyler

14 Henry E. Sigerist: His Interpretations of the History of Disease
and the Future of Medicine   **297**

Elizabeth Fee

List of Contributors   **319**

Index   **323**

ACKNOWLEDGMENTS

THIS BOOK BEGAN with a conference on the history of disease held in 1988 at the College of Physicians of Philadelphia. Organized by the Francis C. Wood Institute for the History of Medicine, a division of the College, it was supported in part by the Milbank Memorial Fund and the Measey Foundation.

Revised versions of seven of the papers given at the conference appeared in a special issue of the *Milbank Quarterly* (Supplement 1, *67*, 1989), edited by Charles E. Rosenberg and Janet Golden (and with the expert help of the *Milbank Quarterly*'s then editor, David P. Willis). These papers—by Steven J. Peitzman, Peter C. English, John Farley, Michael MacDonald, Bert Hansen, John M. Eyler, and Elizabeth Fee—were subsequently revised and are reprinted here with permission of the *Milbank Quarterly*. Additional contributions were solicited from Robert A. Aronowitz, Barbara Bates, Joan Jacobs Brumberg, Ellen Dwyer, Christopher Lawrence, David Rosner and Gerald Markowitz, and Janet A. Tighe. We want to thank all these contributors who patiently tolerated the friction that necessarily accompanies any collective project—as well as those individuals who participated in the original conference and whose criticisms, contributions, and reflections have helped make this a better book: Allan M. Brandt, Ann Carmichael, Caroline Hannaway, Elizabeth Lunbeck, Russell C. Maulitz, Barbara Gutmann Rosenkrantz, and the late Leland J. Rather. We have also benefited from the advice of series editors Judith Walzer Leavitt and Morris Vogel and the comments of two anonymous readers. Finally, we thank Karen Reeds at Rutgers University Press, who has been thoughtful, encouraging, and tolerant.

Charles E. Rosenberg
Janet Golden
Philadelphia
November 1990

INTRODUCTION
Framing Disease: Illness, Society, and History

CHARLES E. ROSENBERG

MEDICINE, an often-quoted Hippocratic teaching explains, "consists in three things—the disease, the patient, and the physician." When I teach an introductory course in the history of medicine, I always begin with disease. There has never been a time that men and women have not suffered from sickness, and the physician's specialized social role has developed in response to it. Even when they assume the guise of priests or shamans, doctors are by definition individuals presumed to have special knowledge or skills that enable them to treat men and women experiencing pain or incapacity, who cannot work and fulfill family or other social obligations.[1]

But "disease" is an elusive entity. It is not simply a less than optimum physiological state. The reality is obviously a good deal more complex; disease is at once a biological event, a generation-specific repertoire of verbal constructs reflecting medicine's intellectual and institutional history, an occasion of and potential legitimation for public policy, an aspect of social role and individual—intrapsychic—identity, a sanction for cultural values, and a structuring element in doctor and patient interactions. In some ways disease does not exist until we have agreed that it does, by perceiving, naming, and responding to it.[2]

In one of its primary aspects, disease must be construed as a biological event little modified by the particular context in which it occurs. As such it exists in animals, who presumably do not socially construct their ailments and negotiate attitudinal responses to sufferers, but who do experience pain and impairment of function. And one can cite instances of human disease that existed in a purely biological sense (certain inborn errors of metabolism, for example) before being disclosed by an increasingly knowledgeable biomedical community. Nevertheless, it is fair to say that in our culture a disease does not exist as a social phenomenon until we agree that it does—until it is named.[3]

And during the past century that naming process has become increasingly central to social as well as medical thought (assuming the two can in some useful ways be distinguished). Many physicians and laypersons have chosen, for example, to label certain behaviors as disease even when a somatic basis remains unclear, and possibly nonexistent—one can cite the instances of alcoholism, homosexuality, chronic fatigue syndrome, and "hyperactivity." More generally, access to health care is structured around the legitimacy built into agreed-upon diagnoses. Therapeutics too is organized around diagnostic decisions. Disease concepts imply, constrain, and legitimate individual behaviors and public policy.

Much has been written during the past two decades about the social construction of illness. But in an important sense this is no more than a tautology, a specialized restatement of the truism that men and women construct themselves culturally. Every aspect of an individual's identity is constructed—so, also, is disease. Although the social-constructionist position has lost something of its novelty during the past decade, it has forcefully reminded us that medical thought and practice are rarely free of cultural constraint, even in matters seemingly technical. Explaining sickness is too significant—socially and emotionally—for it to be a value-free enterprise. It is no accident that several generations of anthropologists have assiduously concerned themselves with disease concepts in non-Western cultures, for agreed-upon etiologies at once incorporate and sanction a society's fundamental ways or organizing its world. Medicine in the contemporary West is by no means divorced from such affinities.

Some of these social constraints reflect and incorporate values, attitudes, and status relationships in the larger culture (of which physicians, like their patients, are part). But medicine, like the scientific disciplines to which it has been so closely linked in the past century, is itself a social system. Even its technical aspects, seemingly little subject to the demands of cultural assumptions (such as, for example, attitudes concerning class, race, and gender), are shaped in part by the shared intellectual worlds and institutional structures of particular communities and subcommunities of scientists and physicians. Differences in specialty, in institutional setting, in academic training, for example, can all play a role in the process through which physicians formulate and agree upon definitions of disease—in terms of both concept formation and ultimate application in practice. In this sense, the designation "social history of medicine" is as tautological as "social construction of disease." Every aspect of medicine's history is necessarily "social," whether acted out in laboratory, library, or at the bedside.

In the following pages I have, in fact, avoided the term social construction. I felt that it has tended to overemphasize functionalist ends and the degree of arbitrariness inherent in the negotiations that result in accepted disease pictures. The social-constructionist argument has focused, in addition, on a handful of culturally resonant diagnoses—hysteria, chlorosis, neurasthenia, and homosexuality, for example—in which a biopathological mechanism is either unproven or unprovable. It invokes, moreover, a particular style of cultural criticism and particular moment in time—the late 1960s through the mid-1980s—and a vision of knowledge and its purveyors as rationalizers and legitimators, ordinarily unwitting, of an oppressive social order.[4] For all these reasons, I have chosen to use the less programmatically charged metaphor "frame" rather than "construct" to describe the fashioning of explanatory and classificatory schemes of particular diseases.[5] Biology, significantly, often shapes the variety of choices available to societies in framing conceptual and institutional responses to disease; tuberculosis and cholera, for example, offer different pictures to frame for a society's would-be framers.[6]

During the past two decades, social scientists, historians, and physicians have shown a growing interest in disease and its history. The attention paid social-constructionist views of disease is only one aspect of a multifaceted concern. Scholarly interest in the history of disease has reflected and incorporated a number of separate, and not always consistent, trends. One is the emphasis among professional historians on social history and the experience of ordinary men and women. Pregnancy and childbirth, for example, like epidemic disease have in recent years become an accepted part of the standard historical canon. A second focus of interest in disease centers on public health policy and a linked concern with explanation of the demographic change associated with the late-nineteenth and early-twentieth centuries. How much credit should go to specific medical interventions for the decline in morbidity and lengthening life spans and how much to changed economic and social circumstances?[7] The policy implications are apparent: What proportion of society's limited resources should be allotted to therapeutic intervention, how much to prevention and social meliorism generally. Third is the rebirth in the past generation of what might be called a new materialism, in the form of an ecological vision of history in which disease plays a key role, for example, in the Spanish conquest of Central and South America.[8] Fourth is the reciprocal influence of demography on a quantitatively oriented generation of historians and of history on a growing number of demographers. For both disciplines, the study of individual disease incidence provides a viable tactic for ascertaining the mechanisms underlying change in morbidity and mortality. Typhoid rates, for example, can tell us something rather more precise about municipal sanitation and public health administration than can aggregate annual mortality figures to which outbreaks of this waterborne disease may have contributed. Finally, and perhaps most widely influential, is a growing interest in the way disease definitions and hypothetical etiologies can serve as tools of social control, as labels for deviance, and as a rationale for the legitimation of status relationships. Logically—and historically—such views have in the past generation often been associated with a relativistic emphasis on the social construction of disease.[9] Such interpretations are one aspect of a more general scholarly interest in the relations among knowledge, the professions, and social power. The more critically inclined among such would-be sociologists of knowledge have seen physicians as articulators and agents of a broader hegemonic enterprise, and the "medicalization" of society as one aspect of a controlling and legitimating ideological system.

Often lost sight of in each of these emphases are, first, the process of disease definition, and second, the consequences of such definitions in the lives of individuals, in the making and discussion of public policy, and in the structuring of medical care. We have, in general, failed to focus on the connection between biological event, its perception by patient and practitioner, and the collective effort to make cognitive and policy sense out of this perception. Yet, this process of recognition and rationalization is a significant problem in itself, one that transcends any single generation's effort to shape satisfactory conceptual frames for those biological phenomena it regards as of special concern.

Where an underlying pathophysiological basis for a putative disease remains problematic, as in alcoholism, for example, we have another sort of framemaking, but one that nevertheless reflects in its style the plausibility and prestige of an unambiguously somatic model of disease. That is, the social legitimacy and intellectual plausibility of any disease must turn on the existence of some characteristic mechanism.[10] This reductionist tendency has been logically and historically tied to another characteristic of our thinking about disease—its specificity. In our culture, the existence of a disease as *specific* entity is a fundamental aspect of its intellectual and moral legitimacy. If it is not specific, it is not a disease, and a sufferer is not entitled to the sympathy, and in recent decades often the insurance reimbursement, connection with an agreed-upon diagnosis. Clinicians and policymakers have long been aware of the limitations of such reductionist styles of conceptualizing disease, but have done little to moderate its increasing prevalence.

Framing Disease

Disease begins with perceived and often physically manifest symptoms. And medicine's historical origins lie in sufferers' attempts to find restored health and an explanation for their misfortune. That search for healing counsel constituted the historical basis for the physician's social role. And an essential aspect of this role developed around the healer's ability to put a name to the patient's pain and discomfort. Even a bad prognosis can be better than none at all; even a dangerous, but familiar and understandable, disease can be emotionally more manageable than a mysterious and unpredictable affliction. It is certainly so from the physician's point of view. Diagnosis and prognosis, the intellectual and social framing of disease, have always been central to the doctor-patient relationship.

The process of framing inevitably includes an explanatory component; how and why did a man or woman come to suffer from a particular ailment? Physicians since classical antiquity have always found intellectual materials at hand with which to explain phenomena they have been asked to treat, imposing some speculative mechanism or another on an otherwise opaque body. The study of an entity or symptom cluster over time indicates the truth of this particular truism.

Physicians have always been dependent on time-bound intellectual tools in seeking to find, demonstrate, and legitimate patterns in the bewildering universe of clinical phenomena they encounter in their everyday practice. In ancient times, for example, references to cooking provided a familiar source for a metaphorical understanding of the body's metabolism, the aggregate functions of which determined the physiological balance that constituted health or disease. Now, at the end of the twentieth century, hypothetical autoimmune mechanisms, or the delayed and subtle effects of virus infections are often used to explain diffuse chronic symptoms. To a physician in the late eighteenth and early nineteenth centuries, as we have suggested, humoral models of balance were particularly important—and used to rationalize such therapeutic measures as bleeding, purging, and the lavish use of diuretics. With the emergence of pathological anatomy in the early nineteenth century, hypothetical frameworks for disease were increasingly fashioned in terms of specific lesions or characteristic functional changes that would, if not modified, produce lesions over time. Fermentation had already provided an experimental basis for metaphors explaining epidemic disease, suggesting the ways in which a small quantity of infectious material might contaminate and bring about pathological change in a much larger substrate (as in the atmosphere, water supply—or a succession of human bodies). The germ theory created another kind of framework for imposing a more firmly based taxonomic order on elusive configurations of clinical symptoms and postmortem findings. It seemed only a matter of time before physicians would be able to understand all those mysterious ills that had puzzled their professional predecessors for millenia; the relevant pathogenic microorganisms need only be found and their physiological and biochemical effects deciphered. This was an era, as is well known, in which energetic physicians "discovered" microorganisms responsible for almost every ill known to mankind.

The major point seems obvious. In crafting an explanatory framework physicians employ a sort of modular construction, using intellectual building elements available to their particular place and generation. But the resulting conceptions of disease and its hypothetical origin are not simply abstract knowledge, the stuff of textbooks and academic debates. They inevitably play a role in mediating doctor-patient interactions. In earlier centuries lay and medical views of disease overlapped to some extent, so that shared knowledge tended to structure and mediate interactions between doctors, patients, and families. Today, knowledge is increasingly specialized and segregated, and laypersons are more likely to accept medical judgments on faith. Diagnostic procedures and agreed-upon disease categories are thus all the more important. They guide both the physician's treatment and the patient's expectations.[11]

Disease as Frame

Once crystallized in the form of specific entities and seen as existing in particular individuals, disease serves as a structuring factor in social situations, as a social actor and mediator. This is an ancient truth. It would hardly have surprised a leper in the twelfth century, or a plague victim in the fourteenth. Nor, in another way, would it have surprised a "sexual invert" at the end of the nineteenth century.

These instances remind us of a number of important facts. One is the role played by laypersons as well as physicians in shaping the total experience of sickness. Another is that the act of diagnosis is a key event in the experience of illness. Logically related to this point is the way in which each disease is invested with a unique configuration of social characteristics, and thus triggers disease-specific responses. Once articulated and accepted, disease entities become "actors" in a complex network of social negotiations. Such negotiations have had a long and continuous history. The nineteenth century may have changed the style and intellectual content of individual diagnoses, but it did not initiate the social centrality of disease concepts and the emotional significance of diagnoses once made.

The expansion of diagnostic categories in the late-nineteenth century created a new set of putative clinical entities that seemed controversial at first and introduced a new variable in defining the feelings of particular individuals about themselves, and of society about those individuals. Inevitably, these often contentious social negotiations evoked questions of value and responsibility as well as epistemological status. Was the alcoholic a victim of sickness or of willful immorality? If sickness, what was its somatic basis? And if a mechanism could not be demonstrated, could it simply be assumed? Was an individual sexually attracted by members of the same sex simply a depraved person who chose to commit unspeakable acts, or a personality type whose behavior was in all likelihood the consequence of hereditary endowment?

Such dilemmas are not simply an incident in the intellectual history of medicine but, more generally, an important—and revealing—aspect of changing social values as well as, of course, a factor in the lives of particular men and women. This style of social negotiation is very much alive today, as physicians and society debate issues of risk and life-style, and as government and experts assess deviance and evaluate modes of social intervention. The historian can hardly decide whether the creation of such diagnoses was positive or negative, constraining or liberating, for particular individuals; certainly the creation of homo-sexuality as a medical diagnosis, for example, altered the variety of options available to individuals for *framing themselves* and their be-havior, its nature and meaning. It offered the possibility, for better or worse, of construing the same behaviors in a new way and of shaping a novel role for the physician in relation to those behaviors.

But this is true not only of such morally and ideologically charged diagnoses. A late-twentieth-century diagnosis of heart disease be-comes, to cite a commonplace example, an important element of an individual's life, to be integrated in ways appropriate to personality and social circumstance. Diet and exercise, anxiety, denial and avoidance, and depression can all become involved in that integration. Once diag-nosed as an epileptic, to cite another example, in centuries before our own—or as a sufferer from cancer or schizophrenia in our generation—an individual became, in part, that diagnosis. In this sense chronic, or "constitutional," illness plays a more fundamental social role (in both economic and intrapsychic terms) than the dramatic but episodic epi-demics of infectious disease that have so influenced the historian's perception of medicine; we have paid too much attention to plague and cholera, too little to "dropsies" and consumption.

From the patient's perspective, diagnostic events are never static. They always imply consequences for the future and often reflect upon the past. They constitute a structuring element in an ongoing narra-tive, an individual's particular trajectory of health or sickness, recovery or death. We are always becoming, always managing ourselves, and the content of a physician's diagnosis provides clues and structures expec-tations. Retrospectively, it makes us construe past habits and incidents in terms of their possible relationship to present disease.

The technical elucidation of somatic disease pictures has steadily added to—and refined—our vocabulary of disease entities. The nine-teenth century saw a host of such developments. The discovery of leukemia as a distinct clinical condition, for example, gave a new sud-denly altered identity to individuals the microscope disclosed as incipi-ent victims. Before that diagnostic option became available they might have felt debilitating symptoms—but symptoms to which they could not put a name. With that diagnosis, a patient became an actor in a suddenly altered narrative. Every new diagnostic tool has the potential for creating similar consequences, even in individuals who had felt no symptoms of illness. Mammography, for example, can suggest the presence of carcinoma in the absence of symptoms. Once the radiological suggestion is confirmed, an individual's life is irrevocably changed.[12] A rather different scenario is acted out in less ominous diseases. Our knowledge of the existence, epidemiological characteristics, and clinical course of chickenpox, for example, constitutes an important social resource. A fevered child suddenly covered with angry eruptions could be extremely alarming to its parents had they not had prior knowledge of that clinical entity called chickenpox and its generally benign and predictable course.

Communities as well as individuals and their families necessarily respond to the articulation and acceptance of explicit disease entities and to an understanding of their biopathological character. Perceptions of disease are context-specific, but also context-determining. For example, when it was recognized in the mid-nineteenth century that typhoid and cholera were discrete diseases spread through the water supply, policy choices were reframed not only in practical engineering terms but in political and moral ones. Vaccination, to cite another example, provided a novel set of choices for philanthropists, government policy makers, and individual physicians. Concepts of disease and its causation and possible prevention always exist in both social and intellectual space.

Individuality of Disease

Disease is irrevocably a social actor, that is, a factor in a structured configuration of social interactions.[13] But the boundaries within which it can play its social role are often shaped by its biological character. Thus, chronic and acute diseases present very different social realities, both to the individual, to his or her family, and to society. In a traditional society, for example, one either survived or died of plague or cholera. Chronic kidney disease or tuberculosis, by contrast, may entail long-term welfare problems for a community and economic and personal dilemmas for particular families. In the case of a chronic disease like tuberculosis or mental illness, for example, institutional programs and policies mediate the complex relationship among patients, families, medical staff, and administrators.

The biological character of particular ills defines both public health policies and therapeutic options. Acute and chronic ills obviously confront physicians, governments, and medical institutions with very different challenges, but acute infections themselves vary, for example, in their modes of transmission and thus have different social connotations. Thus, attitudes toward sexuality and the need to change individual behavior may constrain efforts to halt the spread of syphilis,[14] while the skills of bacteriologists and civil engineers and the decisions of local government may interdict waterborne infections like typhoid and cholera with minimal need to alter individual habits.[15]

Negotiating Disease

The negotiations surrounding the definition of and response to disease are complex and multilayered. They include cognitive and disciplinary elements, institutional and public policy responses, and the adjustments of particular individuals and their families. Involved at all levels is the doctor-patient relationship.

In some cases, society literally—and didactically—acts out such negotiations, for example, when a court weighs a plea of not guilty by reason of insanity, or when a workers' compensation board decides whether a particular illness is a consequence of the claimant's work. In the court, the legal proceedings become a proxy for a debate between competing professional ways of seeing the world, different types and levels of professional training, and conflicting social roles. Recent debates about brown lung and asbestosis are another example of a social negotiation in which interested participants interact to produce logically arbitrary but socially viable, if often provisional, solutions to a dispute. In such cases, agreement upon a definition of disease can provide the basis for mediated compromise and administrative action; conversely, failure to reach a consensus as to the existence, origin, or clinical course of a particular ailment may prolong conflict. Disease can be seen as a dependent variable in such a negotiated situation; yet, once agreed upon, it becomes an actor in that social setting, legitimating and guiding social decision making.[16]

In a more general sense, disease classifications serve to rationalize, mediate, and legitimate relationships between individuals and institutions in a bureaucratic society. This is nicely exemplified in third-party payment schemes, where the inchoate and possibly incommensurable experiences of individuals are transformed into the neatly ordered categories of a diagnostic table—and thus suitable for bureaucratic use. In this sense a nosological table is a kind of Rosetta Stone providing a basis for translation between two very different yet structurally interdependent realms. Diagnoses are rendered literally machine-readable; human beings are not so easily categorized.

CHARLES E. ROSENBERG

Disease as Social Diagnosis

For centuries disease—both specific and generic—has also played another role, that of helping to frame debates about society and social policy. Since at least biblical times the incidence of disease has served as both index of and monitory comment on society. Physicians and social commentators have used the difference between "normal" and extraordinary levels of sickness as an implicit indictment of pathogenic environmental circumstances. A perceived gap between the "is" and the "ought to be," between the real and the ideal, has often constituted a powerful rationale for social action. The meaning of a particular policy stance to contemporaries might well be thought of as the outcome or aggregate of comparisons between what is and what ought to be; the actual is always measured against the presumably attainable ideal.

Late-eighteenth- and early-nineteenth-century military surgeons worried, for example, about the alarming incidence of camp and hospital disease; the frequency of death and disabling sickness in a youthful male population underlined the need for reform in existing camp and barrack arrangements. Social critics in Europe's new industrial cities pointed to the prevalence of fevers and infant deaths among tenement dwellers as evidence of the need for environmental reform; the instructive and unquestioned disparity between rural and urban morbidity and mortality statistics presented a compelling case for public health reform.[17] Between the mid-eighteenth century and the present this telling disparity has always played a role in discussions of public health and social environment.

One could easily cite scores of parallel instances. Disease thus became both the occasion and the agenda for an ongoing discourse concerning the interrelationship of state policy, medical responsibility, and individual culpability. It is difficult indeed to think of any significant area of social debate and tension—ideas of race, gender, class, and industrialization—in which hypothetical disease etiologies have not served to project and rationalize widely held values and attitudes. The debate has hardly ceased, as the recent outbreak of AIDS so forcefully emphasizes.

Unity and Diversity

In a much-quoted essay of 1963, the medical historian Owsei Temkin traced the history of "The Scientific Approach to Disease: Specific Entity and Individual Sickness." He organized his analysis of disease concepts around two distinct yet interrelated orientations. One he termed the "ontological" view of disease: the notion that diseases existed as discrete entities with a predictable and characteristic course (and possibly cause) outside of their manifestation in the body of any particular patient. The other he called "physiological": the view of disease as necessarily individual. Common sense and several centuries of accumulated knowledge tell us that these ways of thinking about disease are separable primarily for analytical purposes; it seems apparent that we do and perhaps must regard diseases as entities apart from their bodily manifestations in particular men and women.[18] At the same time we are well aware that disease as a clinical phenomenon exists only in particular bodies and family settings.

Temkin's distinction parallels another, emphasized perhaps most prominently in recent years by Arthur Kleinman, between illness as experienced by the patient and disease as understood by the world of medicine.[19] Both the Temkin and Kleinman formulations deal with the fundamental distinction between the specific and the general, the personal and the collective. In a sense, of course, these distinctions— ontological versus physiological, disease versus illness, biological event versus socially negotiated construction—are defensible primarily for analytical and critical purposes. In reality, we are describing and trying to understand an interactive system, one in which the formal understanding of disease entities interacts with their manifestations in the lives of particular men and women. At every interface, between patient and physician, between physician and family, between medical institutions and medical practitioners, disease concepts mediate and structure relationships.

Although we have begun to study the history of disease and have cultivated a growing appreciation of the potential significance of such studies, much remains to be done. As I have tried to argue, the study of disease is a multidimensional sampling device for the scholar concerned with the relation between social thought and social structure. Although it has been a traditional concern of physicians, antiquarians, and moralists, the study of disease is still comparatively novel for social scientists. It remains more an agenda for continued research than a repository of rich scholarly accomplishment. We need to know more about the individual experience of disease in time and place, the influence of culture on definitions of disease and of disease on the creation of culture, and the role of the state in defining and responding to disease. We need to understand the organization of the medical profession and the provision of institutional medical care as in part a response to particular patterns of disease incidence and attitudes toward particular ills. This list could easily be extended, but its implicit burden is clear enough. Disease is both a fundamental substantive problem and an analytical tool, not only in the history of medicine but in the social sciences generally.

NOTES

1. Portions of this essay are repeated or adapted from the author's "Disease in History: Frames and Framers," *Milbank Quarterly* 67 (suppl. 1, 1989):1–15 and are reprinted with permission.

2. Disease can and must also be seen as a taxonomy—with individual ailments arranged in some order-imparting structure. For a more general discussion, see Charles E. Rosenberg, "Disease and Social Order in America: Perceptions and Expectations," *Milbank Quarterly* 64 (suppl. 1, 1986):34–55.

3. In the sense I have been trying to suggest, an inborn error of metabolism unknown to a generation's clinicians was not, in fact, a disease but rather an analogy in the realm of pathology to the tree falling in the forest with no ear to hear.

4. The emergence of AIDS and the intractability of certain psychiatric conditions made visible by the deinstitutionalization movement have both played an important role in underlining the need to factor in biopathological mechanisms in understanding the particular social negotiations that frame particular diseases. Physicians and social scientists concerned with such issues necessarily inhabit what might be called a postrelativist moment; neither biological reductionism nor an exclusive social constructionism constitute viable intellectual positions. See Charles E. Rosenberg, "Disease and Social Order," passim.

5. There is, of course, an abundant sociological literature in this area, particularly in relation to psychiatric diagnoses. The work of Erving Goffman has been particularly associated with this emphasis. He also used the "frame" metaphor in his well-known *Frame Analysis: An Essay on the Organization of Experience* (Cambridge: Harvard University Press, 1974) though in a somewhat different context.

6. The very different modes of transmission imply different relationships to relevant ecological and environmental factors.

7. The name of Thomas McKeown has been closely associated with revitalizing this century-old debate; see McKeown and R. G. Record, "Reasons for the Decline in Mortality in England and Wales during the Nineteenth Century," *Population Studies* 16 (1962):94–122; McKeown, *The Modern Rise of Population* (London: Edward Arnold, 1976); McKeown, *The Role of Medicine: Dream, Mirage, or Nemesis* (London: Nuffield Provincial Hospitals Trust, 1976). McKeown's emphasis on the elusive variables that determine tuberculosis incidence has inevitably drawn controversy, but did focus historical and demographic attention on ecological variables in general and contributed to the intellectually and politically related revival of interest in the history of occupational health. See, for example, David Rosner and Gerald Markowitz, eds., *Dying for Work: Worker's Safety and Health in Twentieth-Century America* (Bloomington: Indiana University Press, 1987); Alan Derickson, *Workers' Health, Workers' Democracy: The Western Miners' Struggle, 1891–1925* (Ithaca: Cornell University Press, 1988).

8. Among the most influential works in this area have been A. W. Crosby, Jr., *The Columbian Exchange: Biological and Cultural Consequences of 1492* (Westport, CT: Greenwood Press, 1972); Crosby, *Ecological Imperialism: The Biological*

Expansion of Europe, 900–1900 (Cambridge: Cambridge University Press, 1986; William H. McNeill, *Plagues and Peoples* (Garden City, NY: Anchor Press/Doubleday, 1976).

9. See, among numerous examples, Karl Figlio, "Chlorosis and Chronic Disease in 19th Century Britain: The Social Constitution of Somatic Illness in a Capitalist Society," *Social History* 3 (1978):167–197; P. Wright and A. Treacher, eds., *The Problem of Medical Knowledge* (Edinburgh: Edinburgh University Press, 1982); Elaine Showalter, *The Female Malady. Women, Madness, and English Culture, 1830–1980* (New York: Pantheon, 1985). A recent growth of interest in "imperial" medicine reflects an interest in both the ideological and demographic aspects of disease; see, for example, Roy MacLeod and Milton Lewis, eds., *Disease, Medicine, and Empire: Perspectives on Western Medicine and the Experience of European Expansion* (London and New York: Routledge, 1988); Philip D. Curtin, *Death by Migration: Europe's Encounter with the Tropical World in the Nineteenth Century* (Cambridge and New York: Cambridge University Press, 1989); David Arnold, ed., *Imperial Medicine and Indigenous Societies* (Manchester: Manchester University Press, 1988).

10. This characteristic helps explain the ambiguous status of psychiatry in medicine—and the enthusiasm that greeted recent somatic explanations of behavior and behavior pathology.

11. Contemporary patient advocacy groups may represent in part a response to this asymmetrical distribution of knowledge—and thus power.

12. With today's sophisticated laboratory medicine and screening of populations at risk, we have created an assortment of pre- or protodisease states accompanied by a difficult assortment of personal and policy decisions. Is the middle-aged male with a high cholesterol level a sufferer from disease? What are his personal responsibilities, and those of society on his behalf?

13. It might be objected that the "actor" metaphor is inappropriate, implying volition and autonomy; strictly, only people can be actors. Perhaps disease might be more accurately considered a "script" specifying future behaviors. I prefer the actor metaphor because of its emphasis on the way disease concepts exert influence as independent factors, constraining the options of human actors in social situations.

14. See, for example, Allan M. Brandt, *No Magic Bullet: A Social History of Venereal Disease in the United States since 1880* (New York: Oxford University Press, 1985).

15. The physician's diagnostic situation can reflect another sort of biological reality, the endemic incidence of disease in a particular society. The distribution of sickness constitutes a background against which, and in terms of which, the physician evaluates the comparative plausibility of diagnostic options.

16. Which is not to suggest that the need for decisions in some particular cases precludes conflict in other, parallel instances.

17. Cf. William Coleman, *Death Is a Social Disease: Public Health and Political Economy in Early Industrial France* (Madison: University of Wisconsin Press, 1982); John M. Eyler, *Victorian Social Medicine: The Ideas and Methods of William Farr* (Baltimore: The Johns Hopkins University Press, 1979); Erwin H. Ackerknecht, *Rudolf Virchow, Doctor, Statesman, Anthropologist* (Madison: University of Wisconsin Press, 1965); James C. Riley, *The Eighteenth-Century Campaign to Avoid Disease* (New York: St. Martin's Press, 1987).

18. Temkin himself was careful to note that he employed the terms "physiological" and "ontological" "for brevity's sake." "The Scientific Approach to

Disease: Specific Entity and Individual Sickness," In: A. C. Crombie, ed., *Scientific Change: Historical Studies in the Intellectual, Social and Technical Conditions for Scientific Discovery and Technical Invention from Antiquity to the Present* (New York: Basic Books, 1963), pp. 629–647, reprinted in Temkin, *The Double Face of Janus and Other Essays in the History of Medicine* (Baltimore and London: Johns Hopkins University Press, 1977), pp. 441–455. From the present author's point of view, what Temkin refers to as the "scientific approach" should also be seen as the "bureaucratic approach"—one that lends itself to the functional requirements of large administrative structures.

19. See, for recent expositions, Arthur Kleinman, *The Illness Narratives: Suffering, Healing, and the Human Condition* (New York: Basic Books, 1988) and *Rethinking Psychiatry: From Cultural Category to Personal Experience* (New York: Free Press, 1988); Howard M. Spiro, *Doctors, Patients, and Placebos* (New Haven and London: Yale University Press, 1986); Howard Brody, *Stories of Sickness* (New Haven and London: Yale University Press, 1987).

PART

1

FRAMING DISEASE

1 FROM BRIGHT'S DISEASE TO END-STAGE RENAL DISEASE

STEVEN J. PEITZMAN

Disease begins with perceived symptoms. And each generation of physicians has found ways to explain—and in that sense control—the fear and uncertainty such symptoms may provoke. The pain and dysfunction may not have changed over time, but the framework within which they are explained has changed with succeeding generations.

The condition called "dropsy" provides an excellent case in point. As Steven Peitzman emphasizes in recounting Samuel Johnson's experience of sickness, this was a familiar and ominous symptom, understood in parallel ways by patient and practitioner. Dropsy meant something very concrete to late-eighteenth- and early-nineteenth-century practitioners and their patients. The felt and visible edema— "dropsy" in Johnson's terminology—implied fundamental internal dysfunction, but its precise nature could hardly be determined at that time. Nor could edema arising from a variety of different sources be disaggregated by contemporary clinical skills. Some dropsies seemed to respond to one diuretic or another, for example, some temporarily, some with more lasting results (as in the case of treatment with digitalis). It is hardly surprising that physicians and educated laypersons employed the conceptual tools of a traditional pathology to frame this familiar yet frightening clinical phenomenon. Dropsy seemed to fit neatly into neohumoral models of pathology with their emphasis on physiological balance between intake and outgo; the clinical reality of dropsy, with its accumulation of fluids and responsiveness to diuretics, seemed, in fact, entirely consistent with this speculative model, just as the model helped frame the clinical reality.

But in the course of two centuries since Samuel Johnson's grim illness, the phenomenon called dropsy came to be understood in fundamentally different ways. For example, the London clinician Richard Bright in the 1820s distinguished the portion of this symptom attributable to kidney dysfunction, finding an association between the clinical picture and chemical changes in the urine during life, and a pattern of morbid changes in the kidney at autopsy. Later in the nineteenth and into the twentieth century, clinicians defined and redefined that agreed-upon picture of Bright's disease. Microscopic pathology focused on the fine structure of the lesions characteristically associated with renal disease. In the twentieth century, the interests of a physiologically oriented and self-consciously scientific generation of nephrologists turned to functional criteria, supplanting the anatomical, lesion-oriented conception of the disease so influential in previous decades. Finally, Peitzman argues, Americans have in the past two decades created a very different framework around renal dialysis; most patients never become dropsical at all. Their experience is that of dialysis rather than the illness dialysis is meant to avert. End-stage renal disease (ESRD) is fundamentally an administrative term; it is an automatic trigger for reimbursing providers of dialysis, not a well-defined clinical entity. Yet ESRD is not simply an arbitrary neologism spawned by bureaucratic necessity, but also at one remove the reflection of a real pathology interacting with a specific technology under particular social and political circumstances.

The evolving framework of pathological assumptions describing and explaining "Bright's disease" has been gradually integrated and reintegrated into a series of differently focused explanatory frameworks for the same clinical pictures (the ability of medicine to alter the course of chronic renal disease, as Peitzman emphasizes, remained minor until the dialysis era). It is precisely this process of definition and redefinition that demands scholarly attention because it tells us a great deal about the evolution of medical thought and practice. In another dimension it provides access into the experience of ordinary people during the past century.

—C. E. R.

"BRIGHT'S DISEASE" was the once familiar term for diffuse noninfectious bilateral renal disease, usually marked by albuminuria and sometimes by the symptom complex of uremia. The acronym ESRD, for "end-stage renal disease," entered widespread use as a result of Section 2991 of Public Law 92-603, passed by Congress on October 30,

1972. This unprecedented legislation provided federal financial support to essentially all Americans with a particular chronic disease, kidney failure, so that they could receive treatments to prolong their lives. The treatments were, of course, dialysis and transplantation. Neither Bright's disease nor ESRD referred or refers to any single disease entity defined by either pathological examination or cause.

Knud Faber, in his book *Nosography*, still honored with re-readings sixty years after its publication, said about morbid categories that the clinician "cannot live, cannot speak, cannot act without them."[1] This remains true, and I will return for a closer look to the labels—Bright's Disease, ESRD, and others—placed on categories of sickness. I hope to show how changes in the names used in renal medicine reflected how physicians and others thought about kidney disease. Certainly one theme must be the ways in which first, individual physicians, then, the community of scientific medicine, and finally, government or society, fashioned both conceptions of this disease and the labels given the conceptions. I will also attempt to suggest how some patients thought about their kidney disease, or at least how they experienced it. The patients' contribution to the idea of a disease is more elusive to trace than that of the physicians. Yet even if evidence is fragmentary, the subject warrants exploration and assumes perhaps greater importance near the end of my narrative, when renal patients become dialysis patients. A comprehensive account of the evolution of the pathological categories, etiology, and management of renal disease is a worthy and so far undone task, but here I limit myself to an episodic and selective reconstruction of how successive generations have construed the experience of kidney disorder and created its nomenclature.

Dropsy

The prehistory of Bright's disease was dropsy. From a universe of dropsical patients, Richard Bright of Guy's Hospital in London, beginning in the 1820s, came to recognize a special subset, the forebears of all renal patients.

For hundreds, perhaps thousands of years two forms of serious chronic disease dominated medical practice, inducing in their victims equally grotesque but oddly antipodal transformations. One, consumption, thinned and shrank the body; the other, dropsy, bloated and dulled it. Physician and nosologist William Cullen, in the 1787 edition of his popular *First Lines of the Practice of Physic*, wrote that dropsies were "distinguished from each other according to the parts they occupy, as well as by other circumstances attending them; yet all of them seem to depend upon some general causes, very much in common to the whole." Concerning the frequent association of dropsies with diminished urine Cullen said that "This scarcity of urine may sometimes be owing to an obstruction of the kidneys, but probably is generally occasioned by the watery parts of the blood running off into the cellular texture, and being thereby prevented from passing in the usual quantity." He did not see dropsy as a "cardiac" or "renal disease." His analysis remained more general and symptomatic than local and pathological.[2]

One man's dropsy is especially noted in English literature. In the last year of his celebrated life, writer and lexicographer Samuel Johnson suffered with asthma and dropsy, and told of the experience to his doctors, to James Boswell, and to other friends. On February 11, 1784, he wrote to Boswell: "The asthma, however, is not the worst. A dropsy gains ground upon me; my legs and thighs are very much swollen with water, which I should be content if I could keep it there, but I am afraid it will soon be higher."[3]

Then in late February 1784 a remarkable event relieved Dr. Johnson's pulmonary and peripheral edemas: "Last week I emitted in about twenty hours, full twenty pints of urine, and the tumour of my body is very much lessened, but whether the water will not gather again, He only knows by whom we live and move." Boswell included this mighty postdevotional discharge of fluid in the *Life*, making it unquestionably the most famous diuresis in the history of the English-speaking people. But Johnson's suffering was not over. On July 21 he wrote to one of his physicians: "The water has in these summer months made two invasions, but has run off again with no very formidable tumefaction." Into late summer and early fall, armed with squill and cantharides, Johnson battled the floods, which would rise and fall. On November 4 he reported (to John Ryland) that the "water grows fast upon me, but it has invaded me twice in this last half year, and has been twice expelled, it will I hope give way to the same remedies." Two days later he wrote (to Dr. Brocklesby): "The water encreases almost visibly and the squills which I get here [Lichfield] are utterly inefficacious. My spirits are extremely low. Yet I have recovered from a worse state."[4]

Johnson eventually succumbed on December 13, 1784, and the autopsy revealed both cardiac and renal disease (the right kidney atrophic, the left cystic). But in life Johnson, like many similar patients, had *dropsy*, not Bright's disease or congestive heart failure. The experience of his disease was dreadful: a constant struggle against the drowning of his body by floodwaters from within, a struggle waged with drugs he knew to be inadequate, unpredictable, and noxious.

Bright's Disease

Widespread in private and hospital practice, hideous and lethal when advanced, dropsy captured the attention of some early-nineteenth-century physicians who were remapping diseases on the basis of morbid anatomy and physical examination. In 1827 Richard Bright (1789–1858) published the first volume of his magisterial *Reports of Medical Cases*.[5] Almost half the volume deals with dropsical patients, and the first 24 cases reveal Bright's discovery. Certain dropsied patients had albumin in their urine (detected by coagulation in a candle-heated spoon) and many showed previously unrecognized but striking morbid changes of the kidneys at autopsy. Furthermore, many such dropsical patients experienced characteristic symptoms, including vomiting, headaches, pericarditis, and seizures, amounting to what later would be recognized as uremia. Bright's discovery has rightly been hailed as one of the gleaming achievements of clinical-pathological correlation. Dropsy was ubiquitous, heat-coagulable urine little explored, the uremic symptoms variable, the patients hectically sick, the autopsies undoubtedly odorous and oozing; yet Bright made sense out of all this and established the basis of renal disease.

The detection of albuminuria, as Bright practiced it, was arguably the first practical laboratory aid to diagnosis, but Bright went further by encouraging several physician-chemists to analyze the urine and blood of some of the renal patients.[6] William Prout, John Bostock, and George Owen Rees showed that urea was sometimes detectable in the blood but deficient in the urine, while the albumin was decreased in the serum. (Urea is the major nitrogenous waste of mammals, and its removal is the central function of the kidney.) In 1842 Bright received permission from the managers of Guy's Hospital to assign a number of beds during the summer only to renal cases, for careful prospective clinical and chemical study by a designated team. This hinted at the "metabolic ward," or clinical study unit, of the next century.

Bright's way of understanding the new disease was, however, solidly of the early nineteenth century. Although he recognized that albuminuric renal dropsy could follow scarlet fever, he stressed in all his writings *exposure to cold* as its most important source. Cold could "suppress the insensible perspiration," leading to "sympathy" between the "checked" skin and the kidneys.[7] This language and way of thinking go back at least to Galen and are alien to the physician of the 1990s. Yet for Bright the association of cold with renal disease was not a "theory" or speculation, but indeed a repeated clinical observation: The patients he cared for and reported almost all *did* recount some recent exposure to cold and wet. Cold in time gave way to microbes and then to immunological models as accepted causes of diffuse kidney disease.

A modern physician might expect to find more common ground in Bright's morphological descriptions, enhanced by the magnificent hand-colored mezzotint engravings appended to his *Reports*. In the initial 1827 publication, based on 24 cases, Bright suggested three forms of the deranged kidney structure that accompanies albuminuric dropsy. The first is a kind of softening, with yellow mottling. In the second "the whole cortical part is converted into a granulated texture, and where there appears to be a copious morbid interstitial deposit of an opake white substance." The third "is where the kidney is quite rough and scabrous to the touch externally, and is seen to rise in numerous projections not much exceeding a large pin's head, yellow, red, and purplish. . . . The form of the kidney is often inclined to be lobulated [and there is a] contraction of every part of the organ."[8] These categories and descriptions hold little meaning for the nephrologist of the 1990s, who sees other striking differences—mainly ignored by Bright—when viewing the colored plates. There are two bases for the difficulty. First, Bright relied on a sort of macroscopic tissue pathology of the 1820s derived from Xavier Bichat and (more directly) from Bright's colleague at Guy's, Thomas Hodgkin. Second, the renal physician of today rarely sees or touches a fresh diseased kidney and does not replicate the visual and tactile examinations at autopsy that repeatedly assured Bright of the reality and reproducibility of albuminuric renal disease.

Importantly, Bright allowed that the three forms he distinguished might be only stages of one process. But he seemed to favor three categories. So, from the first publication on the disorder, Bright's disease was not held to be one specific entity, not even by Bright.

Although Richard Bright did not propose the name, the disease or diseases he described quickly became known as *"morbus Brightii,"* "Bright's kidney," "Bright's disease," or *"maladie de Bright."* In fact, I believe that Bright's disease may be the earliest regularly used eponymous name for a disease in English. It is the only one like it in the index to the 1844 edition of Thomas Watson's popular *Lectures on the Principles and Practice of Physic* (unless one counts "St. Vitus's dance"). Watson, however, did not like it. "For this disease," he wrote, "we have no appropriate name. I wish we had. Some call it *granular degeneration* of the kidney, but the epithet granular is not always applicable. It is most familiarly known, both here and abroad, as *Bright's kidney,* or *Bright's disease;* after the eminent physician who in 1837 [sic; should be 1827] first described it, and showed its great pathological importance. These are odd-sounding and awkward terms; but in the lack of better, I must employ them."[9] A British physician named S. J. Goodfellow published a small volume on Bright's disease in 1861 and began by acknowledging that "it has been found very difficult to give an appropriate name to these diseases, one which would embrace, and in some measure define, all the pathognomic symptoms as observed during the life, as well as the anatomical characters found after death. A great deal of thought and ingenuity have been fruitlessly expended in discovering a name. . . . I would advise you to use that of Bright's disease—*Morbus Brightii*, always bearing in mind that it is an arbitrary term for different affections of the kidney."[10]

Why did the naming of this new disease pose a problem and lead to a new way of naming diseases? One obvious answer is the recognition, already noted, that no one morbid appearance characterized Bright's disease. But I would argue for other reasons. Bright's disease required a new sort of naming because it represented a new, nineteenth-century, way of thinking about and defining disease. Patients had Bright's disease if they had at least some of the following: certain symptoms, such as dropsy; certain physical findings, such as a hard pulse or a pericardial rub; certain morbid changes in the kidney, if autopsied; and—perhaps most novel in the 1820s and 1830s—a primordial laboratory abnormality, albuminous urine. The last two—albuminuria and morbid changes of the kidney—only the physician could detect. So this sickness was no longer only the patient's disease; it was also the physician's, Dr. Bright's disease.

Bright recognized that not all these elements were present in every such patient. For example, case 4 of his 1836 hospital report on patients with albuminous urine provided "a strong example of the disease of the kidney passing to its most fatal period, without the slightest symptom of dropsical effusion—a state of things, which, above all, is apt to throw us off our guard."[11] Gradually it also became clear that some patients might progress to renal failure without ever showing substantial albuminuria. Others might display profound anasarca and albuminuria but no kidney defect visible to the naked eye or even under an optical microscope. Bright's disease may be said to have been a way of getting sick through your kidneys. The term was presumably useful, even indispensable, to most physicians, and I suppose (without strong evidence) gradually became comprehensible and useful to patients. A simple and accessible label, "Bright's disease" referred broadly to an ailment sufficiently common that many would have known, or known of, someone who had it.

By 1950 the name Bright's disease evidently lost much of this usefulness; doctors found it old-fashioned. The last two monographs in English I have found with "Bright's disease" as part of the title both appeared in 1948, one by Henry Christian, the other by Stanley Bradley. Thomas Addis, a leading student of renal disease from 1915 to 1949, wrote two books. One appeared in 1931, titled *The Renal Lesion in Bright's Disease*, but the second, in 1948, he called *Glomerular Nephritis*, though its content extended well beyond that particular form of renal disease.[12]

Not until writing this paper did I perceive the vacancy left by the demise of "Bright's disease." No term ratified by both patients and physicians now adequately expresses "getting sick through your kidneys." Without such a joint frame of reference, I often find it difficult to discuss a new diagnosis of chronic renal failure with a patient. The culprit lacks a name and cannot assume the blame; one is left clumsily describing the nature of the vandalism and why the damage cannot be put right.

Before leaving Richard Bright and Guy's Hospital, let me try to evoke the patient's experience of Bright's disease in the nineteenth century, so that I can later draw a contrast with the present day. Here is Bright's description of case XIII of the 1827 *Reports*, Thomas Drudget, a carman hospitalized on December 7, 1826:

> About a fortnight before his admission he was attacked with sickness at the stomach, and shortness of breath; purging then came on, and vomiting: about nine days before admission his face and legs began to swell. The urine had been deficient in quantity the whole time. He complained much of tenderness in the pit of his stomach.

Drudget's urine coagulated to heat. He was treated with cupping over the chest, mercurials, magnesium sulphate, tincture of camphor, potassium supertartrate, jalap, and capsicum, some of which brought temporary improvement. But on the seventeenth he complained to some of his ward mates of a headache.

> About eight o'clock it was observed that he lay in bed making a very singular noise, and on going to him he was in a state of profound apoplectic stertor. Mr. Stocker was immediately called; took away twenty ounces of blood from the temporal artery, gave him ten grains of calomel, and a colocynth injection. He had one or two fresh attacks, accompanied with so much convulsion that he could scarcely be held in bed.

Bright ordered more bloodletting, an enema, and a catharides plaster on the neck, but to no avail: another case entered the *Reports*. The *sectio cadaveris* showed an intracranial bleed as the terminal cause of death; the kidneys were pale and soft with a "motley granulation."[13]

Edematous and convulsive, poor Drudget may stand in for all the victims of advanced Bright's disease before the invention of dialysis. I might also have cited a patient that Robert Tyson, an American authority on Bright's disease, presented to students on October 20, 1892, at the University of Pennsylvania Hospital. She was a fifty-five-year-old woman whose massive "collection of fluid ruptured the skin. Today the leg is swollen, but is reduced though it still pits on pressure. The fluid still trickles over the skin producing an eczema which is treated by antiseptics." She had albuminuria and granular casts in the urine, and was treated with milk and caffeine. Tyson sometimes used pilocarpine, often his beloved "hot-air bath," and even venesection in refractory cases of Bright's disease.[14] Or I might have described Thomas Addis's patient from the 1930s, a young physicist whose case (possibly a composite) was poignantly and instructively narrated in *Glomerular Nephritis*. Over many years Addis helped prolong this patient's useful life with diet, simple medications, Southey tubes, and encouragement. Eventually, edema and nausea announced the onset of a conclusive stage of the disease, confirmed by changes in urinary sediment, albuminuria, and measures of azotemia. Finally, Addis offered the sedative paraldehyde to ease the death of the young man, edematous, nauseated, and exhausted with terminal uremia.[15] Any of these and countless other recorded cases reveal the patient's experience of renal disease in earlier times, and the physician's struggle to cure or palliate it.

Nephritis in Threes

The names that came to compete with Bright's disease, or designate its subcategories, indicate the way nineteenth-century authorities wished to think about the disease. This became increasingly anatomical. The term "nephritis" existed well before the writings of Bright and his contemporaries, appearing for example in Cullen's *Nosology*.[16] The suffix "itis" even by then denoted inflammation, that is, heat, redness, pain, swelling, loss of function. Bright strongly suggested that his disease might be a state of congestion or inflammation seated in the kidney and elsewhere (e.g., the pericardium). Though not all subsequent physicians and pathologists would agree on the essentially inflammatory nature of Bright's disease, "nephritis" entered use by 1840, and the task of classifiers was reduced to adding appropriate modifiers.

The story of the superceding and competing classifications is far too tedious to explore except with the broadest of strokes. Oddly, nephrologists and pathologists looking at altered kidneys have always favored as much as possible a tripartite organization, either seeking simplicity or emulating Dr. Bright. Virchow in 1858 suggested: "parenchymatous nephritis," "interstitial nephritis," and "amyloid degeneration." George Johnson in 1873 proposed the separation into "acute nephritis" and three chronic varieties: "red granular kidney," "large white kidney," and "lardaceous kidney" (which is the same as "amyloid kidney"). Osler in his influential text favored: "Acute Bright's disease," "chronic parenchymatous nephritis," and "chronic interstitial nephritis," dispatching "amyloid" to its own pathological category. In the twentieth century, the extremely influential monograph *Die Brightsche Nierenkrankheit* by Volhard and Fahr, published in 1914, provided a fresh—but still trinitarian—organization: degenerative diseases, the "nephroses"; inflammatory diseases, the "nephritides"; and arteriosclerotic diseases, the "nephroscleroses." Thomas Addis of Stanford University in the 1920s offered a modification of Volhard and Fahr that gained some popularity: "hemorrhagic Bright's disease," "degenerative Bright's disease," and "arteriosclerotic Bright's disease."[17] Even today nephrologists seek first to place a new case of renal parenchymal disease into one of three broad categories: glomerular disease, tubulo-interstitial disease, or vascular disease.

The point is that nineteenth-century physicians interested in Bright's disease relied on gross and microscopic pathology to organize their thinking and teaching about the disorders. Increasingly, since the sick kidney accommodatingly sheds bits of itself into the urine, examination of the urinary sediment aided clinicians. Thus, all later nineteenth-century texts on Bright's disease contain extensive discussion of *casts*, often carefully illustrated. (Urinary casts are microscopic cylinders of tissue debris and protein that result from inflammatory and degenerative renal diseases.) Examination of sediment extended histological diagnosis to the living patient, as would renal biopsy beginning in the late 1940s. Chemical examination, such as blood urea measurement, though precociously commenced by Bright's collaborators, lay nearly dormant until the invention of simpler assays in the early twentieth century.

Functional Diagnosis and Bright's Disease

Although pathology remained the underpinning of organized knowledge and diagnosis in Bright's disease, clinicians of course continued to struggle at the bedside with its obvious functional derangements—dropsy and uremia. In the last decade of the nineteenth century and the early part of the twentieth, some investigators subjected Bright's disease to a more subtle and formal sort of "functional diagnosis," using the laboratory. Methods and thinking analogous to those applied to the stomach and heart were applied to the kidney. In the late 1890s Alexander von Koranyi used cryoscopy to measure renal concentrating power in health and in renal disease. As other workers added tests of dye excretion, test meals, and urea loads, the term "renal insufficiency" entered the language of Bright's disease.[18] The "insufficient" kidney lacked normal power and reserve; further weakening later became known as "renal failure," which meant frank retention of urea and other substances usually discharged by the renal filters. These new names indicated the early twentieth-century enthusiasm for interpreting disease in terms of what an injured organ could or could not do, an enthusiasm fed by the invention of novel measuring instruments.

Thomas Addis, mentioned earlier, represents one of several figures transitional between the dominantly anatomic and dominantly functional ways of envisioning renal disease. An appealing and enigmatic figure, Addis (1881–1949) was a Scottish physician with sound chemical training hired as a young man by the new Stanford University medical school in 1911. There he took up, in clinic and laboratory, a lifelong study of Bright's disease. He titled both his Harvey Lecture of 1928 and his monumental book of 1931 (with pathologist Jean Oliver) *The Renal Lesion in Bright's Disease.* By "renal lesion" he meant both disordered structure *and* the amount of lost function. His scheme for pathologic classification gained validity from his method of standardizing and quantifying the urine sediment examination; that is, the classification and the "Addis count" were inseparable in his mind, and the clinical usefulness of the one depended upon the other.

"Renal lesion" as loss of working renal mass Addis estimated by a timed measurement of the urea content of blood and of urine, expressed as the "urea ratio" (urine urea per unit time/plasma urea concentration, or UV/P). Now this ratio may claim physiologic meaning as the urea clearance. But Addis thought of this number as the amount of "still functioning renal tissue." In a May 15, 1934, letter to Alfred Cohn about a patient, Addis reported "her old 'ratio' figures which show that she then had 69% of the amount of renal tissue proper for her size." He did *not* state that her urea clearance was 69% of normal. Indeed, Addis occasionally decried the introduction of too much physiology into medicine and, in 1931, urged a return to the "straight and narrow road of morphology."[19]

But physiologists such as Homer W. Smith and A. Newton Richards in the 1930s and 1940s built a language of clinical renal physiology based largely on the mathematical idea of clearance and its application to the measurement of glomerular filtration rate. Physiologic analysis increasingly displaced structural thinking as a medical subspecialty called "nephrology" matured.[20] Eventually, the measurement of creatinine succeeded that of urea in importance, and the nephrologist of the 1990s mainly uses serum creatinine as the cardinal indicator of renal health. An elevated creatinine value in a patient *is* "renal failure" and translates—if thought about more—into a decreased glomerular filtration rate. It does *not* call to mind an anatomical image of shrunken renal tissue, as did the urea ratio for Addis, who frequently saw such kidneys at autopsies. For the nephrologist of the late twentieth century, serum creatinine does, however, have associations and is more than an abstract number. To the renal fellow called to see a new consult in the emergency room, a creatinine of "ten" (milligrams per deciliter) conveys far more than twice the urgency of a creatinine of "five." It means "decide about dialysis now," instead of "call me back when the workup is underway." As with the tone of a clarinet, qualitative changes occur as one goes up the scale of serum creatinines.

ESRD—Disease of Entitlement?

The introduction of dialysis and renal fellows into this discussion leads to the last term to be investigated. ESRD—end-stage renal disease—is, perhaps, the strangest of all labels applied to those with seriously sick kidneys.

In 1948 Henry Christian published a small book on renal disease he could still title *Bright's Disease*. In the same year appeared Thomas Addis's *Glomerular Nephritis*, which movingly recounts one patient's decline and death from terminal uremia. One wonders if either author had noted the appearance in 1947 of an obscure, slim, paperbound monograph called *New Ways of Treating Uremia*, by Willem Kolff.[21] Kolff was the most successful of several workers in the 1940s who devised practicable artificial kidneys (hemodialysis). In its first decade or so of regular use, hemodialysis mainly aided patients with acute and reversible renal failure. Then, in the early 1960s, engineer Wayne Quinton and physician Belding Scribner perfected the arterial-venous shunt, which allowed repeated dialytic treatments and the indefinite life-support of patients with advanced irreversible renal failure.[22] (Scribner first become fascinated with renal disease during an elective course for senior medical students with Thomas Addis at Stanford.)

In 1972 Congress included within Public Law 92-603 a provision by which "Chronic Renal Disease [is] Considered to Constitute Disability" (Sec. 2991). The law provided federal financial support for almost all Americans requiring chronic dialytic treatment. Within a few years the "End-Stage Renal Disease Program" under Social Security was formed to implement and regulate this support. "End-stage renal disease" was defined in 1974 as "that stage of renal impairment that cannot be favorably influenced by conservative management alone, and requires dialysis and/or kidney transplantation to maintain life or health."[23] Although clinicians may have used the phrase "end-stage renal failure" or its like before the passage of PL 92-603, with the initiation of the federal program ESRD became a disease defined by eligibility to receive funding for a particular treatment—a disease of entitlement.

The typical patient with ESRD today is a 55 year old with diabetes complicated by retinopathy and nephropathy. When the patient's creatinine reaches the range of 5 to 7 (milligrams per deciliter), surgical preparations are made for chronic hemodialysis or peritoneal dialysis. The patient then has ESRD. Dialytic treatments may begin when the patient reports morning nausea or occasional vomiting, perhaps skin itching or some other early uremic symptom. There are exceptions, of course, and some patients do present feeling very ill with advanced uremia and edema. But ESRD is not the same as "uremia," since most patients usually begin the dialytic treatments soon enough to avoid all but the early and tolerable symptoms. Nor is ESRD equal to "chronic renal failure (CRF)," since many patients with creatinines of 4 or 5 are said to have CRF but still not ESRD.

ESRD, a legislative phrase, *can* lay claim to being a disease. It has become something patients can *have*, although relatively few use the term. One who did, a dialysis patient in Georgia, wrote: "Possibly the name is most frightening, End Stage Renal disease. It sounds like the end of the world. That's the way I felt [upon first learning of it]. I cried and cried."[24] ESRD is a disease nephrologists can write books about, such as *End-Stage Renal Disease, An Integrated Approach*.[25]

Moreover ESRD, like other diseases, encompasses a set of characteristic signs and symptoms. These are a mixture of late consequences of renal failure, rarely seen until dialysis lengthened the lives of renal patients, and complications of the treatment itself. Examples include chronic anemia, renal bone disease, aluminum toxicity, dialysis-induced cramps and hypotension, and amyloidosis—but, above all, clotting and infections of catheters and shunts for dialysis.[26] These last complications harry the days (and some nights) of dialysis patients and their nurses and doctors. Today, since ESRD treatment is started before most patients suffer from engulfing edema or florid uremia, the experience of ESRD is mainly the illness of dialysis. The treatment becomes the sickness. When renal patients write about their experiences—and a fair number do—they write most about *being a dialysis patient*, about the hours on the machine, about their remarkable ability to cope and prevail. Here are some examples:

> Next I spent almost five years on home hemodialysis with my mother as "nurse," two years on CAPD, then back to the machine, in-center. I had a parathyroidectomy in 1976 and had my "closest call" that year with a case of pulmonary edema.
>
> I personally am convinced that the physical aspects are very uncomfortable, and I could gladly do without nausea, cramps, and hypotension so bad I have to crawl into bed after a sudden drop in pressure—because if I stood up I would pass out.
>
> Home dialysis, like many things in life, had its high and low moments. There were times of sharing friendship and thankfulness for extended life and its meaning and pleasures. There were also times of despair, crisis, pain and weariness.
>
> I suppose the longer an individual is on dialysis, the more he incorporates the process into his life. When I am at dialysis I am more willing to take an active role, and when I am away from the unit, the thought of dialysis has an easier time escaping my thoughts.
>
> The first time I saw the kidney machine I thought it looked like a big washing machine. I said "I hope you all aren't going to put me in that machine!"[27]

The physician's experience of renal disease also has changed. Relying on serial laboratory measurements, early dialysis, and potent diuretics, the nephrologist today only rarely encounters fulminant uremia, or dropsy bursting through the skin. Indeed the renal fellow only a few times in a year enjoys the palpable fulfillment of seeing the dialysis machine staunch seizures, penetrate coma, defeat severe uremia. The chronic dialysis patient has been called a "marginal man," suspended between the worlds of the sick and the well, beneficiary and victim of a half-perfected technology. The renal physician, oddly, shares with ESRD patients an incomplete, unresolving medical encounter, one of struggle more with the woes of chronic dialysis than with the woes of uremia. The nephrologist has even lost touch, literally, with the diseased kidney itself, the old granular kidney of Dr. Bright. ESRD patients seldom receive autopsies, and the renal doctor now sees kidneys only microscopically, a biopsied bit at a time, or views shadowy reniform images on an ultrasound screen.

Be assured that I am not calling for more uremic seizures and deaths, or a return to Southey tubes. Clearly the dialysis machine *is* lifesaving. It is the routinization and scale of that lifesaving, with its daily imperceptibility, that are both extraordinary and numbing.

I have been dealing with names. The naming of diseases, I argue with Knud Faber and many others, remains essential to the modern physician who must select a diagnosis and recommend treatment. To most patients, the name of the illness matters less than getting well. But a name sometimes contains meaning. Samuel Johnson wrote in 1784: "My diseases are an Asthma and a Dropsy, and, what is less curable, seventy five."[28] "End stage renal disease," wrote the dialysis patient in Georgia two hundred years later, "it sounds like the end of the world." "Dropsy," "Bright's disease," "nephritis," "renal failure," "ESRD": each name is to some extent elusive, ambiguous. None refers to a well-defined etiological or pathological entity. These terms suggest to me an increasing complexity in the encounter of renal patient and of physician. Evident and clear to both, "dropsy" was disease as symptom. "Bright's disease" was more the physician's abstraction, with two of its key elements—albuminuria and renal structural change—invisible to the patient. Yet I believe "Bright's disease" to patient and doctor talking together came to make sense as "getting sick through the kidneys." It was a simple, accessible label for a broadly defined illness that was sufficiently prevalent for most persons to know of someone who had it. The still-evolving histologic nomenclature of "nephritis" and "glomerulonephritis" represents a language of the renal specialist, too esoteric for patient or general doctor. Finally, "ESRD" is a diagnosis often uncovered by autoanalyzer, defined by need for dialysis, and formally bestowed by government.

Each term has had its use, its particular reality, and its message. Each reflects in some way the experience of the sickness, felt and questioned by the patient. Each reflects as well an experience of the disease observed, contemplated, classified, and treated by the physician—with good intent but with the imperfect medicines and machines of the day.

ACKNOWLEDGMENTS

Research contributing to this paper was supported by grants from the American Philosophical Society and the National Library of Medicine (LM 04906). Support also came from the Department of Medicine and the Humanities Program of the Department of Community and Preventive Medicine of the Medical College of Pennsylvania. The author is grateful for the assistance over many years of the staff of the Historical Collections of the Library of the College of Physicians of Philadelphia.

NOTES

1. Knud Faber, *Nosography*, 2d ed. (New York: Paul Hoeber, 1930), p. 211.

2. William Cullen, *First Lines of the Practice of Physic*, 4 vols. (Edinburgh: C. Eliot, 1787), 4:250–287.

3. Samuel Johnson, *The Letters of Samuel Johnson*, ed. R. W. Chapman, 3 vols. (Oxford: Clarendon Press, 1952), 3:133. Johnson's word "asthma" does not, of course, imply its twentieth-century meaning, a reversible, often allergic, disease of increased spasticity and inflammation of airways. For Johnson, it simply meant attacks of suffocating shortness of breath.

4. Ibid., 3:136, 185, 246, 247. Squills and cantharides are archaic remedies used as diuretics. For a brief account of Dr. Johnson's medical history and autopsy, see Benjamin S. Abeshouse, *A Medical History of Dr. Samuel Johnson* (Norwich, CT: Eaton Laboratories, 1965).

5. Richard Bright, *Reports of Medical Cases Selected with a View of Illustrating the Symptoms and Cure of Disease by a Reference to Morbid Anatomy*, 2 vols. (London: Longman, Rees, Orme, Brown & Green, 1827–1831).

6. Steven J. Pietzman, "Bright's Disease and Bright's Generation—Toward Exact Medicine at Guy's Hospital," *Bulletin of the History of Medicine* 55 (1981):307–321. Bright's test was simple and surprisingly reliable. A small amount of urine was placed in a spoon and heated over a candle flame. Albumin, present only in urine from unhealthy kidneys, formed a coagulum or precipitate when heated. Certain acids also serve to detect proteinuria.

7. For Bright's etiological thought, See Steven J. Peitzman, "Richard Bright and Mercury as the Cause and Cure of Nephritis," *Bulletin of the History of Medicine* 52 (1978):419–434.

8. Bright, *Reports*, pp. 67–69.

9. Thomas Watson, *Lectures on the Principles and Practice of Physic* (Philadelphia: Lea and Blanchard, 1844), p. 779.

10. S. J. Goodfellow, *Lectures on the Diseases of the Kidney Generally Known as "Bright's Disease" and Dropsy* (London: Robert Hardwicke, 1861), pp. 1–2.

11. Richard Bright, "Cases and Observations Illustrative of Renal Disease Accompanied with Secretion of Albuminous Urine," *Guy's Hospital Reports* 1 (1836):338–379.

12. Stanley Bradley, *The Pathologic Physiology of Uremia in Chronic Bright's Disease* (Springfield, IL: Charles Thomas, 1948); Henry Christian, *Bright's Disease* (New York: Oxford University Press, 1948); Thomas Addis and Jean Oliver, *The Renal Lesion in Bright's Disease* (New York: Paul Hoeber, 1931); Thomas Addis, *Glomerular Nephritis* (New York: Macmillan, 1948). The alternative name "Bright's kidney," briefly in use in the early nineteenth century, may be seen as a revealing synecdoche—the organ for the whole disease—that succinctly recognized the individual pathologist and his work in that period.

13. Bright, *Reports*, pp. 29–33. The polypharmacy employed was typical of treatment at the time. The various remedies mentioned included some thought of as "depleting" or "antiphlogistic" (anti-inflammatory) and others used as laxatives and diuretics.

14. John M. Swan, "Case Reports on a Great Variety of Ailments Observed at Philadelphia General Hospital and University Hospital, 1890–1893" (unpaginated manuscript notebooks), case 60, Historical Collections, Library of the College of Physicians of Philadelphia; James Tyson, *A Treatise on Bright's Disease and Diabetes* (Philadelphia: Lindsay and Blakiston, 1881), pp. 120–123, 137–147. Tyson's hot-air bath had "A tin pipe two or three inches in diameter with an expanded extremity under which a spirit-lamp is placed, while the other end is placed under the bed clothing. . . . An ordinary rain-spout may be used" (p. 121). The intent, of course, was to induce perspiration, so that "the skin may be made to substitute the action of the kidney." Moreover, suppression of perspiration, a "check on the skin" induced by a chill, persisted as a central idea in constructing the etiology of Bright's disease into the late nineteenth and even early twentieth centuries.

15. Addis, *Glomerular Nephritis*, pp. 289–314. Southey tubes were small silver tubes used to drain tissue edema, particularly from the lower legs. Although the case Addis described may be a composite, recently uncovered case records of Dr. Addis's outpatient practice reviewed by the author confirm the treatment as outlined in *Glomerular Nephritis*. These case records are in the Archives, Lane Library, Stanford University School of Medicine, Palo Alto, CA.

16. William Cullen, *Nosology, or a Systematic Arrangement of Diseases* (Edinburgh: W. Creech, 1800).

17. Carl Bartels, "Diseases of the Kidney," vol. 15 of *Cyclopedia of Practice of Medicine*, ed. H. Von Ziemssen. American edition, ed. A. H. Buck (New York: William Wood, 1877); Tyson, *Bright's Disease*, pp. 79–84; George Johnson, *Lectures on Bright's Disease* (London: Smith Elder, 1873); William Osler, *The Principles and Practice of Medicine*, 7th ed. (New York: D. Appleton, 1909), pp. 686–703; Franz Volhard and Theodor Fahr, *Die Brightsche Nierenkrankheit* (Berlin: Springer, 1914); Thomas Addis, "A Clinical Classification of Bright's Diseases," *Journal of the American Medical Association* 85 (1925):163–167.

18. See Faber, *Nosology*, pp. 112–171; Arthur M. Fishberg, *Hypertension and Nephritis* (Philadelphia: Lea and Febiger, 1930), pp. 39–54.

19. For the work and life of Addis, see particularly his *Glomerular Nephritis* (New York: MacMillan, 1948), a summation of almost three decades of his clinical and investigative work, and S. J. Peitzman, "Thomas Addis: Mixing Rats, Patients, and Politics," *Kidney International* 37 (1990):833–840. The letter quoted is in the Alfred Cohn Papers, Library of the American Philosophical Society, Philadelphia, PA. The call for a return to morphology is from Addis and Oliver, *Renal Lesion*, pp. 4–5.

20. S. J. Peitzman, "Nephrology in America from Thomas Addis to the Artificial Kidney," in *Grand Rounds: One Hundred Years of Internal Medicine*, ed. Russell C. Maulitz and Diana E. Long (Philadelphia: University of Pennsylvania Press, 1988), pp. 211–241.

21. Willem Kolff, *New Ways of Treating Uremia* (London: Churchill, 1947).

22. There exists no complete technical and social history of hemodialysis. A useful source, however, is Patrick T. McBride, *Genesis of the Artificial Kidney* (n.p.: Baxter Healthcare Corporation, 1987). Baxter Healthcare is the successor to the Travenol company, an early manufacturer of dialysis filters and equipment. MacBride's book is available from Baxter, which produced it as a professional service. Mr. MacBride also contributed a historical introductory chapter to *Clinical Dialysis*, ed. A. R. Nissenson, R. N. Fine, and D. E. Gentile (Norwalk, CT: Appleton-Century-Crofts, 1984), pp. 1–22.

23. "Proposed Rules [End-stage renal disease]," *Federal Register* 39 (1974): 35819.

24. Connie Jones, *Even in Heaven They Don't Sing All the Time: Experiences of Kidney Patients and Their Families* (Atlanta: National Kidney Foundation of Georgia, 1984), p. 34.

25. W. J. Stone and P. L. Rabin, *End-Stage Renal Disease: An Integrated Approach* (New York and London: Academic Press, 1983).

26. Chronic renal failure causes an anemia, which erodes the strength and energy of dialysis patients; however, as of 1989 this complication could be treated effectively with recombinant human erythropoietin, the hormonal stimulant to the bone marrow normally produced by healthy kidneys. Dialysis patients remain subject to destructive bone disease caused by the inability of dialysis to fully correct calcium and phosphorus imbalances. Aluminum toxicity, from dialysate water in some cases, or from "phosphate binders" given orally, can lead to yet another form of bone disorder. Better understanding of these complex interactions and improved water preparation will decrease the frequency of such complications. Transient falls in blood pressure during hemodialysis treatments, along with rapid osmotic changes in the patient's blood, can cause uncomfortable but only rarely dangerous drops in blood pressure. Patients feel these drops as headaches, nausea, or dizziness. Some patients suffer miserably with muscle cramps during or shortly after the dialytic treatments; the genesis of these remains uncertain.

27. The passages are from (in order): Connie Jones, *Even in Heaven They Don't Sing All the Time: Experiences of Kidney patients and their Families* (Atlanta: National Kidney Foundation of Georgia, 1984), p. 29; Lou Sand, "A Patient's Opinion," *Journal of Nephrology Nursing* (May–June 1986):109–110; James D. Campbell and Anne R. Campbell, "The Social and Economic Costs of End-Stage Renal Disease," *New England Journal of Medicine* 299 (1978):386–392; Mark Levi, "After All These Years," *Journal of Nephrology Nursing* (September–October 1984):109–111; M. Rosenberg, *Patients: The Experience of Illness* (Philadelphia: Saunders, 1980). See also Lee Foster, "Man and Medicine: Life without Kidneys," *Hastings Center Report* 6, no. 3 (June, 1976):5–8. These are but a sampling of narratives written by patients with ESRD. They tend to appear in nursing journals and foundation newsletters more often than in mainstream medical journals.

28. Johnson, *Letters*, 3:236.

2 EMERGENCE OF RHEUMATIC FEVER IN THE NINETEENTH CENTURY

PETER C. ENGLISH

Each generation of physicians can call upon a different repertoire of framing materials in suggesting an understanding of pathological phenomena; but the phenomena themselves may also change. Peter English reminds us of this complex and elusive aspect of disease history. He argues that rheumatic fever might well not have existed in its nineteenth-century clinical form much before its perceived emergence during that period of enormous social change. Yet, as he also reminds us, medical institutions and ideas were evolving at the same time. The greater likelihood of hospitalization, for example, and the prominence of hospital-based studies in discerning and defining clinical entities almost certainly played a role in framing what physicians came to call rheumatic fever, with its characteristic—and attention-focusing—incidence of cardiac involvement. English is equally well aware that the conceptual and technical tools for correlating appearances after death with symptoms in life did not really exist before the beginning of the nineteenth century. This timing poses an intricate and intractable, yet highly significant, dilemma. How does one make sense of this interactive negotiation over time, this framing of pathophysiological reality in which the tools of the framer and the picture to be framed *both* may well have been changing (and in which the relation between the subgroup called rheumatic fever and *all* symptom-producing interactions between humans and streptococci remained unclear)?

The history of the clinical entity designated rheumatic fever illustrates not only a conceptual evolution but a necessarily related—and parallel—evolution in representation. The first clinical descriptions were discursive narratives of individual doctor-patient interactions (which came to include a revelatory epilogue in the form of postmortem findings) so characteristic of the late eighteenth and early nineteenth centuries. By midcentury, representation of the disease took the form of numerical aggregates summarizing hospital experience. Walter B. Cheadle's depiction of rheumatic fever as a linked cluster of symptoms constituted still another step in the evolution of efforts to represent—and in representing to legitimate—this elusive clinical experience. Put another way, all these efforts sought to generalize a valid clinical entity from among the disparate symptoms of an extraordinary variety of cases, no one of which might stand as ideally typical with entire precision. In retrospect it is hardly surprising that Cheadle's work should have appealed to contemporary practitioners. It provided a viable—if schematic—compromise between a satisfyingly unified clinical entity and its protean manifestations in hospital ward or consulting room. In more recent generations, of course, laboratory findings and a variety of imaging techniques have became central to a new style of conceptualizing and representing this elusive ailment.

—C. E. R.

CONGESTIVE HEART FAILURE of a child from rheumatic heart disease, so common a century ago, is fortunately now rare. To the medical students I teach, rheumatic fever is a disease of history. They still memorize T. Duckett Jones's 1944 criteria for diagnosis (as they do Robert Koch's postulates), but nearly all have lacked the opportunity to apply them—that is, until the last several months when the unexpected return of rheumatic fever in several isolated epidemics surprised both pediatricians and epidemiologists.

So complete seemed the demise of rheumatic fever that pediatrician-epidemiologist Leon Gordis's 1985 T. Duckett Jones Lecture before the American Heart Association was entitled, "The Virtual Disappearance of Rheumatic Fever in the United States: Lessons on the Rise and Fall of Disease."[1] That same year Milton Markowitz entitled his Lewis W. Wannamaker Memorial Lecture "The Decline of Rheumatic Fever: Role of Medical Intervention."[2] In 1986, the 86th Ross Conference was dedicated to the clinical problem of "Management of Pharyngitis in an Era of Declining Rheumatic Fever." Clinicians over fifty will recall that as late as the 1950s and 1960s rheumatic fever was a major health problem that touched in a central way many of the basic medical sciences (bacteriology, immunology, pathology, epidemiology, genetics, endocrinology, pharmacology), most of the clinical services (internal medicine, pediatrics, surgery, psychiatry), as well as many of the medical technologies (culturing, immunological assay, erythrocyte sedimentation rate, electrocardiogram, and Xray). The clinical entity that Walter Butler Cheadle defined in London in the 1880s and Jones clarified sixty years later in Boston is now virtually gone, despite its recent recrudescence.

As striking as the decline and reappearance of rheumatic fever is that the disease, as Cheadle understood it in 1890, may very well not have existed a century earlier. How, then, did it emerge? I argue that biological, technological, clinical, institutional, possibly even geographical and climatic elements of rheumatic fever's "ecology" changed in the nineteenth century in a way that focused attention on damage done to the heart as the clinically most important facet of the disease, and it was this cardiac emphasis that precipitated and shaped the clinical recognition of rheumatic fever.

In the late eighteenth century, rheumatic fever was imbedded in the diagnostic category of "rheumatism," a broadly defined group of illnesses characterized by fevers, aches and pains of the limbs, and debility.[3] Rheumatism was a routine medical diagnosis in the eighteenth century. I believe that the biological nature of a portion of rheumatism may well have changed in the late eighteenth century so that the heart, especially tissues of the pericardium and the endocardium, became inflamed, an injury that had not commonly occurred previously.

This biological alteration must remain to a degree speculative. What is without question, however, is that the late eighteenth century witnessed the beginning of a clinical appreciation of cardiac involvement in ills characterized as "rheumatism." In 1812 William Charles Wells, a Charleston, South Carolina, native who trained in Edinburgh, later remained loyal to Great Britain after the Revolution, and practiced medicine in London, definitively linked heart disease with rheumatism.[4] Nevertheless, it is clear from reading late-eighteenth- and early-nineteenth-century accounts of patients suffering from rheumatism that a few practitioners were becoming aware of the cardiac connection before Wells. For example, Gerhard van Swieten reported in the mid-eighteenth century that "sometimes, when the pain in the limbs ceases, there arises an anxiety in the breast, a palpitation of the heart, and intermitting pulse."[5] While it is certain that van Swieten understood the occasional involvement of the heart, he did not believe it was a common element of rheumatism. William Cullen in the 1760s called attention to the "frequent, full and heard pulse" in some rheumatic patients.[6] Corvisart, in *Organic Diseases of the Heart and Great Vessels* (case 37) described one patient with rheumatic pains who on autopsy had vegetations on the mitral valve.[7] In the first extensive statistical analysis of rheumatism, John Haygarth noted that pulse records of 93 patients with acute rheumatism showed 55 with a heart rate greater than 96 beats per minute.[8]

Haygarth did not make the connection between heart damage and rheumatism, and an elevated pulse rate does not always mean heart disease (fever alone can raise the pulse). But his observations indicated that physicians were beginning to look at the cardiovascular system. Twelve of Haygarth's patients died. This detailed case history indicates that three died either with severe chest pain or with shortness of breath. While the exact pathological cause of death is not certain, it is possible to speculate that pericarditis and/or congestive heart failure contributed. Giovanni Morgagni, in his *On the Seats and Causes of Diseases*, supports the paucity of cardiac involvement in rheumatism. Case histories and discussions of patients with mitral valve disease indicated that none clearly suffered from rheumatism. And Morgagni, when analyzing rheumatism, could recall only two deaths with this diagnosis.[9]

Wells did not claim priority for linking heart disease with rheumatism. Rather, he credited David Pitcairn, a prominent British physician, with the initial association in 1788. Pitcairn had failed to publish his observations, so Wells considered his paper as recording Pitcairn's idea, to which he added seven cases of his own. Wells also remarked that Matthew Baillie, a Scottish pathologist, had made the initial pathological investigation of the death from heart disease of a patient with rheumatism:

The muscular parietes of the heart being generally very thin in proportion to the enlarged size of its cavities, the heart has little power to propel an increased quantity of blood into the more distant branches of the arterial system. At times there is much difficulty of breathing; and there is a purplish hue of the cheeks and lips. . . . The causes which produce a morbid growth of the heart are but little known; one of them would seem to be rheumatism attacking this organ.[10]

In a footnote, Baillie confirmed that "Dr. Pitcairn has observed this in several cases." Baillie's description contained no prior reference to rheumatism, so it is not possible to understand how he arrived at his conclusion that rheumatism was responsible.

Wells also called attention to the experience of David Dundas, sergeant-surgeon to the king, who in 1809 reported 9 patients with heart disease and rheumatism he had seen in thirty-six years of practice. Most had suffered chest pains, anxiety, an increased pulse, ascites, pleural fluid, or peripheral edema following one or more attacks of rheumatic fever. Seven of the 9 were under twenty-two years of age, and 7 died, usually after a period of several months. He autopsied 6 and found the heart enlarged in most; pericardial fluid surrounded one heart, and in several others the pericardium adhered to the surface of the heart.[11]

Wells encountered his first rheumatic patient with heart disease in 1798. Enlightened by Baillie's description, he suspected heart disease because the eighteen-year-old boy complained of an "oppression in his chest." Wells consulted with Pitcairn, who confirmed his diagnosis. The boy subsequently died, and at autopsy Wells discovered an enlarged heart. Wells observed his second case of rheumatism associated with heart disease four years later. This time his patient, Martha Clifton, recovered after eleven weeks.

Is it possible that Pitcairn, Wells, and Dundas simply observed what others had plainly overlooked in the past? I think not. Rather, it is at least possible that cardiac involvement was new at the end of the eighteenth century. Let us imagine that there was a clinical spectrum of heart involvement in rheumatism:

1. Heart not involved
2. Heart involved but patient asymptomatic
3. Heart involved, patient initially asymptomatic but develops significant (and symptomatic) heart disease years after a bout with rheumatism
4. Heart involved, patient symptomatic (congestive heart failure, pericarditis, chest pain)
5. Death

Even an astute observer could miss the first three categories at the onset of rheumatism. A careful physician, however, would not miss the last two. Even Haygarth, van Swieten, and Cullen—who did not associate rheumatism with heart disease—nevertheless did remark on these "unusual" symptoms. In the eighteenth century, patients with rheumatism were not discomfited, except from their joint pains, and did not die. A close look at individual case reports from Wells and Dundas leaves little doubt that these early patients had cardiac symptoms.

Example 1. In the beginning of August, shortly after remaining some time in a cold cellar, she was seized with pains, swelling, and redness of her joints, and fever. These symptoms lasted only ten days. Immediately upon their ceasing, her heart began to beat with considerable violence. Her right hypochondrium soon after became painful, and about the same time she began to complain of a pain in the tops of her shoulders. The palpitation of the heart, which had never ceased from its first appearance, was distinctly felt in every part of the thorax, to which my hand was applied. In the arteries, only a shaking was perceivable, which could not be divided into distinct pulsations. The strokes of the heart were one hundred and ninety in a minute; she frequently complained of a great and indescribable anxiety in her chest. The external jugular veins were swollen, and alternately rose and fell. After her death, the following are the principal morbid appearances, which, as I was afterwards informed, were observed: The whole of the internal surface of the pericardium was attached to the heart.[12]

Example 2. The patient complains of great anxiety and oppression at the praecordia; has generally a short cough, and a difficulty of breathing, which is so much increased by motion or by an exertion, as to occasion an apprehension that a very little additional motion would extinguish life. There is also frequently an acute pain in the region of the heart, but not always.[13]

The observation that took Wells and Dundas years to repeat became commonplace in the nineteenth century. For example, P. M. Latham, a physician at St. Bartholomew's Hospital, tabulated that 13 percent of all patients admitted between 1836 and 1840 with the diagnosis of acute rheumatism suffered from pericarditis.[14] The proportion of patients with pericarditis as part of their rheumatism grew to 22 percent at Middlesex Hospital between 1853 and 1859[15] and to 24 percent at Guy's Hospital between 1870 and 1872.[16] The prevalence of endocarditis in some hospitals was even higher. Following Wells's lead, a trickle of practitioners began to examine the heart in all cases of rheumatism, with or without cardiac complaints. For example, Laennec listed "gouty or rheumatic affections" as an occasional cause of pericarditis.[17] Such case reports increased to the point where a Parisian medical student could sustain in 1824 a twenty-four-page thesis for graduation from medical school, entitled "Considérations sur la Rhumatisme de Coeur."[18]

These initial reports resulted from external inspection of patients followed by confirmatory autopsies. The stethoscope, technical breakthrough, changed this. While Laennec's introduction of the stethoscope in 1816 has been well studied by historians, a great deal less is known about its reception by practitioners. The acquisition by practitioners of skills that correlated sounds at the bedside with structural changes at the autopsy table took time. The best historical accounts demonstrate that the stethoscope received a slow but steady welcome, especially from Paris-trained physicians. Initially, sounds emanating from the lungs received attention, and only by the 1830s did clinicians begin to sort out which abnormal heart sounds came from a particular chamber or valve of the heart.[19]

Using the stethoscope, Jean-Baptiste Bouillaud, a controversial Parisian clinician, argued forcefully that there was a "constant coincidence either of endocarditis or of pericarditis with acute articular rheumatism. . . . In auscultating the sounds of the heart in some individuals still laboring under, or convalescing from acute articular rheumatism, I was not a little surprised to hear a strong, full, saw or bellows sound. . . . such as I had often met in chronic or organic induration of the valves, with contractions of the orifices of the heart."[20]

What the stethoscope permitted Bouillaud to discover was a greatly increased number of patients with asymptomatic heart disease (categories 2 and 3 above) and to locate more specifically which part of the heart was affected. In a short monograph devoted entirely to the subject, Bouillaud outlined a systematic approach to examining the heart of patients with rheumatism. Like other members of the "Paris school," he used percussion and auscultation extensively and followed unsuccessfully treated cases to the autopsy room. Hospital-based, he saw enough cases to estimate that nearly half the people with acute rheumatism suffered from pericarditis, endocarditis, or both, either symptomatic or asymptomatic. By midcentury, then, the technology and experience were available to diagnose whether the pericardium or endocardium was involved; the stethoscope was not, of course, so helpful in determining myocardial damage.

Less speculative were the distinct shifts in clinical thinking about acute rheumatism in the nineteenth century brought about by the very complexity of the disease. Each patient with acute rheumatism experienced the disease differently. Various joints were swollen, inflamed, and painful, but rarely in any perceptible pattern. Some patients had rashes, others fever, but in no common sequence. Some behaved peculiarly; some were mildly discomfited; others died rapidly in great pain. In some families, several were sick at one time; in other families, several generations suffered. Some were sick for days, others for weeks or months. Some had one bout, others more than ten. Some were noticeably sick from other illnesses before rheumatism struck (scarlet fever, for example), others not. Some responded to therapy, others got better unaided. Still others died despite all remedial measures.

Rheumatic fever required physicians to sort out common and useful patterns in a disease that was complex in its presentation. This sorting out occurred in three distinct phases, clearly reflected in clinical reporting. Early in the century, practitioners wrote about and stressed elements peculiar to individual cases. Schematically, the format was similar to the case reports of Wells and others.

The focus was on joint involvement, but other manifestations, such as cardiac damage, were added to the picture in a supporting role.[21] In midcentury, hospital-based physicians in Great Britain and France, with access to hundreds to patients with rheumatism, were able to describe a "statistically average" case. George William Fleetwood Bury's analysis of 476 patients admitted to the Middlesex Hospital with the diagnosis of rheumatism between 1853 and 1859 showed that 253 suffered heart disease (138 with endocarditis, 35 with isolated pericarditis, 71 with both) to which, the numbers indicated, women and younger patients of both sexes appeared the most susceptible.[22] At Guy's Hospital, medical registrar Philip Henry Pye-Smith assembled the records of 300 hospitalized patients with rheumatism and an additional 100 outpatients seen during 1870 to 1872. He showed that fully half were under twenty years of age.[23] What is apparent from these tables (only three from hundreds) is that the literature had shifted from individual cases to statistical averages based on large numbers and that heart damage had surfaced as the most significant clinical manifestation of the disease.

The final third of the nineteenth century witnessed yet another shift in clinical thinking about rheumatic fever. "Individual" case histories showed the immense variability of acute rheumatism; "statistically" analyses demonstrated in a general way how rheumatism affected populations. Neither approach was entirely helpful to the practitioner confronted with sick patients.

What emerged in the 1880s was the concept of the "typical" case that incorporated common elements yet allowed for variability. This approach, which clustered elements from many cases, permitted physicians to make a certain diagnosis in a disease that was most uncertain in its presentation. Pioneering in this strategy was Walter Butler Cheadle, physician to the Hospital for Sick Children, Great Ormond Street.

The question Cheadle addressed in his Harveian Lectures was how a practitioner could recognize the "various manifestations of the rheumatic state." His solution was to describe a "typical" case, or rather ten of them. These were not statistical averages. Typical cases did not last 25.5 days, exhibit fevers of 102°F., or have a 56 percent chance of heart disease.

First, Cheadle identified the common elements: "The claims of endocarditis, of pericarditis, of pleurisy, of tonsillitis, of exudative erythema, of chorea, of subcutaneous nodules, will hardly, I think, be seriously disputed."[24] What he did next was pioneering. He claimed that each element was separate and could appear in nearly any combination, in almost any order. In children, the variation was even more extreme. He called this variation "phases in the rheumatic process or series." Each member of the series—for example, chorea or carditis—also had causes other than rheumatism, but rheumatic fever was one of the most common if not the most common predisposing cause.

Cheadle's clinical organization of thinking about acute articular rheumatism largely settled the difficult task of diagnosis. Leading physicians quickly adopted Cheadle's scheme, and nearly all subsequent discussions at meetings and lectures, in textbooks, and in scholarly papers showed clearly that his method swiftly reached the practitioner as well as prominent workers in the field. Indeed, no one has improved upon his general approach. Even T. Duckett Jones's criteria, which medical students began to memorize after 1944, must be understood as simply introducing mathematical precision to Cheadle's earlier clinical triumph. A diagnosis of rheumatic fever is indicated if a patient has two of the major manifestations or one major and two or more minor manifestations.

Clinical thinking shifted its emphasis from the joints to the heart. Individual cases showed that carditis was part of rheumatism; statistical averages showed that most morbidity and mortality resulted from heart disease; the typical case showed that the crucial element in rheumatism was heart involvement.

Closely associated with shifts in clinical thinking about rheumatic fever were institutional changes. For the most part, individual practitioners in the 1800s described individual case histories from experiences at the patient's bedside, normally at home. This practice accounted for the relatively small number of and long period between cases, as we saw in the practices of Wells and Dundas. The "statistical average" and "typical" cases were characteristically hospital-based.

Hospital observations, which normally included hundreds of patients, reinforced the conclusion that damage to the heart was the clinical event of most concern. Almost certainly, this reflected some selective bias. Contemporary case reports make clear that only the sickest, and of course the poorest, patients were admitted to hospitals; "sickest" usually meant pericarditis or congestive heart failure. Less common reasons for admission were extreme joint pain or poorly controlled chorea. In other words, nineteenth-century hospital practice tended to concentrate patients suffering from heart disease in the hands of hospital physicians, who were frequently leaders, authors, and educators.

Deaths from rheumatism invariably came from heart complications, a valid finding with the autopsy techniques available at midcentury. Pericarditis and endocarditis are easily seen with the naked eye, requiring no special stains or a microscope. One example of this emerging cardiac prominence came from St. Bartholomew's Hospital where heart disease (valvular disease, mitral stenosis and aortic regurgitation, and pericarditis) ranked second only to tuberculosis as the leading cause of death from 1830 to 1872.[25] While causes other than rheumatic fever could account for each of these pathological entities, acute rheumatic fever certainly played a large role.

Climate and geography also may have contributed to the rise of rheumatic fever. As is obvious from this account, much of the writing on this disease came from Britain or Northern France—northern latitudes with a generally cold and damp climate. Was this just coincidence? In 1881 August Hirsch, in his *Handbook of Geographical and Historical Pathology*, found evidence of rheumatic fever at all latitudes and in all climates but confirmed the generally held view that rheumatism was more prevalent in northern, cold, and damp locations.[26] Studies in the early twentieth century reconfirmed this uneven geographical distribution. In 1924 James Faulkner and Paul Dudley White compiled published statistics from twenty-eight hospitals around the world; those in the coldest and wettest climates reported the most rheumatic fever.[27] Also in 1924, Tinsley Harrison and S. A. Levine similarly noted the geographical variation: At the Peter Bent Brigham Hospital in Boston, 1.85 percent of all admissions were for rheumatic fever, compared with 0.73 percent at Johns Hopkins in Baltimore and 0.3 percent at Charity Hospital in New Orleans.[28] That rheumatic fever occurred most frequently in a region of the world that both introduced the stethoscope and organized the practice of medicine in hospitals could only have reinforced the clinical and pathological recognition of heart damage.

No discussion of the rise of rheumatic fever can be complete without mentioning the streptococcus. We know that the virulence of the streptococcus, owing in part to the phage-induced M protein, changes with time, sometimes abruptly, and that in general streptococcal diseases (childbed fever, erysipelas, scarlet fever) were more invasive a century ago. Could it be that the rise of rheumatic fever resulted solely from biological changes within the streptococcus? This hypothesis holds a certain attraction. It would explain the relatively sudden appearance and brisk rise of rheumatic fever as the course followed by any new infection moving through a previously nonimmune or "virgin soil" population, and its decline as due to immunity developing within an "experienced" population. Unfortunately, I do not believe that it was so simple. Rheumatic fever is not an infection in the usual sense; rather, it is a host response following a relatively innocuous infection. While it was true that people died of overwhelming streptococcal sepsis following childbed fever, erysipelas, or scarlet fever, patients did not succumb to streptococcal infection in rheumatic fever. I think it more likely that a change in the streptococcus induced a host response that damaged the heart. This peculiar response differs markedly from the host's ability to destroy the streptococcus or to be overwhelmed by bacterial infection (as is the case with childbed fever, erysipelas, and scarlet fever).

The shift in clinical focus to the heart was not without its ironies. Fever and joint pain were the symptoms that frequently brought patients to doctors. Yet, it was just these obvious complaints that physicians were asked to ignore in favor of potential dangers, often unperceived by the patient, that had to be detected through new technological devices. This disparity between what was clinically apparent and what was pathologically relevant did not die easily. A look at how clinicians referred to the disease illustrates the point. Early in the nineteenth century they called it "rheumatism" or "acute rheumatism" (following William Cullen), using a term that clearly focused on joints. Well into the twentieth century (long after heart damage had emerged as the key problem), English writers used "acute *articular* rheumatism," and French physicians "rheumatisme *articulaire* aigu," again with emphasis on the joints. Only later did "rheumatic fever" and "rheumatic heart disease" attain predominance.

A similar pattern occurred in therapeutics. In the early nineteenth century, physicians treated fever and joint pain. Both patient and physician were satisfied if these bothersome symptoms were ameliorated.

This pattern of gauging successful treatment did not cease with the introduction of salicylates after 1874. Fifty years later physicians still debated whether salicylates benefited carditis, in addition to lowering fever and easing joint pain, in part because there were no carefully crafted clinical studies measuring the value of salicylates in reducing cardiac damage. All investigations had determined dosing and benefits to the joints and fever.

I would argue that rheumatic fever arose in the late eighteenth century as the result of distinct biological changes (organism and host) that led to cardiac damage. Clinicians appreciated this alteration through the assimilation of technological changes (stethoscope and autopsy), refinements in clinical thinking ("the typical case"), the concentration of rheumatic invalids in hospitals, and quite possibly the serendipitous influence of geography and climate.

A final comment on the role of heart disease in the resurgence of rheumatic fever in the last two years: Although rheumatic fever has many clinical components, what catches the practitioner's eye are transient, migratory arthritis, acquired heart disease, and chorea. Of these, the arthritis is pathologically insignificant. Indeed, temporary arthralgia and arthritis can accompany many conditions, and are generally dismissed. Not so with chorea and acute heart disease. Both are dramatic and press the clinician into action. In the recent outbreaks of rheumatic fever, 91 percent of afflicted children in the Utah epidemic suffered carditis;[29] in Columbus, 50 percent;[30] in Pennsylvania 60 percent.[31] In only one report (northeast Ohio) was the percentage much lower (30 percent).[32] While no report singled out the presence of carditis as influential in the initial diagnosis of rheumatic fever, I suspect that acute heart disease is what caught the clinician's attention, in a way that may recapitulate the prominent role of heart involvement in the clinical recognition of rheumatic fever in the nineteenth century.

NOTES

1. Leon Gordis, "The Virtual Disappearance of Rheumatic Fever in the United States: Lessons in the Rise and Fall of Disease," *Circulation* 72 (1985):1155–1162.

2. Milton Markowitz, "The Decline of Rheumatic Fever: Role of Medical Intervention," *Journal of Pediatrics* 106 (1985):545–550.

3. John Swan, *Entire Works of Dr. Thomas Sydemham New Made English From the Originals* (London: St. John's Gate, 1746); William Cullen, *First Lines of the*

Practice of Physic, for the Use of Students in the University of Edinburgh, 2d ed. (Philadelphia: Steiner and Ast, 1781); Gerhard van Swieten, *Commentaries Upon Boerhaave's Aphorisms Concerning Knowledge and Cure of Diseases* (Edinburgh: Charles Ellis, 1776).

4. William Charles Wells, "On Rheumatism of the Heart," *Transactions of the Society for the Improvement of Medical and Chirurgical Knowledge* 3 (1812):373–424. Harry Keil, "A Note on Jenner's Lost Manuscript on Rheumatism of the Heart," *Bulletin of the History of Medicine* 7 (1939):409–411; Keil, "Dr. William Charles Wells and His Contribution to the Study of Rheumatic Fever," *Bulletin of the History of Medicine* 4 (1936):789–816.

5. van Swieten, *Commentaries*, vol. 13, p. 32.

6. Cullen, *First Lines*, p. 156.

7. Jean Nicolas Corvisart des Marets, *Organic Diseases of the Heart and Great Vessels*, [1806], 1984.

8. John Haygarth, *A Clinical History of Diseases: (I) A Clinical History of Acute Rheumatism* (1805; reprint ed. Oceanside, NY: Dabor Science Publications, 1977).

9. Giovanni Morgagni, *On the Seats and Causes of Diseases* (1761; reprint ed. New York: New York Academy of Medicine, 1960).

10. Matthew Baillie, *The Morbid Anatomy of Some of the Most Important Parts of the Human Body*, 2d ed. (London: J. Johnson, 1797).

11. David Dundas, "An Account of a Peculiar Disease of the Heart," *Medico-Chirurgical Transactions* 1 (1809):37–46.

12. Wells, "On Rheumatism," pp. 373–424.

13. Dundas, "Account," pp. 37–46.

14. William Selby Church, "An Examination of Nearly Seven Hundred Cases of Acute Rheumatism, Chiefly with a View to Determining the Frequency of Cardiac Affections, and Especially Pericarditis, at the Present Time," *St. Bartholomew's Hospital Reports* 23 (1887):269–287.

15. George William Fleetwood Bury, "A Statistical Account of Four Hundred Cases of Acute Rheumatism Admitted to the Wards of Middlesex Hospital during the Years 1853–1859," *British and Foreign Medico-Chirurgical Review* 28 (1861):194–198.

16. Philip Henry Pye-Smith, "Analysis of the Cases of Rheumatism, and Other Diseases of the Joints, Which Have Occurred in the Hospital during Three Consecutive Years, with Remarks on the Pathological Alliances of Rheumatic Fever," *Guy's Hospital Reports* 19 (1874):311–356.

17. Rene Théophile Hyacinthe Laennec, *A Treatise on the Diseases of the Chest* (London: T. & G. Underwood, 1821).

18. Joseph Irénée Itard, "Considérations sur le Rhumatisme du coeur" (Thèse Présente et Sontenue a la Faculté de Médecine de Paris, le 13 juillet, 1824).

19. Audrey B. Davis, *Medicine and Its Technology* (Westport, CT: Greenwood Press, 1981); Dale C. Smith, "Austin Smith and Auscultation in America," *Journal of the History of Medicine and Allied Sciences* 33 (1978):129–149.

20. Jean-Baptiste Bouillaud, *New Researches on Acute Articular Rheumatism in General, and Especially on the Law of Coincidence of Pericarditis and Endocarditis with This Disease* (Philadelphia: Haswell, Barrington, and Haswell, 1837).

21. Alfred Stille, "Case of Rheumatism, Endopericarditis, Pleurisy, and Double Pneumonia, with Autopsy," *Medical Examiner* 3 (1840):21–25; Richard

Bright, "Cases of Spasmodic Disease Accompanying Affections of the Pericardium," *Medico-Chirurgical Transactions* 22 (1839):1–19; J. Kingston Fowler, "On the Association of Affections of the Throat with Acute Rheumatism," *Lancet* 2 (1880):933–934; Thomas Barlow, "On Subcutaneous Nodules with Fibrous Strictures, Occurring in Children, the Subjects of Rheumatism and Chorea," *Transactions of the International Medical Congress* 4 (1881):116–128; Archibald Edward Garrod, "On the Relation of Erythema Multiforme and Erythema Nodosum to Rheumatism," *St. Bartholomew's Hospital Report* 24 (1888):43–54.

22. Bury, "Statistical Account, pp. 194–198.

23. Pye-Smith, "Analysis," pp. 311–356.

24. Walter Butler Cheadle, "Harveian Lectures on the Various Manifestations of the Rheumatic State, as Exemplified in Childhood and Early Life," *Lancet* 1 (1889):821–827, 871–877.

25. Thomas Rogers Forbes, "Mortality at St. Bartholomew's Hospital, London, 1839–1872," *Journal of the History of Medicine and Allied Sciences* 38 (1983):432–439.

26. August Hirsch, *Handbook of Geographical and Historical Pathology* (1881).

27. James Faulkner and Paul Dudley White, "The Incidence of Rheumatic Fever, Chorea, and Rheumatic Heart Disease, with Especial Reference to Its Occurrence in Families," *Journal of the American Medical Association*, 83 (1924): 425–426.

28. Tinsley R. Harrison and Samuel Albert Levine, "Notes on the Regional Distribution of Rheumatic Fever and Rheumatic Heart Disease in the United States," *Southern Medical Journal* 17 (1924): 1914–1915.

29. L. George Veasy et al., "Resurgence of Acute Rheumatic Fever in the Intermountain Area of the United States," *New England Journal of Medicine* 316 (1987):421–427.

30. Don Miller Hosier, Josefa Maria Craenen, Douglas William Teske, and John James Wheller, "Resurgence of Rheumatic Fever," *American Journal of the Diseases of Children* 141 (1987):730–733.

31. Ellen F. Rashkow Wald et al., "Acute Rheumatic Fever in Western Pennsylvania and the Tri-State Area," *Pediatrics* 80 (1987):371–374.

32. Blaise L. Congeni et al., "Outbreak of Acute Rheumatic Fever in Northeast Ohio," *Journal of Pediatrics* 111 (1987):176–179.

3   PARASITES AND THE GERM THEORY OF DISEASE

JOHN FARLEY

Each generation of medical scientists seeks to understand the mechanisms underlying the manifestations of disease; each is limited by the intellectual framing materials available to it. In his discussion of parasitology and its tangential relationship to the germ theory, John Farley illustrates the more general truth that particular framing options were not equally available to would-be framers. In the first half of the nineteenth century the parasite model seemed to have little relevance to most human ills, and the study of parasites was intellectually segregated. Thus, in the formative decades of the germ theory, the opportunity to create a unified understanding of infectious disease causation was ignored. In fact, such unification was not a real possibility.

In the late nineteenth and early twentieth centuries, parasitology was also institutionally segregated, in schools of tropical medicine, public health, and agriculture (paralleling the marginalization of tropical medicine away from the central foci of mid-twentieth-century medicine in Europe, England and North America). That is, intellectual and institutional history made certain options immediately accessible, leaving others relatively unavailable to physicians and biologists seeking understanding of particular ills. The slowness to generalize models of parasitism to human disease was as much the consequence of a particular history as was the often unthinking and mechanical application of other models—most conspicuously in the parallel history of the bacteriological theory of infectious disease where would-be bacteria hunters announced scores of premature or inaccurate discoveries of disease agents.

—C. E. R.

"THERE HAS ARISEN," noted Dr. August Hirsch in his *Handbook of Geographical and Historical Pathology* published in the early 1880s, "a prospect of adding to this department of pathology." The "department" was that of "parasitic diseases," and the prospective addition was "infectious diseases." Such diseases, he estimated, may well be of "a parasitic nature, or that there occur in them organisms of the lowest rank of organic development—the micrococci and bacilli." He was arguing that the so-called germ theory of disease (bacterial and later viral) then making rapid inroads should be termed the "parasitic theory of disease," and that bacteria should be considered, along with worms and protozoa, as belonging to a group of disease-causing parasites. Hirsch, like many others at that time, stood on the brink of accepting a generalized parasitic theory of disease.[1]

Such a theory did not soon emerge. Parasitic organisms, the worms and protozoa, stood apart from infectious agents such as bacteria, with the result that those who studied parasites seemed to have little to offer those concerned with infectious diseases. Important parasite-related discoveries made in the mid-nineteenth century did not enter into debates over the etiology of infectious diseases and had little if any impact on the genesis of the germ theory. Similarly, after the germ theory became generally accepted, parasitologists continued to exclude bacteria and viruses from the organisms they studied, and bacteriologists did not embrace the study of parasites. The two groups remained intellectually and institutionally unwed. I shall examine the history of parasitology in the nineteenth and early twentieth century for reasons that the world of parasites did not expand to include bacteria and viruses, and, consequently, a generalized parasitic theory did not develop.

The professional study of parasitic organisms predated the era of Pasteur and Koch, essentially originating in Germany at the beginning of the nineteenth century. In 1810, for example, Carl Rudolphi, who began his career at the University of Griefswald in Sweden after completing an M.D. thesis entitled "Observations on Intestinal Worms," moved to the University of Berlin where he acquired a reputation writing on the natural history of these strange creatures. His *Entozoorum Synopsis*, published in 1819, included nearly 1,000 species of parasitic worms. These worms, called Helminthes, were usually divided into five groups: nematodes, acanthocephalans, flukes, tapeworms, and so-called Cystica, later shown to be the bladderlike larval stages of some tapeworms.

In those early years certain pathologies were associated with the presence of these parasitic worms, but the worms were generally understood to be the symptoms not the cause of disease. They were, in other words, assumed to be spontaneously generated within the host. As Johannes Müller noted in his account of Rudolphi's life, "What he [Rudolphi] says in favour of *Generatio Aequivoca* [spontaneous generation], is still "almost the only recorded expression of opinion on which the defence of this doctrine can be made to rest."[2]

As I have explained elsewhere,[3] the arguments in favor of the spontaneous generation of parasitic worms were very persuasive. In the late eighteenth century, for example, Marcus Bloch's essay (for a prize awarded by the Danish Academy of Sciences) argued that parasitic worms were obviously destined to live only in very particular locations within a specific host organism. In the lawful, ordered, mechanical world of the eighteenth century, such parasites could not possibly arrive at these locations by chance; they had to be generated there.[4] No other conclusion was possible.

But the most telling arguments for spontaneous generation were published at the beginning of the nineteenth century by Johann Bremser, a friend and colleague of Rudolphi.[5] Parasitic worms, he argued, are members of a peculiar group of organisms that occur nowhere but inside the bodies of other animals. How then do they arrive at these locations? Are they produced inside the body, he asked, or do they or their eggs arrive there from outside? In his answers to these questions one is immediately struck by a parallel with debates over the contagious or noncontagious nature of some infectious diseases. Do these diseases, such as yellow fever and cholera (later in the century), likewise arise locally by a sort of spontaneous generation, or are they brought in from the outside by contagions?[6] In both cases a broad consensus favored spontaneous generation of both infectious diseases and parasites.

As a general rule, most infectious diseases seemed to remain localized within particular areas of a city, just as parasitic worms lived only in specific areas of the body. That both the diseases and the parasites arose there rather than being carried in from the outside seemed, at that time, to offer the best explanation for their distribution. Also, parasites, like certain infectious diseases, had a limited geographic range. Any explanation based on the external transmission of contagions or eggs made little sense. Why would humans in eastern Europe, for example, be infected with the bothriocephalid tapeworm while those in the west carried taeniid tapeworms, if worm eggs were the source of infection? What would stop eggs, or contagions, from being carried across Europe from one region to another? Likewise, the contagion theory offered no basis for understanding why some diseases were similarly limited in range, or why some seemed to arise "spontaneously" in certain areas with no known previous contact with a possible source of contagion.

There were also other problems. If a parasitic worm arose from eggs passed from another worm of the same species in another host animal, Bremser asked, how are the eggs passed? People rarely ate food spoiled by the feces of a neighbor, he pointed out, and how would the eggs survive outside the body long enough to make contact? And how are eggs passed between animals that do not drink much, or do not eat each other, still less each other's feces? How are we to explain, Bremser wondered, the fact that herbivores carry as many worms as carnivores, and the occurrence of hydatid cysts within the muscles of vegetable-eating ruminants?[7] In both cases the answer seemed obvious: Both the diseases and the worms arose locally because of organic decay, whether in miasmatic swamps or diseased body tissues. The following statement, made with reference to hydatid cysts, could equally well have been used to describe the origin of infectious diseases:

> Humidity, abundance and the bad or vegetable quality of the nourishment of an animal, are unequivocal means of producing acephalocysts (hydatid cysts) . . . irritation in fact, of a specific kind, has been excited by which a state favourable to their development has been produced.[8]

In an earlier paper dealing with the origin of these parasitic worms, I remarked that, without the concept of a vector or intermediate host, contagionists could never hope to explain the outbreak of some of these infectious diseases. Nor, without such a concept, could those opposed to spontaneous generation explain the origin of parasitic worms within the body. "The discovery of intermediate hosts for parasitic worms later in the century," I wrote, "was, then, of great significance for the understanding of contagious diseases."[9] I no longer believe this last statement to be true; most of the nineteenth-century concepts drawn from the study of parasites seem inapplicable to the understanding of infectious disease.

The Discovery of Intermediate Hosts

By the early nineteenth century, naturalists had come to realize that animals could develop in many different ways. Insect life cycles, involving nymphs or larvae and pupae, had long been known to differ from what was usually seen as the normal egg-to-adult pattern. By this time many benthic invertebrates also were known to produce larval stages that differed markedly from the adult animal and required complex metamorphoses in order to attain the adult form.[10] Then, in 1842, the Danish naturalist Japetus Steenstrup described yet another developmental process: *Generationswechsel*, or the alternation of generations. In this case an immature stage, rather than behaving like a larva by transforming directly into a subsequent stage in the life cycle, as in insects, actually seemed to reproduce so as to generate more than one member of the next stage. The alternation of generations, Steenstrup wrote, occurs by "an animal giving birth to a progeny permanently dissimilar to its parent, but which itself produces a new generation, which either it-self or in its offspring, returns to the form of the parent animal."[11]

Jellyfish provided the best-known example of such a life cycle. The minute free-swimming stage hatching from the egg did not transform directly into a jellyfish but developed instead into a small sessile "poly-piform creature," quite unlike the parent jellyfish. This creature did not then transform into a jellyfish as a larval stage might have done, but instead budded off numerous tiny jellyfishlike forms, which detached from the polyp to grow eventually into true jellyfish. But the most im-portant examples of such a life cycle, as far as this article is concerned, occurred in the trematodes or flukes. Flukes were then known to develop from free-living larval stages called "cercariae." But, Steenstrup asked, "whence come then the free swimming cercariae?" They arose in large numbers, he concluded, from saclike bodies within the tissues of snails, which in no way resembled the adult flukes. But Steenstrup, intent on interpreting the nature of this alternation of generations, was not interested in parasites per se, and remained unaware that he had intro-duced two key concepts into the world of parasitology: *the intermediate host* (in this case a snail, in which a parasite develops but only to an im-mature stage) and *the parasitic life cycle,* in which the parasite not only changes its host but also its form.

In the 1850s a series of important feeding experiments with tape-worms, associated mainly with Friedrich Kuechenmeister, Carl von Siebold, and Pierre-Joseph van Beneden, revealed that the so-called Cystica was not a separate animal taxon but consisted of the larval stages of terrestrial tapeworms.[12] This discovery was extremely signifi-cant, not only because it proved that tapeworms too had complex life cycles and intermediate hosts, but because the Cystica, lacking re-productive organs and found in muscles or the body cavity with no known outlet to the outside world, had traditionally been seen as the most obvious examples of spontaneously generated worms. Thus, what Steenstrup had described earlier in flukes was true also for tape-worms: Neither group arose by spontaneous generation, but passed from host to host in a complex life cycle involving great changes in body form.

In theory, at least, the concepts of a complex life cycle and intermedi-ate host might have been suggestive to contagionists striving to understand the distribution and behavior of infectious diseases; for ex-ample, host animals might convey contagions from place to place. But the concepts did not carry over into the medical world. Contagionists, to my knowledge, never proposed that either or both concepts might help to explain seeming anomalies in a contagionist interpretation of infectious disease. Of course, since few contagionists believed that liv-ing organisms caused infections, they had no reason to make the connection.

But another reason made ideas unlikely to flow from parasitology to medicine at that time. All parasitic animals were thought to belong to a *single* taxon: the Entozoa, or Helminthes. The word "parasite" was not widely used; they were helminths or entozoans, and their study was termed "helminthology," and less often "entozoology" or "parasitology."[13] Parasitism, therefore, was not seen as a life style shared by a wide variety of animal groups. The concepts of helminth life cycles and intermediate hosts were not transferable; they seemed applicable only to a single animal taxon.

The Beginning of Parasitology

This state of affairs was to last until the 1880s. By then, as Rudolph Leuckart had discovered, parasites included more than one animal taxon. Leuckart can, therefore, be seen as the "father of parasitology," as distinct from helminthology. He first became interested in helminths during the 1850s while on the medical faculty at the University of Giessen. In 1869 he moved to the University of Leipzig where, eleven years later, he opened perhaps the world's first parasite laboratory in a new zoological institute built for him. Sometime during his career at Leipzig, he recognized the coccidians to be parasitic members of the unicellular protozoans and placed them in the class Sporozoa. As a result of this discovery Leuckart could claim that animal parasites consisted not only of helminths, "but numerous other creatures that sometime resemble so completely certain free-living animals . . . that an independent mode of existence has been actually ascribed to them."[14] Parasites now included three groups of animals: worms, protozoans, and arthropods. As the title of Leuckart's book illustrates, these discoveries resulted in the increasing use of the words "parasite" and "parasitology" to describe a group of organisms and a new discipline.

At approximately the same time Patrick Manson, a relatively unknown physician working for the customs service in China, uncovered the role of bloodsucking flies in the life cycle of filarial nematode worms.[15] He almost, but not quite, discovered the distinction between biological vectors and intermediate hosts.

A few years before, Dr. Timothy Lewis, a member of the British Army Medical Service stationed in India, had discovered six minute active "snake-like" nematode worms, or filariae, in a single droplet of human blood.[16] "It is an almost universal law in the history of the more dangerous kinds of Entozoa," Manson remarked in 1877, "that the egg or embryo must escape from the host inhabited by the parent before much progress can be made in development." After that, he noted, the embryo either lives independently for a short while or is swallowed by another animal.[17] Manson found that the filariae were picked up by the female mosquito during its blood meal and then metamorphosed in the mosquito's stomach into what he considered to be adult nematode worms. "There can be little doubt as to the subsequent history of the filaria," he concluded. "Escaping into the water in which the mosquito died, it is through the medium of this fluid brought into contact with the tissues of man, and then either piercing the integuments, or, what is more probable, being swallowed, it works its way through the alimentary canal to its final resting place."[18]

Manson was close to observing the role of vector hosts but was misled by the model of cestodes (and trematodes to a lesser extent), which are transmitted from host to host through the food chain, to the natural conclusion that humans become infected by ingesting nematodes rather than through reinfection by biting flies. His use of the word "nurse" to describe the mosquito in the title of his first paper illustrates that he was thinking of the cestode-trematode model. Steenstrup himself had coined the word *Ammen* ("nurses") for what we would call the intermediate host. Manson's acceptance of the cestode-trematode food-chain model was also reinforced by the generally held assumption that a female mosquito took only one blood meal before depositing her eggs, after which she died.

Although the parasitic web had extended to include the protozoans by the 1880s, the trematode-cestode life-cycle model remained in place. The model had guided Manson's filarial work, and now the filarial work provided the model for his malarial theory, first proposed in 1894. Neither the malarial disease nor the malarial organism can be directly communicated from person to person, Manson noted; they "can be acquired only indirectly either through the air, the water, by food, or by another unknown way." Since escape from the body was necessary to transmit any parasite, and no trace of the parasite appeared in physiological or pathological discharges, Manson concluded that the same mechanism applied to the malarial protozoan as to the filarial worms studied earlier.

If this be the case, the mosquito having been shown to be the agent by which the filaria is removed from the human blood vessels, this, or a similar suctorial insect must be the agent which removes from the human blood vessels those forms of the malarial organism which are destined to continue the existence of this organism outside the body. It must, therefore, be in this or in a similar suctorial insect or insects that the first stages of the extracorporeal life of the malarial organism are passed.[19]

Manson constantly drew analogies with the helminths and other sporozoans. He noted that the malarial parasite must escape from the body in the same fashion as the larval stages of tapeworms. Both must be eaten, and, by analogy with filarial worms, mosquitoes seemed the most obvious villains. Likewise, he assumed that humans acquired the malarial parasites in the same manner as they acquired their parasitic worms: by eating them. The mosquito, Manson argued, "seeks out some dark and sheltered spot near stagnant water. At the end of about six days she quits her shelter, and, alighting on the surface of the water, deposits her eggs thereon. She then dies, and, as a rule, falls into the water alongside her eggs."[20] People became infected with the malarial parasite, Manson assumed, either by swallowing contaminated water or by inhaling dust containing resistant cysts of the parasite or created by decomposed infected mosquitoes. Again, Manson's views reflected prevailing notions about mosquitoes. In answer to the Italian parasitologist Amico Bignami's claim that a mosquito bite passed the parasite to man, Manson repeated that "the habit of the mosquito is to bite once only."[21]

In Texas cattle fever also, as Theobald Smith and F. Kilborne noted in their famous paper, the cattle tick was at first assumed to pass the disease organism in a helminthlike manner. "Hitherto we had supposed," they wrote, "that the cattle tick acts as a carrier of the disease between the Southern cattle and the soil of the Northern pastures. It was believed that the tick obtained the parasite from the blood of its host, and in its dissolution on the pasture a certain resistant spore form was set free, which produced the disease when taken in with the food."[22]

The error was finally revealed in 1898 by Ronald Ross.[23] Mosquitoes, he found, not only removed the malarial parasite from the blood but acted as true vector hosts. The parasites developed in the mosquitoes and eventually migrated to the salivary glands from which they could be injected back into the blood of the human host; people could acquire their parasites by inoculation as well as by feeding. The discovery that female mosquitoes produced more than one batch of eggs and took more than one blood meal added considerable credence to the vector theory.[24] Thus, arthropod vectors were a particular kind of intermediate host: The parasite developed in them and they carried the parasite both to and from the final host.[25]

Smith and Kilborne did not make that fundamental discovery, as often claimed. They could not decide whether a tick accidentally carried spores of the protozoan parasite responsible for Texas cattle fever or was a necessary host in which the parasite developed and then became "localized in certain glands of the young tick," from which it would be discharged into the blood of the cattle. Being unable to locate the parasite inside the tick, they remained uncertain. "Further investigations," they urged, "are necessary before the probable truth of one or the other of these hypotheses can be predicted with any degree of certainty." Smith and Kilborne faced the same difficulty as Manson. Just as mosquitoes were believed to bite only once, so the tick was presumed to spend its life on a single host, making host-to-host transmission by successive biting seem improbable.

But whether we ascribe the final breakthrough to Ross or to the Italian workers, certainty eventually came. By 1900 parasitic helminths and protozoa were known to develop in intermediate or vector hosts and to be transmitted to humans through either infected food or the bites of bloodsucking arthropods.

The period between Manson's "half-vector" filarial-worm discovery in the late 1870s and the realization by 1900 that a vector was a true biological host (that is, picks up the parasite, nurtures it through some developmental stages, and then transmits it) coincided, of course, with Koch's famous discoveries of the bacterial source of anthrax, tuberculosis, and cholera. It witnessed the growing belief, by such eminences as Hirsch, that bacteria cause many infectious diseases and could be added to the list of parasitic organisms. A generalized parasitic theory of disease seemed imminent.

It did not occur. Instead, a belief in two distinct sorts of disease developed: those caused by bacteria and those caused by parasites. The latter were not contagious; their life cycle inevitably involved passage through an intermediate or vector host in which obligatory stages of development took place. By contrast, bacterial diseases were usually contagious, although they might also be transmitted by arthropods in a purely accidental and mechanical fashion. As Victoria Harden has explained, the association between obligatory vector hosts and parasitic organisms was so strong that once a vector was discovered the disease was assumed to have a protozoal (or helminthic) cause; vice versa, once bacterial organisms were suspected, vector hosts were discounted.[26] As noted by Charles Stiles, who had been trained in parasitology in Rudolf Leuckart's laboratory:

> We may lay down two general biologic rules. . . . The first rule, to which at present a few exceptions are known, is that diseases which are accidentally spread by insects are caused by parasitic plants, particularly by bacteria. The second, to which no exceptions are as yet known, is that those diseases which are dependent upon insects or other arthropods for their dissemination and transmission are caused by parasitic animals, particularly by sporozoa and worms.[27]

The implications of these "laws" were profound. When yellow fever, long thought to have a bacterial cause, was shown in 1900 to be transmitted by the bite of mosquitoes, its cause became suspected of being protozoan. Similarly, it was thought that the plague, being caused by bacteria and harbored by the rat, could not be transmitted to humans by the bite of infected fleas. Fleas might carry the bacteria, according to Stiles, but the organisms could infect people only through a wound or scratch.

Patrick Manson also made a sharp distinction between bacterial and parasitic diseases in his address at the opening of the London School of Tropical Medicine.[28] The tropical external climate, Manson argued, influences the distribution of pathogens, restricting many to the tropics and bringing about diseases with "a limited climatic range." But, he added, coming to the crux of his argument, such climatic rules do not affect bacterial pathogens. They are cosmopolitan precisely because they live in the human body and rarely come under the influence of the external climate. Instead, "transmitted directly from host to host, they can be acquired in any climate when suitable social conditions occur." On the contrary, parasitic diseases are climatically localized, often because they passed through intermediate or vector hosts (mosquitoes, flies, snails, etc.) native to the tropics but not to temperate zones. The discovery by Arthur Looss, Professor of Parasitology at the University of Cairo and another student of Leuckart, that the larval stages of the hookworm passed directly into human skin without an intermediate host does not seem to have undercut the acceptability of the two-disease rules. Indeed, Looss saw the discovery as an anomaly. "I admit that this claim might arouse doubts," he wrote, "since it is an event which, until then, was without analogy in the history of parasites."[29] He later put the validity of the two-disease theory in question again when he claimed that the life cycle of the schistosome worms lacked an intermediate host. Although he received support, this too was perceived as an anomaly.

These reactions to Looss's "anomalous" discoveries support Harden's claim that parasitologists seemed to have been the most fervent supporters of the two-disease theory, probably related to the growth of parasitology as a distinct discipline in twentieth-century Britain and the United States. In both countries parasitology and bacteriology pulled apart; the latter became essentially a medical discipline while the former occupied niches outside the mainstream of medicine. Since intermediate and vector hosts were a unique part of the parasitologist's field, a claim that parasitic and bacterial disease were significantly different may well have been part of an attempt to delineate and legitimate a new discipline. And so, once again, as in the nineteenth century, parasitologists had little to offer investigators of infectious diseases.

The Parting of the Ways

Johannes Müller, writing about his predecessor at the faculty of medicine in Berlin, made the extraordinary claim that the success of Germans in helminthology reflected a lack of overseas possessions; whereas the British looked outward to the flora and fauna of empire, the Germans were forced to look inward. "The limitation caused by our geographical position," he wrote, "has imparted to our spirit a certain direction towards what is concealed, and has made us much the greater in the investigation of a world of concealed inhabitants of our native creatures, viz. the Entozoa."[30] He turned out to be quite wrong; it was, in fact, the expansion of the British Empire at the very end of the nineteenth century that drew the British into helminthology which, together with protozoology and medical entomology, became synonymous with a new postgraduate medical field of study—tropical medicine.

In March 1898, the British Colonial Office forwarded a memo to the War Office, Foreign Office, India Office, General Medical Council, Seamen's Hospital Society, and all twenty-six British medical schools. Joseph Chamberlain, the Colonial Secretary, according to this memo was "anxious to do anything in his power to extend the benefits of medical science to the natives of tropical colonies and protectorates, and to diminish the risk to the lives and health of those Europeans who . . . are called upon to serve in unhealthy climates." He requested, therefore, that all British medical schools provide instruction in tropical medicine.[31] Three months later, however, Chamberlain seems to have acquiesced to Patrick Manson's view that only one school of tropical medicine was needed, to be housed at the Seamen's Hospital at Greenwich. By July 1898, the foundations of the new school were being laid.

Tropical medicine became the main impetus for the emergence of parasitology as a discipline in Britain.[32] The London School of Tropical Medicine established lectureships in helminthology and protozoology, and the Liverpool School of Tropical Medicine, which opened at the same time, established lectureships and chairs in parasitology and entomology. Bacteriology was ignored.

Interuniversity jealousies were mainly responsible for this rather curious curriculum of the London School of Tropical Medicine. In November 1898, two weeks after the medical schools had been apprised of the decision to build a separate postgraduate school at Greenwich, King's College Medical School forwarded its belated reply to the original Colonial Office memo. Basically, the letter extolled the virtues of its bacteriological laboratory and supported the idea of special training in tropical medicine, which must, they noted, "of necessity include a thorough practical training in bacteriological methods."

The council trust that the Government would be willing, in any arrangements they may eventually make, to recognize instruction given in the bacteriological laboratory of King's College as a qualifying course in Tropical Medicine.[33]

Bacteriology became a thorny issue. In a letter to Chamberlain, Manson argued that, if the King's College proposal to recognize its course as a qualification in tropical medicine were agreed to, then

"every medical school or bacteriological laboratory in London and
throughout the country would have an equal claim to be regarded as
affording a qualifying course of study."[34] To justify the existence of a
single special school, through which all medical officers had to pass,
clearly necessitated avoiding duplication of the curricula of British
medical schools. What better way than to omit bacteriology and
hygiene from the curriculum of the new school? By 1900, training in
bacteriology was available at most British medical schools; indeed, it
was the most obvious sign that a school was progressive and promoting
the spirit of modern medical science. Thus, the London School of
Tropical Medicine, put its greatest emphasis on protozoology and helminthology while downplaying bacteriology and hygiene. Michael
Worboys has expressed it well: "In an important sense tropical medicine was defined initially by what an orthodox medical degree left out."
It had left out parasitology.

But this decision also gave institutional meaning to the growing idea
that bacterial and parasitic diseases were in fact not alike. Manson's address at the opening of the London School not only distinguished one
from the other but claimed that the study of parasitic diseases should be
the sole function of parasitologists working in the new field of tropical
medicine. The presence of vector and intermediate parasitic hosts produced diseases with limited geographical range as opposed to those
produced by cosmopolitan bacteria. The parasitic diseases limited to
the tropics became the focus of tropical medicine in the British Empire.
Indeed, apart from the rather unsuccessful Molteno Institute for Research in Parasitology at the University of Cambridge, the only institute
of parasitology in the British Empire that was not linked to tropical
medicine was founded in 1932 at the MacDonald College of McGill
University in Montreal. Supported by the Canadian National Research
Council, the Quebec Department of Agriculture, and the Empire Mar-
keting Board in London, its mission was to investigate parasites of
economic importance to the swine, sheep, and poultry industries, although during the Second World War the faculty of the institute was
called upon to offer classes toward a McGill diploma in tropical medicine.[35] But Canada was not only a member of the British Empire, it
shared the North American continent with the United States where
parasitology, as at McGill, often had a veterinary and agricultural
focus.

In the United States, land grant colleges, not schools of tropical medicine, became the major forum for parasitology; its practitioners were
mainly zoologists, not physicians. In these land grant colleges parasitology either had a veterinary and agricultural emphasis[36] or, as "medical
zoology," was tailored to the needs of a new and increasingly important
set of undergraduate students called "premeds." American parasitology was nurtured in Nebraska and Illinois; its father was Henry
B. Ward, who spent two years in Germany studying at Göttingen,

Freiburg, and eventually under Rudolf Leuckart at Leipzig. In 1893 he came to the University of Nebraska at Lincoln, where he began to include the study of parasites in his undergraduate zoology classes. The University of Nebraska Catalogues show that by 1904 Ward was presenting a class called "Parasites of Man." It was clearly aimed at premedical students, who were required to spend their first two years at Lincoln before transferring to the Omaha Medical College.

Reflecting the shift of American medical education away from proprietary schools towards university-based schools of medicine, the Omaha College had become the College of Medicine of the University of Nebraska. At the same time, the introductory zoology class, now also increasingly geared for premedical students, was retitled "Introduction to Animal Biology and Medical Zoology." By 1910 the division between medicine and agriculture had become complete; the two parasitology classes were called "Medical Zoology," for premedical students, and "Animal Parasites," for zoology students. By then, however, Ward had gone. Denied the deanship in medicine at Omaha, he had accepted the chair of zoology at Illinois where he built up the first American graduate school in parasitology, trained the first generation of American parasitologists, and founded the *Journal of Parasitology*.[37]

But, as in Britain, American parasitology also served the needs of tropical medicine. The Department of Medical Zoology in the Rockefeller-funded Johns Hopkins University School of Hygiene and Public Health became one of the most active research centers in American tropical medicine.[38] Divided into the three classical divisions—protozoology, helminthology, and entomology—it became the department with the strongest links to the International Health Division of the Rockefeller Foundation, which dominated tropical medicine between the two world wars, and whose members often circulated through the department to gain necessary expertise in tropical diseases. As in the London School of Tropical Medicine, parasitology at Johns Hopkins became primarily a postgraduate medical study, although some of its degrees were also available to science graduates.

For most of the nineteenth century the world of parasites was quite separate from the world of infectious diseases. Diseases were not deemed to be caused by organisms, and parasites were restricted to a single and very peculiar taxon of parasitic worms. Thus, the discovery that these worms were not spontaneously generated and that their life cycles involved passage through intermediate hosts played no role in the debate over the contagious or noncontagious nature of infectious diseases. In the last two decades of the nineteenth century, however, accumulated knowledge could have bridged the barrier. Parasitism seemed to be a way of life common to a wide variety of organisms, which could embrace the bacteria, now viewed to be the cause of many infectious diseases. Intellectual options were available to build around a parasitic theory of disease. In the words of Hirsch, all belonged to a "single department of pathology." But no fruitful interchange seems to have taken place, and no such single department of pathology was born. Instead, institutional differences came to play a central role; parasitology and the medical discipline of bacteriology distanced themselves from each other with arguments that each dealt with a different set of diseases. Parasitology in Britain and the United States became established as a discipline outside the mainstream of medicine, segregated from the modern medical field of bacteriology. It concentrated on describing nonbacterial parasites and their life cycles that were of veterinary, agricultural, or imperial importance. Thus twentieth-century parasitology came to resemble nineteenth-century helminthology; hermatically confined by cultural circumstances, neither managed to share its resources with the medical search for an understanding of diseases.

ACKNOWLEDGMENTS

This work is part of a broader study of the history of tropical medicine, which has been generously supported by grants from the Social Sciences and Humanities Research Council of Canada.

NOTES

1. August Hirsch, *Handbook of Geographical and Historical Pathology.* English trans. C. Creighton (London: Sydenham Society, 1885), vol. II., p. 279.

2. Johannes Müller, "On the Life and Writings of the Late Professor Rudolphi." *Edinburgh New Philosophical Journal* 25 (1838):221–242.

3. John Farley, "The Spontaneous Generation Controversy (1700–1860): The Origin of Parasitic Worms." *J. Hist. Biol.* 5 (1972):95–125. See also Farley, *The Spontaneous Generation Controversy from Descartes to Oparin* (Baltimore: Johns Hopkins University Press, 1977), Ch. 3.

4. Marcus Bloch, *Abhandlung von der Erzeugung der Eingeweiderwürmer und den mitteln wider dieselben* (Berlin: S. F. Hesse, 1782).

5. Johann Bremser, *Ueber lebende Würmer im lebenden Menschen* (Vienna: C. Schaumburg, 1819).

6. The arguments for the noncontagious nature of these diseases were described by Noah Webster in 1799 in *A Brief History of Epidemic and Pestilential Diseases with the Principal Phenomena of the Physical World Which Precede and Accompany Them*. Webster's work was discussed by Charles Winslow, *The Conquest of Epidemic Disease* (Princeton: Princeton University Press, 1943). Reprint (New York: Hafner, 1967), pp. 214ff.

7. Hydatid cysts was the name given to what was later found to be the bladderlike larval stage of certain tapeworms.

8. B. Phillips, "Cysts," in R. B. Todd ed., *The Cyclopaedia of Anatomy and Physiology* (London: Longman, 1835), vol. I, p. 790.

9. Farley, "The Origin of parasitic worms."

10. The confusion generated by these discoveries has been discussed by Mary Winsor, "Barnacle Larvae in the Nineteenth Century," *Journal of the History of Medicine*, 24 (1969):294–309.

11. Japetus Steenstrup, 1842, *On the Alternation of Generations; or, the Propagation and Development of Animals through Alternate Generations*. English trans. G. Busk (London: The Ray Society, 1845). For a detailed examination of the alternation of generations concept, see Mary Winsor, *Starfish, Jellyfish, and the Order of Life* (New Haven: Yale University Press, 1976) and John Farley, *Gametes and Spores: Ideas about Sexual Reproduction* (Baltimore: Johns Hopkins University Press, 1982).

12. These experiments are discussed in Farley, "The Origin of Parasitic Worms" and *The Spontaneous Generation Controversy*.

13. A good example was T. S. Cobbold, *Entozoa: An Introduction to the Study of Helminthology, with reference more Particularly to the Internal Parasites of Man* (London: Groombridge & Sons, 1864).

14. Rudolf Leuckart, 1876, *Die menschlichen Parasiten und die ihnen herruhrenden Krankheiten*, 2 vols. English trans. W. Hoyle, *The Parasites of Man and the Diseases Which Proceed from Them* (Edinburgh: Pentland, 1886).

15. Patrick Manson's papers were published in the obscure *Medical Reports of the China Imperial Maritime Customs, Shanghai*. Some have been partially reprinted in B. H. Kean et al., *Tropical Medicine and Parasitology: Classical Investigations* (Ithaca: Cornell University Press, 1978). His first paper on the filarial worms, "On the Development of *Filaria sanguinis hominis* and on the Mosquitoes Considered as a Nurse," was completed in 1877, published in China, read before the Linnaean Society in 1878, and published in *Journal of the Linnean Society* 14:303–311. The mature worms of this group of nematodes release large numbers of minute prelarval stages, called "filariae," into the peripheral blood stream, where they are picked up by biting flies.

16. Timothy Lewis, "On a Haematozoon Inhabiting Human Blood, Its Relation to Chyluria and Other Diseases," in *Eighth Annual Report of the Sanitary Commissioner with the Government of India*, 1872. Partially reprinted in Kean, *Tropical Medicine*, p. 379.

17. Manson, "Further Observations on *Filariae sanguinis hominis*." Partially reprinted in Kean, *Tropical Medicine*, pp. 387–392.

18. Manson is often assumed to have suspected the mosquito only after noting a periodicity of the filariae in the bloodstream—they appear first at sunset, reach a peak about midnight when the mosquitoes bite, and thereafter decrease so that by midmorning few if any appear in the blood. But he noted this periodicity after discovering the role of the mosquitoes; at the time he had merely observed the temporary absence of the filariae from the blood, but "was not aware . . . of any law governing this." See Patrick Manson, "Additional notes on *Filaria sanguinis hominis* and filaria disease." Reprinted in Kean, *Tropical Medicine*, p. 394.

19. Manson, "On the Nature and Significance of the Crescentic and Flagellated Bodies in Malarial Blood," *British Medical Journal* 2 (1894):1306–1308. Reprinted in Kean, *Tropical Medicine*, pp. 55–59.

20. Manson, "The Goulstonian Lectures on the Life-history of the Malarial Germ outside the Human Body," *British Medical Journal* 1 (1896):716.

21. Amico Bignami, "The Inoculation Theory of Malarial Infection: Account of a Successful Infection with Mosquitoes," *Lancet* (1898):461–463, 1541–1544. Manson to Ronald Ross, October 12, 1896, reprinted in P. H. Manson-Bahr and A. Alcock, *The Life and Work of Sir Patrick Manson* (London: Cassell, 1927).

22. Theobald Smith and F. Kilborne, "Investigations in the Nature, Causation, and Prevention of Southern Cattle Fever," *8th and 9th Annual Reports, Bureau of Animal Husbandry* (Washington, DC: Government Printing Office, 1893). Reprinted in "Theobald Smith: 1859–1934," *Medical Classics* 1 (1937):372–598.

23. Manson, "The Mosquito and the Malarial Parasite," *British Medical Journal* 2 (1898):849–853. The story of Ross's discovery has been told in G. Harrison, *Mosquitoes, Malaria and Man: A History of Hostilities since 1880* (New York: Dutton, 1978).

24. Manson-Bahr & Alcock, *Life of Manson*, attributed this discovery to Thomas Bancroft in 1899, a year after Ross's findings. If so, one can assume that Bignami's claim may have influenced Ross and Manson more than either was ready to admit.

25. Purists will argue that the mosquito is the final host of the malarial parasite, in that the parasite reaches maturity in the insect, and man acts as the intermediate host. However, the concepts of "adult" and "maturity" make little sense in the context of single-celled organisms.

26. Victoria Harden, "Rocky Mountain Spotted Fever Research and the Development of the Insect Vector Theory, 1900–1930," *Bulletin of the History of Medicine* 59 (1985):449–466.

27. Charles Stiles, "Insects as Disseminators of Disease," *Virginia Medical Semi-Monthly* 6 (1901):54. Quoted in Harden, "Insect Vector Theory."

28. Manson, "Introductory Address," included in a bound series of addresses entitled *London School of Tropical Medicine and Hospital for Tropical Diseases: Miscellanea 1899–1927*, Archives, London School of Hygiene and Tropical Medicine.

29. Arthur Looss, "Ueber das eindringen der Ankylostomalarven in die menschliche Haut," *Centralbl. f Bakteriologie und Parasitenkunde* 29 (1901):733. Partially reprinted in Kean, *Tropical Medicine*, pp. 309–312.

30. Müller, "On Professor Rudolphi."

31. Memo to General Medical Council and British Medical Schools, March 11, 1898, *Miscellaneous Papers, Printed for the Use of the Colonial Office*, CO 885/7/119. I discuss the origins of tropical medicine in *Bilharzia: A History of Imperial Tropical Medicine* (New York: Cambridge University Press, 1991).

32. Michael Worboys, "The Emergence of Tropical Medicine," in G. Lemaine et al., eds., *Perspectives on the Emergence of Scientific Disciplines* (London:

Mouton, 1976); "The Emergence and Early Development of Parasitology," in K. S. Warren and J. Z. Bowers, eds., *Parasitology: A Global Perspective* (New York: Springer-Verlag, 1983).

33. King's College to Colonial Office, Nov. 22, 1898, *Miscellaneous Papers, Colonial Office*, CO 885/7/119.

34. Manson to Colonial Office, December 26, 1898, ibid.

35. T. W. M. Cameron, "The Institute of Parasitology," *McGill News*, Autumn, 1940.

36. C. Schwabe, "A Brief History of American Parasitology: The Veterinary Connection between Medicine and Zoology," in K. S. Warren and E. Purcell, eds., *The Current Status and Future of Parasitology* (New York: Josiah Macy, 1981).

37. H. van Cleave, "A History of the Department of Zoology in the University of Illinois," *Bios* 18 (1947):75–97. Also the H. B. Ward papers, University of Illinois Archives, Urbana, Illinois. His students included W. W. Cort, Ernest Faust, George LaRue, Harold Manter, Justus Mueller, Horace Stunkard, and others.

38. Elizabeth Fee, *Disease and Discovery: A History of the Johns Hopkins School of Hygiene and Public Health* (Baltimore: Johns Hopkins University Press, 1987).

4 "DEFINITE AND MATERIAL": Coronary Thrombosis and Cardiologists in the 1920s

CHRISTOPHER LAWRENCE

The history of medicine often seems to be the history of a progressive revelation in knowledge, a gradual but inexorable inching closer and closer to an understanding of the natural world. And as we have seen, a key component of that real world was the nature and course of discrete disease entities. Not surprisingly, historians of medicine, especially practitioners in a particular field, have been concerned with elucidation of the disease entities that now loom so prominently in everyday practice. And often, as Chris Lawrence suggests, they indulge in the luxury of reflecting on the obtuseness of their predecessors who had for so long missed the seemingly so obvious. One such entity was coronary thrombosis.

Elaborated in the decade after World War I and generally accepted as a coherent and unambiguous clinical reality by the early 1930s, the concept of coronary thrombosis, Lawrence contends, was in fact neither clear nor obviously coherent. Patients had long complained to physicians of one or several of the associated symptoms, and pathologists by the turn of the century had described elements of the characteristic pathology. Yet the unpredictable and diverse configuration of symptoms remained a dilemma to practitioners.

Conventional historical accounts cite electrocardiography as the consolidating key, a technical innovation that sedimented this inherently real and coherent entity out of a mixture of functional, circulatory, and other ills, the low-grade clinical ore in which the pure concept was embedded. Yet, as Lawrence maintains, even the early ECG findings were far from plain and indisputable; the new concept took a gradual and meandering path before it swept all before it—before it was accepted as inevitable and obvious truth. This was not because the new organizing concept was inconsistent with the very real symptoms and signs but because it was not the only pattern in which the evidence could be construed. The boundaries of the new concept, Lawrence argues, mirror a fundamental social reality: the creation of a cardiological subspecialty wedded to a somatic and technical cognitive identity that would clearly define and legitimate the specialty, setting it categorically apart from that of other clinicians. In a more general sense, Lawrence seeks to underline the relationship between the cognitive boundaries enclosing the classificatory concept of coronary thrombosis—together with the somatic style in which those boundaries were maintained and legitimated—and the social identity of the cardiological community itself.

—C. E. R.

A New Disorder and Its Historiography

In 1928 the english physicians John Parkinson and Evan Bedford wrote:

> When a man of advancing years is seized while at rest with severe pain across the sternum which continues for some hours and which is accompanied by shock, collapse, and dyspnoea he has had an anginal attack of no ordinary kind. It is only reasonable to suppose that something definite and material has happened in the heart, and investigation is actually proving that such attacks are the result of acute infarction of the heart muscle from coronary occlusion.[1]

Historians and clinicians generally agree that only toward the end of the 1920s did the bedside diagnosis of acute myocardial infarction (sudden death of heart muscle) consequent on coronary thrombosis (clot formation in the heart's arteries) become a regular clinical event.[2] This condition, "heart attack" as it popularly became known, was soon to be designated one of the leading causes of death in Western societies. By 1930 the disease seemed to have a clear-cut character. For instance, in 1929 Terence East and Curtis Bain, discussing coronary thrombosis in their *Recent Advances in Cardiology*, wrote, "Now that a considerable amount has been written on the subject, a definite clinical entity stands out. The disease has become easy to recognise at the bedside."[3] In retrospect, this generation of physicians were puzzled at their earlier failure to recognize so apparently distinct a condition. In 1946 Sir Maurice Cassidy recalled, "Looking through my notes of patients seen twenty or thirty years ago, I come across occasional cases where I failed to recognise the coronary thrombosis, which now, on paper, is the obvious diagnosis."[4] These two quotes neatly frame a problem addressed by a number of historical studies: Why did physicians before the 1920s fail to describe, discover, or diagnose so obvious a disease? I argue, however, that solving this problem is ahistorical and suggest that the framing of it was part of the process whereby the disease was originally given an identity. Alternatively I outline how the history of the appearance of coronary thrombosis in the 1920s might be written so as to avoid anachronism.

The coronary thrombosis problem, as I will call it, arose because of certain historiographic assumptions. These assumptions are epistemological and are commonly employed in the study of the history of diseases. The most important of them are that diseases are unproblematic natural entities and that they can be described in an objective or value-neutral language. Once so described, diseases are often held to have a manifestly obvious character. This historiographic tradition also assumes that, after a long period of mistaken or partial description, diseases are finally and definitively described by significant individuals. Such individuals are held to have precursors: clinicians who nearly described the disease or who described some feature later deemed to be part of a greater whole. It is now common to trace the history of coronary thrombosis to William Heberden's description of angina pectoris and then recount the lives and work of John Hunter, Edward Jenner, Caleb Parry, and numerous other figures who, according to the author's persuasion, receive varying degrees of emphasis. Leibowitz, in his important study, *The History of Coronary Heart Disease*, paid particular attention to Carl Weigert and his description, in 1880, of pathological changes in the heart associated with coronary occlusion. Many authors acknowledge the work of Adam Hammer, a German physician practicing in America, who, according to Leibowitz, made "the first diagnosis of coronary occlusion . . . during the life of the patient."[5] All these precursors are, in some way, deemed to have been engaged in the enterprise of defining coronary thrombosis.

Thus for the historian writing within this tradition there was by 1900, or even earlier, an overwhelming amount of clinical, pathological, and experimental data pointing to the existence of the syndrome. As Leibowitz put it:

In the course of our investigation it has become evident that observations leading to deeper insight had become increasingly overlapping and more difficult to separate. In fact, coronary thrombosis, as shown already, had been repeatedly presented in the literature by most impressive descriptions.[6]

By framing the question in this way the historian is encouraged to seek out factors that caused the "delay" in the disease's description. The factors usually invoked to explain this delay commonly include the "errors" or "mistaken beliefs" of previous observers. Thus, H. A. Snellen, who delineated the history in much the same way as Leibowitz, noted "it is difficult to imagine the great obstacl :s that prevented the recognition [of coronary thrombosis]."[7] Bruce Fye, in one of the best histories of the disorder, explicitly adopted the delay framework.[8]

Historians usually single out an American, James Bryan Herrick, as *the* describer of coronary thrombosis. For instance, Ralph H. Major, in his *Classic Descriptions of Disease*, declared that it was Herrick who had "cleared up the confusion surrounding this disease."[9] Willius and Keys, in their *Classics of Cardiology*, reproduced a paper of Herrick's published in 1912 and argued that, until Herrick, physicians had mistakenly believed that coronary thrombosis always caused death.[10] Also claiming Herrick's centrality, Leibowitz advanced yet another solution to the coronary thrombosis problem, arguing that Herrick's 1912 paper (Willius and Keys's "outstanding contribution") was ignored by clinicians and that it was a 1919 paper by Herrick that put coronary thrombosis on the clinical map.[11] In this paper, Leibowitz observed, Herrick correlated the damage of infarction with electrocardiographic (ECG) changes, thus demonstrating the reality of the disorder. Here, then, is a technological solution to the problem. Earlier clinicians lacking the ECG could not identify the condition. Similarly, Joel Howell more recently asked why the distinguished English physician Thomas Lewis failed "to recognize . . . application of the ECG in the diagnosis of . . . myocardial infarction" when, in 1918 in America, James Herrick "had described the ECG changes in myocardial infarction,"[12] I suggest below that these claims about the determining role of the ECG are incorrect, and that evidence is wanting that Herrick described what all agreed were "the ECG changes" of infarction.

In contrast to these approaches I argue that the focus of inquiry should be on how communities come to frame or see diseases, indeed, in some cases such as smallpox and Down's syndrome, to see them so clearly that nonmedical individuals can recognize them.[13] What is required is an explanation of how and why perceptions are structured as they are and how and why they change. The establishment of coronary thrombosis, I suggest, was a complex, socially sustained, reclassification procedure. It was not a negative process of removing obstacles but a positive restructuring of clinical and pathological experience. Further, the features held to be characteristic of the disease were not suddenly recognized but were arrived at by a process of negotiation and persuasion over a period of time. In other words, there was disagreement over whether the disease existed and what constituted its significant features. I argue that this is equally true of the ECG changes established by consent over some fifteen years. In the 1920s clinical and pathological investigations of heart disease offered the possibility of alternative interpretations, the construction of alternative diseases.

In giving this account of the appearance of coronary thrombosis, I employ Mary Hesse's network theory and the sociological twist given to it by David Bloor.[14] I suggest that in the reclassification process that gave rise to the disease the cognitive and social interests of cardiologists and academic physicians determined the emergence and form of a new disorder. These interests led to the creation of a syndrome that offered great potential for epidemiological work, had therapeutic possibilities, and could be the focus of pathological, physiological, and technological research. I shall also suggest, however, that in defining coronary thrombosis as a "definite and material" disease these clinicians were also defining their own social identity.[15] The demarcation of coronary thrombosis was a means of signaling their own medically elite status. Conceptual redefinition was part of the process of specialty formation.

It is necessary to note that viewing coronary thrombosis as a disease entity that exists only as part of a classification network, sustained by the interests of a particular group, does not devalue the experience of patients or the knowledge of the medical community. The patient's experience is, in a sense, open-ended. Pain, collapse, or any other symptom is real and terrible enough and the basic stuff of medical classification. But medical knowledge does not simply sort out such events into preexisting natural categories.[16] Medical knowledge in general, and of heart disease in particular, exists *because* of its constructed and interested character, not in spite of it.

Cardiac Pathology 1870–1930

Here I explore concepts of cardiac pathology as evidenced by a number of works used in Britain and America before and during the period when the coronary thrombosis was delineated. The first edition of the pathology textbook of a Charing Cross Hospital physician, Thomas Henry Green, appeared in 1871. Green defined nutrition as the fundamental taxonomic category of pathology. Pathological states were the consequence of increased or decreased nutrition. Green, however, had little to say about the heart. The tissues of the heart, like those of any other organ, could be the victim of nutritional derangements. Inadequate nutrition, he observed, led to "an alteration in the quality of the tissue," giving rise to the disorder "fatty degeneration." This disorder was frequently caused by a "Diminished supply of blood," and, Green noted, "In the heart . . . this form of degeneration is not an infrequent

concomitant of disease of the coronary arteries."[17] Of the clinical manifestations of this disorder Green said nothing.

Pathologists certainly identified at postmortem the condition described by Green and recognized it as a cause, or at least precursor,
of sudden death. In 1875 a man was brought in to St. Bartholomew's
Hospital dead after falling down in the City. The postmortem revealed
only, "The heart very flabby: muscular tissue soft and easily broken. . . .
Under microscope muscular fibres extensively fattily degenerated . . .
extensive calcification of coronary artery." The report concluded, "*Nature
of Disease* Fatty Heart."[18] Such reports seem to have been common.
Green's text, it should be noted, also described, in general terms, thrombosis and consequent local tissue death, or necrosis. However, he gave no
specific account of the condition in the heart.

Green's work went through numerous English and American editions. The seventh edition, in 1889, was revised and enlarged by
Stanley Boyd. Boyd explained how fatty degeneration could be seen in
the heart "as the result of atheromatous changes [degenerative narrow-
ing] in the coronary arteries."[19] Fatty change, he noted, could be quite
circumscribed and sudden, but he gave no indication of the cause.
Boyd, like Green, acknowledged that an arterial thrombosis could be a
cause of tissue death, although he exemplified this with cerebral softening rather than heart-muscle damage.

The eighth edition of Green's text, in 1895, was revised by another
Charing Cross man, H. Montague Murray. Murray elaborated further
on the pathology of fatty degeneration and included a new chapter,
"Diseases of the Circulation." Here he discussed the general problem
of "infarction." Infarcts, or tissue death following stoppage of the
blood supply, were of two types, red, occurring in the lungs, spleen, and
kidney, and white, occurring in the brain, retina, and "the muscular
walls of the heart." Murray then noted that "Cessation of function,
soon follows cessation of nutrition." In the case of the cerebral arteries,
obstruction was followed by loss of consciousness and "plugging of one
of the coronary arteries by sudden paralysis of the heart."[20] Here, at
least, seems to be confirmation of the view that earlier doctors regarded
coronary obstruction as always fatal.

However, Murray's pronouncement needs cautious handling. Contemporaries would almost certainly have read it as meaning a sudden
plugging of one of the two main coronary arteries, not the smaller
branches. Thus, Tanner Hewlett's textbook of general and special
pathology stated that arterial thrombosis was commonest in the cerebral and coronary arteries, adding that "Sudden occlusion of a large
branch [of a coronary artery] may cause sudden death." The text also
cited "anaemic necrosis" as the consequence of small branch occlusion.[21] It did not give symptoms.

The ninth edition of Green appeared in 1900, thoroughly revised by

Murray. It now said the heart was susceptible to white and red infarctions. In the section on diseases of the heart, a new disorder appeared—"Myomalacia cordis"—described as "the term applied by Ziegler to the occurrence of necrosed areas in the myocardium as a result of the local deprivation of arterial blood. . . . Usually due to thrombosis."[22] The necrosed areas could be either dark red or, in the absence of hemorrhage, yellow. Similarly, Beattie and Dickson's textbook of 1908 described "Infarction of the heart wall," most commonly caused by thrombosis in diseased coronary arteries. Infarcts were "irregular in shape and firm, and the central part . . . either yellowish or dull white."[23]

The pathological conception of infarction as contributory to the heart's decline is quite clear in these texts, but not all explicitly described and named myocardial infarction. Thus, the fourth edition of Sims Woodhead's *Practical Pathology*, in 1910, instructed the student on how to recognize such conditions as cloudy swelling, fatty infiltration and degeneration, brown atrophy, and fibroid degeneration of the heart.[24] A contemporary pathologist, of course, might have identified any of these as a sequel to infarction.

By the beginning of the second decade of the twentieth century, therefore, the literature on the diseased appearances of the heart was extensive. It included descriptions of cardiac infarction immediately following thrombosis and the appearances after healing. Yet it identified no special clinical symptoms or signs as associated with new or old infarction, except for occlusion of a major trunk when sudden death followed. The physiological literature described the possibility of immediate death after the sudden occlusion of *healthy* coronary arteries. However, physiological investigation had also shown that dogs with suddenly occluded coronaries could survive. In 1909 a much-cited paper concluded that "Either main trunk of the left coronary artery may be ligated [in dogs] without seriously disturbing the heart."[25]

Accounts of cardiac infarction in pathological texts changed little over the next two decades. Green's text, for example, was hardly altered. Suddenly, however, a remodeling appeared in the early 1930s. The fifteenth, "largely rewritten" edition of Green's text, in 1934, contained a new section, "Coronary Disease," with a subsection, "Cardiac Infarction,"[26] describing the new clinical syndrome. However, the text contained nothing new on pathology. William Boyd's new 1932 textbook contained five pages on "Coronary Artery Occlusion," stating that the "essential pathology" of the disorder was "an infarct of the heart, an anaemic or ischaemic necrosis." Infarcts Boyd described as "pale yellow in colour often surrounded by a red zone."[27] Boyd also gave a detailed account of the pathological change in the infarct from day to day.

In a morbid anatomical sense, this was nothing new. Pathologists had long given descriptions of new and old infarcts, acute and chronic cardiac degeneration, and the effects of reduced blood supply to the heart. Clinicians, however, were being asked to reorganize their clinical experience. From their everyday marshaling of symptoms they were being asked to isolate a new clinical disease based on an old pathological category, myocardial infarction. The next questions, therefore, are: How had clinicians organized their experience before this time, and how did they employ the pathological categories I have described in order to explain it?

Clinical Perceptions

The first edition of Thomas Savill's *System of Clinical Medicine* appeared in 1903. It included a chapter on diseases of the heart and its sac, the pericardium. Savill distinguished the three cardinal signs of heart disorder as breathlessness, dropsy (ankle swelling), and cyanosis (turning blue). In patients over 35 years of age with breathlessness but no other cardiac signs, among the many obscure disorders physicians should suspect was "Coronary Obstruction (*i.e.,* diminution of the calibre of the coronary arteries by atheroma, calcification, or other disease)." Savill here was referring to chronic obstruction, a condition that "can never be more than suspected during life." His classification of acute heart diseases also included angina pectoris, which he said afflicted males past middle life and was characterized by constricting pain in the chest, sometimes radiating down the left arm, often after exertion, which left the patient pallid and sweating. Attacks could last from two minutes to two hours or more. In a certain proportion of cases "death closes the scene." The pulse could vary from normal to feeble and could also be irregular. Angina, he explained, was symptomatic when an organic cause could be detected, idiopathic when it could not. "After death it is said that no structural disease of the heart and arteries may be found, though far more frequently the heart walls are found to be degenerated, flabby, or fatty." This was the fibroid or fatty degeneration well described in the pathology textbooks. Similarly, he observed that "Atheroma of the coronary arteries [may lead to] . . . thrombosis, and thus to a more or less localized degeneration of cardiac muscle,"[28] again a condition described in the pathology texts but not regarded by Savill as a distinct *clinical* disease. The cause of the pain in angina, Savill averred, was a damaged heart laboring to sustain output against some factor that suddenly increased the peripheral resistance to blood flow.

The third edition of Wheeler and Jack's *Handbook of Medicine and Therapeutics* appeared in 1908. It identified angina as a symptom caused by any condition interfering with the heart's nutrition. At postmortem "the walls and valves [of the heart] often show extensive morbid changes. The coronary arteries are extensively diseased."[29]

Angina might also be accompanied by no visible pathology, in which case, the authors said, embarrassment of the heart muscle followed peripheral vasoconstriction (sudden reflex narrowing of the body's small arteries). It is important to note here the primacy of the clinical category; failure to find heart damage at postmortem did not overturn a clinical judgment of angina. The fourth edition, in 1912, added a new section, "Diseases of the Myocardium," and stated that thrombosis in large vessels in patients with prior coronary damage could cause an "anaemic infarct, which may lead to aneurysm or rupture of the heart."[30] Editions of this text remained virtually unchanged through the 1920s. The first edition of Wheeler and Jack to appear in the 1930s (1932) included a short new section, "Coronary thrombosis," a condition that "must," the authors warned, "be differentiated from angina pectoris."[31] Savill's text had already introduced this new disorder two years previously. However, Savill still linked angina to infarction following thrombosis and did not specify any unique or distinctive pathological changes associated with the new disorder.[32]

More specialized works in the early twentieth century had a similar perspective to that of the general textbooks. The fourth edition of William Broadbent's *Heart Disease*, in 1906, indicated sclerosis or narrowing as the commonest affection of the coronary arteries. Sclerosis could drastically reduce the bloodflow to the heart, causing failure and consequent breathlessness. Sclerosis could also lead to thrombosis, which could produce wedge-shaped infarcts in the myocardium, usually "anaemic, or a pale yellowish colour." Thrombosis in a major trunk could cause sudden death; a nonlethal infarction could give rise to heart failure and breathlessness and was one of the causes of angina. Broadbent thought it commoner than generally supposed, although it had "no characteristic physical signs or symptoms." Fatty degeneration could also occur, presenting with breathlessness, palpitation, and angina.[33]

Thus, the early-twentieth-century clinician held that the heart could fail quickly or slowly, and a sudden failure could supervene on a chronic failure. Ailing coronary arteries were often to blame, producing chronic degeneration and acute pathological change. However, clinicians seem to have argued that so individual were the manifestations—one patient might be breathless, another in pain—and so poor the correlation—one patient's pain might be due to degeneration and/or infarction and another's accompany no apparent pathology—that the identification of exact clinicopathological entities was not feasible.

Part of the reason for the primacy of clinical categories and doubts about being able to correlate clinical signs and symptoms with distinct pathological changes lay in the significance then accorded to what was known as functional heart disease. Functional diseases were undetectable organic disorders of the nerves presumed to cause the observed symptoms. Such disorders featured prominently in late Victorian and Edwardian medicine.[34] They were not merely a catchall diagnostic wastebasket but an important category of disorders to which the heart was regarded as especially prone. Any physician confronting a cardiac case had to consider functional possibilities. For instance, as Savill noted in 1914, "A feeling of discomfort or constriction, or a sense of suffocation, is a symptom frequently present when the action of the heart is deranged by functional or structural diseases—oftener perhaps by functional."[35] Angina was often attributed to a functional disorder, for which cardiac degeneration could be a predisposing cause. Importantly, functional disorders exhibited ordinary clinical manifestations (Were palpitations a nervous or an organic cardiac disorder?) and shaded imperceptibly (as in angina) across the boundary between organic and inorganic phenomena. In no sense were functional derangements illegitimate diagnoses, or the preserve of nerve specialists and beyond the domain of the general physician. In 1903 the American author of a text on heart disease noted that the term functional may be problematical but was "sanctioned by usage." He added that

> It would no doubt be more in accord with the pathology of these cases to relegate these so-called cardiac neuroses to the domain of neurology, where they properly belong; but the symptoms calling attention to the heart are so often the dominant ones that they mislead the patient into the belief he has heart-disease. Indeed, the correct interpretation of the sensations is often puzzling to the physician, and hence it is customary to consider these cases in works of this kind.[36]

Physicians did of course speak of distinct clinicopathological entities, rheumatic disorders, for instance, when dealing with heart diseases, and some argued that various other cardiac symptoms also might signify such entities. For one, Clifford Allbutt claimed angina was an aortic root disease.[37] Indeed, *conceptualizing* a clinicopathological syndrome around coronary thrombosis was also perfectly possible. In 1887 the Glasgow physician John Lindsay had written that "fibrous transformation of the heart wall" had a "relationship . . . to obstructed coronary circulation" and

> one of the objects which I have at present in view is an attempt to establish that in [this disorder] . . . we have a diseased condition which has as much right, from a clinical as well as from a pathological point of view, to be considered an independent affection of the heart as fatty degeneration.[38]

Nothing could make clearer the point that *no cognitive obstacles prevented coronary thrombosis from being recognized as a disease entity.* Establishing it as a routine diagnosis was quite a different matter.

Angina Pectoris

How then did early-twentieth-century clinicians reorganize their experience to include a new disease, coronary thrombosis? Part of the answer lies in an area of clinical experience later claimed to have been misconstrued: the perception of angina pectoris, or intermittent crushing chest pain.

In the early twentieth century clinicians declared angina to be relatively rare and puzzling, as in the frequently cited 1910 Lumleian Lecture by William Osler. Osler thought the "disease," as he called it, uncommon, a consultant in active practice seeing only 10 or 15 cases a year. Angina could present in varying degrees of severity and its causes could range from functional to demonstrably organic with every shade in between. In its mildest form, pseudoangina, it was "emotional and muscular" in origin. A less mild form was related to neurotic, vasomotor, or toxic factors. In its severest form its clinical presentation was varied indeed. Some patients had chronic angina for years, others recovered completely after one or two attacks. In the worst cases, "status anginosis," death occurred during the first or second episode. Only three of Osler's 225 cases of severe angina died in the first attack. At postmortem of severe cases Osler found that the coronaries were usually at fault, and complete blocking of a branch was very common in instances of sudden death. Some patients, he observed, may "live for some time," the infarct being recognized by consequent pericardial friction. Yet he alluded to no simple clinicopathological correlation or the possible value of distinguishing a particular anginal episode as a distinct disease. All senior clinicians, he stated, could furnish cases of patients with narrowed and occluded coronaries who had no symptoms, and of others who suffered angina without pathology. As far as the attacks themselves were concerned, the symptoms had no prognostic value. A mild attack could be followed by a severe one, or a "man may come out of a state which seems absolutely desperate."[39]

Clinicians on both sides of the Atlantic marshaled pathological evidence to show the protean nature of angina. In America in 1918 E. R. LeCount published a series of 60 postmortems of sudden deaths from "difficulties with the circulation of the blood in the coronary arteries." Thirty-four had "fibrous myocarditis," in which the heart was badly scarred and had other degenerative lesions. The other 26 were associated with obstruction, usually thrombosis, at the root of the coronary artery. Most had large areas of infarction. "It may be assumed," he observed, "that none of these patients suffered from angina pectoris."[40] LeCount concluded that his cases showed that sudden obstruction could cause death (to which everyone agreed) and that severe coronary and/or myocardial disease was not necessarily associated with angina but could itself cause sudden death. LeCount's article, which made no argument for a new clinicopathological syndrome, was later used to support the view that there was such a syndrome.

This was also the case with numerous postmortems Sir James Mackenzie cited to indicate the range of pathologies underlying angina. Mackenzie, the elder statesman of English physicians interested in the heart, held angina to be a symptom of what the heart could, or could not, do. Death occurred when the heart could no longer function because of the absence of a suitable blood supply. The coronaries were, therefore, central to his view of angina. However, he threw cold water on the value of morbid anatomy, as it was then practiced, for revealing anything useful about the living but sick heart. He said that "I have long wished that we could be able to correlate the functional efficiency of the muscle with its morbid state, so as to know whether 'friability' after death has any relation to functional inefficiency."

Mackenzie's cases included all the types described by Osler, from those with regular pains ending in a spectacular episode to those having cardiac pathology without symptoms. In case one, for example, numerous episodes of pain induced by exercise were succeeded by a final massive attack. Postmortem found a branch of the left coronary artery calcified and "obliterated by clot . . . the heart muscle supplied by it had undergone necrosis."[41] The caption accompanying the figure explained that "This change was evidently due to progressive narrowing of the vessels and the eventual obliteration of the main artery."[42] The pathological process was deemed to be continuous. In the second case, anginal pain had disappeared, only to be succeeded by breathlessness. The pain eventually returned, and the patient finally dropped dead on the street, the postmortem revealed "old standing sclerosis of the [coronary] artery, leading to fibrous changes, and in addition a recent obstruction of one of the vessels causing acute degeneration of a portion of the heart wall."[43] Mackenzie's catalog went on through the spectrum of symptoms, pain and dyspnea, signs of heart failure, and the pathology of sclerosis, fibrosis, and infarction. There was no simple rule. Each patient presented and progressed in an individual way.

Early in this century then, angina was quite frequently discussed. However, its clinical and pathological meaning embraced a spectrum, from localized organic changes to functional disorders. Part of the process of categorizing coronary thrombosis was effecting a change in the meaning of the term angina—eventually, anything that was not diminished blood supply to cardiac muscle (nearly always linked with coronary sclerosis) was not angina.[44] Relatedly, coronary thrombosis and myocardial infarction with chronic scarring had been well described at the turn of the century as both independent and associated pathological states. These pathological states, however, were not considered the basis of a distinct disease. Clinicians did not start from the pathological entities and organize their clinical experience to conform to them. Rather, things ran in the other direction. Clinical experience began with pain, dyspnea, and heart failure, for which these pathological conditions might sometimes, but not always and not consistently, account. Myocardial infarction subsequent to coronary thrombosis did not define a clinical disorder; rather it was one among a number of conditions—functional and organic—invoked to explain a misbehaving heart. This organization of clinical experience was broken during the twenties by creation of the distinct clinical entity—coronary thrombosis.

Making Coronary Thrombosis

I suggested above that clinicians earlier viewed the symptoms now attributed to several distinct heart diseases as a continuum and interpreted pathology in accordance with this view. How then did a new syndrome become an important fact of their clinical experience during the 1920s? The significant factor, as suggested at the beginning, was a slow negotiation of the features of the disease as part of a reclassification process. In Mary Hesse's network theory, knowledge production is regarded as a classificatory activity, the framing of laws to relate different sorts of objects. Such laws structure our expectations. Just as the objects of classification are organized by the formation of laws, however, so laws in turn are organized into higher systems, or networks. In consequence, any new object may be assigned to a different part of the network according to which of its properties are deemed significant. Thus, we can imagine a classificatory universe in which the discovery of the whale creates the problem of whether to allocate it to the category of mammals or fish. As Bloor noted "Resemblance alone pulls us in opposite directions, so the choice cannot be resolved by experience." More generally

> The lesson to be learned from this example is that the functioning of a name or predicate cannot be explained fully by similarity relations but depends on the laws into which they enter. Conversely the laws will depend on conventions about the boundaries of the classes they relate. This reciprocal dependence: the discretion it allows, and the choices it forces upon us, are completely general phenomena. They do not just apply to new and problematic cases which—like the whale in the example—have yet to be assigned to a class. The principle applies to existing classifications and can lead to their reassessment. New contingencies can always prompt retrospective revisions.

Nothing, it follows, has privilege in the networks. Even cases regarded as exemplars of a class may be reclassified. Thus, just as alloys and salt were once both held to be compounds because both were an intimate blending of constituents, so adoption of the law of constant proportions meant that an apparently paradigmatic member of the class—alloys—had to be dropped. Bloor concluded:

> The general picture is now emerging. The appropriate verbal description of an object is always a matter of context as well as direct confrontation. There is no "direct" verbal rendering of experience; it is always mediated through a network of laws. Like analogies and metaphors, which are but special cases, laws repair the fragmentary character of experience; they highlight real or imagined patterns in it, and prompt and justify threshold adjustments. They act as selective filters while permitting us to impute an inner nature to things. . . . All the elements of this network of classifications are equally open to negotiation, and equally the outcome of such processes.

It is important to recognize that a network is not a "free-floating system of thought"; classification decisions "are made with reference to the world and in the light of experience."[45] However, since classification is a simplifying procedure involving a choice about which properties of an object are significant, the possibility of straightforward correspondence between classification and an indefinitely complex reality is lost. It is here, drawing on Mary Douglas, that Bloor gave the argument its sociological turn. In brief, he argued that we must look to the network users' social interests to understand which features of an object—which resemblances—are chosen when the object is classified. Similarly—for it is part of the same process—it is to social interests we must look to discover which parts of the network are protected, and kept stable, and which are selectively manipulated.

It is now possible to investigate how aspects of the earlier classification were manipulated and clinical experiences assigned to the new category of coronary thrombosis. The process was quite complex, since a pathological state—coronary thrombosis—was raised to the status of a new classifying element (or law, in Bloor's sense) for clinical experience. Clinicians were then asked to place patients in a new category defined by immediate signs and symptoms. Patients who might once have been assigned to such categories as, for example, further evidence of chronic heart degeneration, anginal attack, first episode of heart failure, and non-organic functional attack, were now to be distinguished as having a definite and nameable *disease*. What is striking is how the new classification system was defended despite what now appears as striking disagreement over the salient classifying features.

Framing the new disease involved an assertion that it had, and always had had, an existence. Thus, its existence was grounded in its having a past. All previous cardiac investigation showed that the disease indeed had a history. Herrick presented an impressive quantity of historical evidence that his predecessors had come within an ace of recognizing the syndrome.[46] Joseph Wearn of Boston, writing in 1923, traced relevant investigation back to Harvey and then showed that the meanings of clinical experience for his forebears were the same as his own—for instance, "It is to Osler . . . that we are indebted for the first real correlation of the clinical and pathologic findings."[47] By creating a history for the disease it was possible to diagnose it retrospectively in the works of the great clinicians. In 1925 J. W. McNee of Baltimore claimed that he was able, with difficulty, to "sift out [from Clifford Allbutt's book] the clinical accounts which probably refer to the special group of cases described here."[48] In 1926 White discovered it in 18 of the 160 cases of angina described by Mackenzie.[49]

Argument that the disease had a real history and that great clinicians had nearly, but not quite, described it, made it necessary to show why they had been misled. One explanation imputed false belief, notably Herrick's view that before him it was commonly held that "this condition was almost always suddenly fatal."[50] However, other explanations were possible. Wearn suggested that previous failure to spot the disease originated in a mistaken estimate of incidence. Formerly, he said, it was "classed among the rarities of medicine . . . most of the textbooks of medicine . . . dismiss it with a brief paragraph." Wearn, however, asserted the disorder's relative frequency: "As a matter of fact the condition is not a very rare one; it merely goes unrecognized during life."[51] Not all clinicians agreed. Herrick, for example, said that "Patients with this condition do not present themselves very often."[52] Here then were the clinical origins of the coronary thrombosis problem inherited by historians. What are now harmless historical explanations for the absence of the disorder in the earlier literature were, in the 1920s, ways of bringing the disease into existence. Its previous nonexistence, in spite of notable delineations, was deemed to be a failure to recognize it.

These assertions about fatality and incidence were significant beyond demonstrating why the disease was formerly unrecognized. They were also *positive* assertions about the character or qualities of the disorder that enabled it to be recognized—people survive it, and it may or may not be common. These were classifying features. It is striking, however, that despite agreement that the disorder existed consensus was markedly lacking as to its symptoms and signs. Authors did agree that it was more common in males than in females and more prevalent at ages over forty, but propositions about the social class most at risk were more contentious. In Britain it was suggested that this class was underrepresented in the public wards of a hospital, which primarily tended the poor; that is, the middle and upper strata were presumed to be the most susceptible but did not appear in records. The same was also commonly said of America, but, Wearn observed, "The idea has gained prevalence . . . that the disease is almost always limited to the upper classes," whereas his cases included machinists, housewives, and, he added with some satisfaction, "one man who gave as his occupation 'Christian Science Healer.'"[53]

If incidence, age distribution, and class were no sure initial guide to recognition, what aids did the physician have? Again it is striking how authorities differed. Herrick, an experienced clinician, declared that "No simple picture of the condition . . . can be drawn."[54] Henry Christian of Boston, however, recorded that the condition was "easily diagnosable" and that his fourth-year medical students "adopt or eliminate [the diagnosis] with a very considerable degree of accuracy."[55] J. W. McNee claimed "the clinical picture is so characteristic that it could scarcely be missed."[56] However, the symptoms ascribed to the condition varied widely. The only consistent agreement on symptoms in the literature was that many but by no means all cases exhibited severe, persistent pain in the abdominal area or chest.

Other symptoms of this "easily diagnosable condition" were variously described. Herrick stressed the importance of pain, but acknowledged a report of a painless case.[57] He also noted that pain was frequently preceded by other symptoms. Wearn, in 1923, reported to the contrary that they were infrequent. Wearn also distinguished three sorts of presentation: seizures of pain, shortness of breath, and instances in which antecedent heart failure masked the onset. Three of the four patients Wearn placed in this last group he described as "irrational."[58] The three patients had dyspnea and cyanosis but no pain. Pain, said Parkinson and Bedford, was "usually sternal, most often in the middle or lower third."[59] McNee wrote that "The distribution of the pain is irregular and peculiar."[60] Wearn said pain could occur in the epigastrium, over the heart, under the sternum, or at all of these sites and "in other regions"; it could radiate to the arms, nipples, and "various other places." He described the pain as characteristically persistent, but added that it "may be intermittent with periods of complete absence."

Diagnosis, Wearn asserted, depended on physical examination since "there are several signs in this condition which occur with remarkable constancy."[61] Parkinson and Bedford, however, held that "Physical examination at this early stage reveals little."[62] In 1920 Whittington Gorham considered pericardial friction a valuable sign, and Savill, in 1930, held it was "usually present."[63] But Christian maintained that pericardial friction was "unfortunately rarely present."[64] Wearn and McNee thought the heart generally enlarged;[65] so did Christian, but he remarked that it "scarcely helps."[66] Parkinson and Bedford stated that the heart may appear normal.[67] Christian said blood pressure was "practically always lowered,"[68] but Wearn reported "wide variation."[69] Gibson reported "no constancy in the effects of occlusion" and a "fast or slow" pulse that could suggest cardiac failure but "nothing else."[70] Irregular heart rhythm (arrhythmia), according to Christian, was "present in a majority,"[71] and Henderson reported the pulse "is rapid, soft and irregular,"[72] while Parkinson and Bedford maintained that "usually the rhythm remains normal."[73] Also, Parkinson and Bedford "rarely" heard murmurs (abnormal heart sounds), but over a third of

Wearn's patients had nonrheumatic murmurs.[74] Wearn reported fever was commonly present, but "Other patients had subnormal temperatures."[75] Slight fever and a raised white cell count, said Christian, were "almost invariably found."[76] McNee agreed but concluded they "are chiefly important from their liability to lead the clinician astray,"[77] while Parkinson and Bedford thought fever "a useful diagnostic sign."[78]

As these by no means atypical examples imply, the clinical definition of coronary thrombosis was a complex temporal process in which some clinicians tried to persuade colleagues to reclassify their clinical experiences, to denominate as a new collective entity certain findings previously classified in other ways. By accounting for the description of the syndrome in this way, rather than characterizing it as a simple accretion of knowledge, I argue that no previous unproblematic natural category determined the description of the disease. This is not to say, however, that knowledge of heart disease did not increase. It certainly did. However, it was knowledge gained by classifying heart disease in one particular way rather than another, thus precluding alternative approaches.

Alternative classifications were indeed possible. During the twenties, rather different, pathologically driven organizations of the clinical world were suggested. In 1925 M. H. Nathanson, of the Department of Medicine at the University of Minnesota, proposed, on the basis of postmortem work, the category coronary artery disease as the basis for organizing clinical experience. Whereas Osler and Mackenzie had written first about clinical events for which coronary disease was a possible pathology, Nathanson regarded this as regrettable: "It is unfortunate that various types of cardiac diseases are named and classified by the clinician without regard for the underlying structural changes found in the heart at necroscopy." Postmortem, he argued, showed coronary disease as "one of the frequent causes of death in individuals above the age of forty years." However, in light of the fact that patients with sclerosed coronaries often had been free of angina, while patients with clean coronaries often had complained of chest pain, angina "cannot be accepted as the constant clinical expression of coronary disease."

Although Nathanson acknowledged Herrick's work on coronary thrombosis, he lamented that it had distracted attention from the wider entity. He also cited the responsibility of Mackenzie's work for failure of the pathological project to make headway. The aim, Nathanson said, was "to construct a clinical picture of coronary sclerosis from the analysis of the clinical data of a series of fatal cases which came to necroscopy." The symptoms of Nathanson's entity included attacks of pain and respiratory distress and clinical data related to heart size and the presence or absence of congestive failure. Nathanson claimed he had found that "certain clinical manifestations are quite constant for the group as a whole," and identified four clinical subgroups. Twenty-three of Nathanson's cases exhibited thrombosis at necropsy and clearly fitted

Herrick's clinical classification, but the same was true of forty cases of "coronary sclerosis without a thrombosis." Herrick's clinical category, in other words, encompassed both thrombotic and nonthrombotic cases; that is, "the clinical features of coronary thrombosis are essentially similar to those of coronary sclerosis." Nathanson concluded, "Since coronary thrombosis is constantly associated with sclerosis of the vessels, it is most reasonable to regard a thrombus not as an entity, but merely as one of the end results of coronary disease."[79] Mackenzie might have agreed, although for entirely different reasons.

In fact, because Nathanson's system was not chosen we have no way of knowing how useful—therapeutically, for example—it might have been. On similar grounds some physicians, while acknowledging coronary thrombosis as a pathological entity, continued to be reluctant to elevate the disease to a simple clinical status. They questioned the value of so organizing clinical experience, either because it seemed to have no particular beginning or end, or because what mattered most was the patient's incapacity. This was Mackenzie's problem, echoed by Paul White in 1931 in a rather different way in the first edition of his *Heart Disease*:

> Coronary thrombosis responsible for an acute myocardial infarct large enough to be readily recognized clinically has not been wholly divided off in this book from the discussion of coronary disease in general, as is often done, since such separation is artificial and rather misleading. Such coronary thrombosis is but one phase of coronary disease, although a striking one. Sometimes thrombosis and even extensive infarction, if slow in development, may occur with a different clinical course from that ordinarily labelled coronary thrombosis, and even clinically recognizable acute myocardial infarction may assume a variety of pictures.[80]

The ECG

By the early 1930s, textbooks described coronary thrombosis but variously emphasized different symptoms and signs. By arguing that a description of the disorder was only slowly arrived at, I have tried to suggest that coronary thrombosis was not "discovered" or "described," in the common medical sense of the words, but was "made" by clinicians gradually coming to agree on the conventions defining the entity whose existence they affirmed. I will return to the reasons for this shift in medical perception. It is first necessary to assess the technologically determinist argument that the ECG was largely responsible for the definition and acceptance of the syndrome. This claim was made explicitly by Leibowitz and is implicit in Howell's view that after Herrick had described "ECG changes, the syndrome of myocardial infarction was rapidly accepted in the United States."[81] I maintain that ECG signs of coronary thrombosis were, like all other signs and symptoms, negotiated. ECG signs were the subject of dispute in the 1920s. Only *after* agreement had been reached was the ECG credited with the objective definition of the disease.

The ECG had come into limited use in the early twentieth century. By the end of the first decade it was agreed that it recorded the heart's electrical activity, depicting a characteristic complex (P, QRS, and T waves) that mirrored the temporal sequence of the heart's muscular contractions. An inverted T wave, usually upright, was held to be a sign of severe heart disease. Today it is taken as highly suggestive of old infarction. In 1915, at a symposium of American physicians, Herrick asked if there was a "definite explanation of the inversion of the T wave."[82] No mention of coronary thrombosis was reported. He was told that the inverted T had no diagnostic specificity. The physician Alfred Cohn said it was not "necessarily . . . significant" but "it is seen so regularly in so very many serious lesions that it is usually regarded as a bad prognostic sign."[83] In 1918 Herrick's colleague F. M. Smith published a paper describing ECG changes following ligation of the coronary arteries of dogs, displaying records taken at intervals from five minutes to thirty-eight days after ligation. The findings included the fact that "changes in the T-wave following the ligation of any branch of the left coronary artery were among the most constant." Smith noted that usually "immediately following the ligation the T-wave became more prominent. . . . Within twenty-four hours it became sharply negative."[84] Within four weeks the wave became positive, then, in one or more leads, negative again. Howell states that Herrick, also in 1918, described "the typical ECG changes of myocardial infarction."[85] This claim does not seem warranted. Herrick, in his papers of 1918 and 1919, said that Smith's work showed that there "seems to be a fairly consistent variation in the electrocardiogram following the lesion of a particular branch of the coronary."[86] Herrick's paper included a reproduction of two of Smith's ECG recordings, taken 2 and 34 days after ligation of branches of the left coronary of a dog. Both showed inverted T waves. He also reproduced two ECGs showing inverted T waves, taken 41 and 178 days after a clinical episode diagnosed as coronary thrombosis, in a patient found at postmortem about six months after the original attack to have an organized clot in the coronary artery and a "gristly" left ventricle.[87] Herrick described no other ECG signs, certainly none occurring at or around the time of the presumed infarction. American physicians, it might be suggested, would hardly have been amazed by this paper or the pictures. They would all have been familiar with such a patient, a man suffering severe chest pain, followed by a six-month history of extreme shortness of breath, and with various grave cardiac signs, who had inverted T waves on an ECG.

In 1920 another American physician, Harold Pardee, published a paper often cited as a classic work on the ECG signs of coronary thrombosis. Pardee's paper described a patient with anginal pain and various other symptoms who eventually died. He argued, on clinical evidence alone (since the man did not come to postmortem), that the patient had suffered coronary thrombosis during a particularly severe attack of substernal pain. He reported that the significant feature of the ECG twenty-four hours after obstruction was "the extreme height of the T wave and the fact that this wave starts from a point on the QRS group well away from the base line."[88] The raised ST segment, as this change became known, is now taken as a possible sign of very recent infarction. Pardee stated that his observation was in line with the results of two other workers, Eppinger and Rothberger, and with those of Herrick's colleague Smith.[89] Pardee reproduced one of Smith's curves, taken 5 minutes after experimental obstruction of the artery in a dog, which (to us) clearly shows a raised ST segment, although Smith had commented only that the T wave became "markedly elevated" and "more prominent."[90] Later, Pardee observed, the T wave in his patient became inverted.

Since the patient did not come to postmortem, Pardee did not correlate the signs of immediate and old infarctions with pathological evidence. He did, however, generalize on the typical sign of old infarction, "the QRS group is usually notched and usually shows left predominance: the T wave starts from a point near but not directly on the base line, and in one lead . . . quickly leaves it in a sharp curve . . . the T wave is usually turned downward in two of the three leads." He presented five ECGs with these signs. Of the five patients, three had suffered "typical attacks of anginal pain," and one had had "precordial discomfort which was felt on walking."[91] None of these four, in other words, had presented with any sort of clinical episode definitively identified as coronary thrombosis. The fifth case was Herrick's, the only one to have had a postmortem. Significantly, Pardee acknowledged that an inverted T wave was not exclusive to old, acute infarction but could be produced by slow degeneration of heart muscle following gradual narrowing of the coronary lumen. Pardee's work thus contained the suggestion that Herrick's "consistent variation" of the ECG in infarction had other possible causes.

Pardee's paper may now have the status of a classic, but it is hardly surprising that it did not transform overnight the myocardial literature of the 1920s. Evidence is wanting that Herrick's and Pardee's "signs" were soon adopted as definitive.[92] Authors of the time displayed great uncertainty about the ECG signs of coronary thrombosis. Wearn found abnormalities of the QRS complex in three of his ten cases and in all ten "some alteration of the T waves in at least one of the leads," but no consistency. "Many were isoelectric while others were inverted or diphasic, but in one record only was the T-wave found to come off the R-wave as described by Pardee." This deflection occurred only in a single lead on just one day. "[I]t is obvious", said Wearn, "that no one form of electrocardiogram is characteristic of this condition."[93]

Thus, by the midtwenties the ECG signs of infarction had not been agreed on. "It is evident," wrote McNee in 1925, "that there is no special electro-cardiogram characteristic of coronary thrombosis but many deviations from the normal may be found."[94] Although historians have declared the syndrome to have been defined by the ECG, and by Herrick's work in particular, it is quite clear that many authors at the time held traditional bedside techniques to be the best guide to diagnosis and the ECG untrustworthy. "Cardiac infarction," Henry Christian wrote, "stands out as a clean-cut clinical entity easy of recognition. . . . Simple bedside methods of examination suffice to make the diagnosis in most cases." This view, as we have seen, was hardly supported by the practice of other clinicians. Christian agreed that "In many patients the electrocardiogram does show changes in the form of the ventricular complexes which are almost pathognomonic." He did not, however, describe them, and added, "it is to be remembered that the electro-cardiogram may be entirely normal in patients with cardiac infarction." The "history and other findings" were paramount.[95] In Britain things were no easier. In 1925 Alexander Gibson reported, "The electro-cardiogram has been explored as a means of getting greater accuracy [of diagnosis]."[96] However, he made no reference to the work of Herrick or Pardee but cited abnormalities of the R waves described by Carter, and A. N. Drury's refutation (see below).

The problematic status of ECG evidence of infarction in the early 1920s suggests why the absence of an endorsement of Herrick's work by Sir Thomas Lewis cannot fairly be described as one of his "failings."[97] Howell has argued that Lewis, trained as he was by Mackenzie, used the ECG as a sophisticated polygraph, that is, a simple amplifier, to investigate the heart's rhythm but not the state of its muscle and thus failed to see the significance of changes of shape in the complexes.[98] More convincingly, Howell has also suggested another reason that Lewis did not acknowledge Herrick's work. Lewis was interested only in technologies that provided information about clinical signs accessible to the trained but unaided senses. An inverted T wave, of course, had no clinical equivalent.[99] On evidential grounds, however, it was perfectly reasonable in the 1920s for Lewis not to commit himself to the view that myocardial infarction induced characteristic ECG changes. Many other leading physicians had not done so.

Indeed it would still have been possible at that time to challenge claims that a distinct syndrome existed. Some in Lewis's circle even saw convincing evidence that there were no typical ECG changes. For example, in 1921 A. N. Drury, one of Lewis's co-workers, published a paper in Lewis's journal *Heart*[100] refuting the work of Edward Carter and of Oppenheimer and Rothschild. In 1916 Oppenheimer and Rothschild had reported consistently abnormal ECGs in a number of patients. They described the QRS complex as abnormally long and the

R wave as notched. In many instances, they claimed, these changes followed sclerosis and closure of the left coronary artery with consequent fibrosis of the left ventricular wall. They called the ECG pattern "arborization block."[101] These ECG changes were quite different from those described by Pardee, who had suggested that a similar pathological condition ("coronary obstruction [which] may arise gradually by a narrowing of the lumen of the vessel") could cause what he regarded as typical ECG features.[102] In 1918 Edward Carter in Cleveland reported a series of cases of cardiac sclerosis with ECG changes similar to those described by Oppenheimer and Rothschild. He found prolongation of the QRS complex with "bizarre notching," inversion of the T wave in lead I, and inversion of the P wave.[103]

A. N. Drury published his work to refute these conclusions, addressing the question: Was there a characteristic ECG appearance in patients with myocardial damage? He presented the case of a man who for two years had experienced precordial pain, palpitation, and dyspnea, which was "latterly brought on by slight effort." An ECG was taken in 1919. The man was then admitted to the hospital the following year with cardiac failure and died shortly afterward. A postmortem showed "Extensive fibrosis of the muscular tissue at the apex of the left ventricle," and "The descending branch [of the left coronary artery] was stenosed by a fibrotic lesion, probably of a few weeks' standing, which rendered the vessel almost impervious." He had had a myocardial infarction following a coronary thrombosis, a diagnosis Osler, Mackenzie, and Herrick would presumably have agreed on. The first ECG of 1919 was "normal."[104] The second curve, taken on admission to the hospital a year later, showed auricular flutter and no other abnormality. None of the features described by any of the American authors were found.

It is arguable therefore that Lewis's apparent "failure" to endorse Herrick's work was a straightforward acknowledgement that, in the 1920s, experimental workers and clinical observers had given different and conflicting accounts of the relation between muscle damage and ECG traces. Only if we assume that the consensus of the 1930s (that there is a syndrome with characteristic ECG features) existed in the early 1920s does Lewis's noncitation of Herrick or Pardee look like a failure.[105] In fact Lewis had worked on coronary ligation and had suggested that muscular damage would be likely to produce ECG changes and commented that "it cannot be expected that the curves will be exact duplicates of what are now regarded as normal."[106] As early as 1913 Lewis had noted that inverted T waves were pathological in humans. His *Clinical Electrocardiography* in that year stated:

[I]nversion of *T* in these leads (I and II) is always pathological, or nearly always pathological. . . . A good deal of attention has been paid to this deflection in electrocardiograms, and it is a helpful sign in prognosis. Certainly in my own patients, it has often been associated with signs or symptoms of ill omen; and often the patients who have presented it have been short lived. This experience accords with that of other observers.[107]

The question is complex, however, for both Howell's work and my own suggest that Mackenzie and Lewis did have particular ways of viewing heart problems such as angina that were repudiated rather than extended in the 1920s.[108]

Only by the end of the twenties was consensus building that coronary thrombosis induced characteristic ECG changes. Parkinson and Bedford noted of the ECG that "occasionally it provides the only objective sign of a cardiac lesion."[109] The characteristic feature they referred to was the plateau in the RT interval (Pardee's sign) and deep inversion of the T after a *few weeks*. This, the characteristic sign of 1928, was nothing like the supposedly "classic" sign reported by Smith and Herrick in 1918. Smith had reported inversion of the T wave after *24 hours* and that the T wave in a patient of Herrick's "ran a course similar to that of the dogs."[110] By 1930 this claim was no longer acceptable. In a much-cited paper Barnes and Whitten observed the "electrocardiographic changes observed by Smith, in dogs, do not parallel the phenomenon found by us in the human being following infarction."[111]

The point about consensus here is important. I do not wish to be read as implying that an entirely arbitrary relation exists between ECG changes and coronary thrombosis, that physicians could have picked any old change they liked and agreed that it represented infarction. When Smith and Herrick, or Pardee, or Parkinson and Bedford, or any other figures began their work, only certain selections or classifications were possible, just as it was with the clinicopathological entity itself. Interpretative flexibility was extremely limited but did exist, as evidenced by Pardee's revising the once unproblematic reports of Smith and Herrick and having his own reports subsequently reinterpreted by Barnes and Whitten. The work of all these figures was open to alternative interpretation, characterized in each instance by the increasing creation of order from relative disorderliness.

As Latour and Woolgar put it in 1979, "Scientific reality is a packet of order, created out of disorder by seizing on any signal which fits what has already been enclosed and by enclosing it, albeit *at a cost*."[112] The cost in classifying heart disease as coronary thrombosis was leaving the door open only to certain ways of investigating the heart (choosing and ordering certain sorts of signals) and closing it to others (Mackenzie's program, for example). The decision to classify constrained future decisions.

As was the case for other symptoms, the opinion that associated particular ECG changes with infarction had to be promoted by persuasion in the face of plenty of counterevidence and competing claims. For example, ECG changes were sometimes absent despite pathologically proven infarction and sometimes present when postmortem disclosed no evidence of infarction. In the construction of the ECG changes of myocardial infarction we see a typical instance of the production and "stabilization" of certain sorts of facts as facts of nature and their subsequent deconstruction (when no longer considered facts of nature) into local, social circumstances.[113] Smith and Herrick's fact—that a certain inversion of the T wave after 24 hours indicated human infarction— became Barnes and Whitten's artifact—that it was the curve for a dog.[114] To regard the ECG as an unequivocal producer of medical "facts" is to remove the social processes producing those facts.

Heart Attack

I have thus far not indicated the impetus that might have instigated the "making" of the new disease, coronary thrombosis, in the 1920s. Here my account becomes somewhat more conjectural. We must now consider the interests of the physicians who created and manipulated the classification network. Such interests need not be immediately social, in the sense of material. They may have been cognitive. The creation of coronary thrombosis seems to have been an extension and modification of a new pathophysiological approach to heart disease that developed in the previous decade. I have argued elsewhere that, in Britain, the men who took this approach based it on physiological grounds and self-consciously established themselves as experts in heart disease.[115] Many were leading figures in what they called the "new cardiology" and in the promotion of academic clinical medicine. Their goal was to correlate symptoms, such as pulse irregularity and palpitations, with demonstrable changes in physiological properties, such as cardiac rhythmicity and irritability, and, it was hoped, with underlying organic disorder. New cardiology had been particularly successful in the study of rheumatic heart disease.[116] The definition of coronary thrombosis represented an extension of this approach to degenerative disease, and to other sorts of symptoms, such as breathlessness and pain.

In 1919, in a review of past and present cardiac history, the "new cardiologist" Edgar Lea pictured what he regarded as the cardiologist's dilemma. In a chapter perhaps significantly entitled "The Loneliness of the Cardiologist," Lea lamented that the "divorce" between cardiac symptoms and the ability to correlate them with "objective evidence" was greater than in any other branch of medicine. He claimed that neurologists were in a particularly enviable position in this respect. He also stated that the "confused" relation between organic and functional disorders of the heart was a problem "of a kind not met with to anything like the same extent in other spheres [of medicine]." He recounted, however, that cardiological investigation had recently found that some disorders, such as paroxysmal tachycardia (intermittent rapid heart rate), "are now to be strictly considered as essentially evidence of cardiac organic disorder." This and other examples, he said, showed that the progress of cardiology lay in reducing functional disorders to demonstrable organic or physiological changes. Functional symptoms that could not be so reduced should be relegated to another sphere, beyond the interest of the cardiologist. Cardiology, he said, should be creating an organic base for "those multifarious complaints—pain, breathlessness, faintness, palpitation—the common symptoms which the laity fully recognise as meaning 'heart,'" Cardiology would then be rid of the "anaemic and pampered old spinster who consults doctor after doctor for her heart."[117] He questioned whether the term "functional" had the value in cardiology that it did in neurology.

Functional disease disappeared from the cardiological literature between 1910 and 1930. For instance, although texts still said tachycardia could be caused by "nervousness" or "neurosis," they nowhere suggested that it was a concern of the cardiologist except as cases to be excluded from the field of interest.[118] A consensus was also forming that angina was due solely to "acute deficiency of the coronary circulation."[119] Any chest pain that could not be so identified *was no longer angina*. The cardiologists were defining their field as constituted by demonstrably organic and pathophysiological disorders. They were ridding it of functional diseases, the cardiac neuroses, the protean disorders that sprawled across the whole of medicine, which any general practitioner of thirty years before would have felt competent to handle.

All the men defining coronary thrombosis were interested in promoting academic medicine and were leaders in the field, but many also wanted to make cardiology a separate specialty.[120] For instance, John Parkinson "Throughout the whole of his professional life . . . worked untiringly to establish cardiology as an acknowledged specialty within medicine."[121] Largely because these men were at the head of their profession, they were able to bring the disease into existence.[122] By publishing, teaching, and lecturing they were able to persuade other medical men to adopt their perspective.

The proposition that giving status to the heart specialist and establishing academic medicine involved creating a thoroughly organic discipline free of "functional" disorders is only part of the history of the appearance of coronary thrombosis. It explains the reason for the manipulation and organization of one area of cardiac knowledge but not the way in which it was manipulated. It indicates only that these men were striving to create more concrete organic syndromes, or organic bases for symptoms, in certain areas of their experience but not why they created that particular disorder. Why coronary thrombosis rather than some other clinicopathological complex (which, as we have seen, might have been justifiable from the evidence)? Part of the answer, of course, is like that given in the discussion of the ECG. Available choices were relatively limited by the direction these physicians had taken and were ultimately determined by local clinical circumstances: the use of existing diagnostic technology (especially the ECG), potential therapeutic opportunities (the clot after all eventually became the object of therapy), relating symptoms (angina) to specific organic changes, and epidemiological possibilities (coronary thrombosis could be fairly well defined for purposes of quantification).[123]

We can add a further explanatory layer. David Bloor's sociological account of network theory is an attempt to demonstrate in practice the claim by Durkheim and Mauss that "the classification of things reproduces the classification of men."[124] Such a natural reproduction of social categories can be seen in the creation of academic medicine and cardiology. At the turn of the century general physicians routinely treated heart and functional disorders. No rigid social boundaries marked off doctors treating heart disease any more than natural boundaries did heart diseases. During the second decade of the century the physicians determined to create an academic medicine and cardiology as a specialty attempted to designate as organic (or at least pathophysiologically demonstrable) a range of cardiac phenomena, notably pulse changes. In doing so they made finer distinctions between cardiac and noncardiac disorders. This process continued into the 1920s, when more recalcitrant symptoms, for example, angina and breathlessness, were molded into a concrete disease entity, coronary thrombosis. Anything not construable as organic or physiological was no longer cardiological. Lines were being drawn around a discipline and the diseases it studied. A definite (natural) entity reproduced the material arrangement of men.

Epilogue

The status of coronary thrombosis as a disease entity has recently been questioned by some cardiologists who prefer to see it is an "incident." Are we watching the beginnings of the disappearance of so definite, material, and natural a category as a heart disease? If so, how will future generations explain this disappearance?

ACKNOWLEDGMENTS

This paper was originally given in the Science Studies Unit at the University of Edinburgh, and I should like to thank the members of the Unit for their valuable remarks on that occasion. In addition, I am specifically indebted to Mike

Barfoot, David Bloor, John Henry, Peter Fleming, Charles Rosenberg, and Steven Sturdy for comments on an earlier draft. Joel Howell deserves special thanks for his helpful and thoughtful reply to my discussion of his work. Without Jacqui Canning at the word processor the paper could never have been finished.

NOTES

1. John Parkinson and D. Evan Bedford, "Cardiac Infarction and Coronary Thrombosis," *Lancet ii* (1928):4.

2. I shall use the pathological term "coronary thrombosis" to denominate a relatively simultaneous complex of clinical and pathological events— symptoms, clinical signs, thrombosis in a coronary artery, and myocardial infarction—each of which might occur separately. This usage is historically sanctioned and should not cause confusion as long as the proviso is clear. I shall indicate when I am using the term solely as pathological description. I appreciate that my definition of infarction is not quite correct.

3. C. F. Terence East and C. W. Curtis Bain, *Recent Advances in Cardiology* (London: J. and A. Churchill, 1929), p. 4.

4. Maurice Cassidy, "Coronary Disease," *Lancet ii* (1946): 588. The case is the same as highlighting how an idea "suddenly" occurred to a scientist as opposed to diffusing the idea into a social network. See Bruno Latour and Steve Woolgar, *Laboratory Life: The Social Construction of Scientific Facts* (Beverly Hills, CA: Sage, 1979), p. 170.

5. J. O. Leibowitz, *The History of Coronary Heart Disease* (London: Wellcome Institute of the History of Medicine, 1970), pp. 129–30, 137.

6. Ibid., p. 147.

7. H. A. Snellen, *History of Cardiology* (Rotterdam: Donker Academic Publications, 1984), p. 167.

8. W. Bruce Fye, "The Delayed Diagnosis of Myocardial Infarction: It Took Half a Century!" *Circulation* 77 (1985):262–271.

9. Ralph H. Major, *Classic Descriptions of Disease*, 3d ed. (Springfield, IL: Charles C. Thomas, 1945), p. 434.

10. Frederick A. Willius and Thomas E. Keys, *Classics of Cardiology*, 2 vols. (New York: Henry Schurman, 1941), vol. II, pp. 815–816; James B. Herrick, "Clinical Features of Sudden Obstruction of the Coronary Arteries," *Journal of the American Medical Association* 59 (1912):2015–2020.

11. Leibowitz, *Heart Disease*, pp. 149–150.

12. Joel D. Howell, "Early Perceptions of the Electrocardiogram: From Arrhythmia to Infarction," *Bulletin of the History of Medicine* 54 (1984):92, 97.

13. See the excellent case study of the description of Down's syndrome in the context of Victorian anthropology and degenerationism in Lilian Zihni, "The Relationship between the Theory and Treatment of Down's Syndrome in Britain and America from 1866 to 1967," University of London, Ph.D. thesis, 1989.

14. Mary Hesse, *The Structure of Scientific Inference* (London: Macmillan,

1974); David Bloor, "Durkheim and Mauss Revisited: Classification and the Sociology of Knowledge," *Studies in the History and Philosophy of Science* 13 (1982):267–297.

15. Parkinson and Bedford, "Cardiac Infarction," p. 4.

16. Things are, of course, not quite as simple as this because the experience and expression of the raw data of medicine, for instance, pain, are themselves socially determined. For two complementary accounts of pain as a socially determined experience, see Helen King, "The Early Anodynes: Pain in the Ancient World," and Wendy Savage, "The Management of Obstetric Pain," in Ronald D. Mann, ed., *The History of the Management of Pain* (Carnforth, Lancs.: Parthenon, 1988), pp. 51–62, 187–200.

17. T. Henry Green, *An Introduction to Pathology and Morbid Anatomy* (London: Henry Renshaw, 1871), pp. 29, 31, 35.

18. Rodney Finlayson, "Ischaemic Heart Disease, Aortic Aneurysms, and Atherosclerosis in the City of London, 1868–1982," *Medical History*, suppl. 5 (1985):167.

19. Green, *Introduction*, 7th ed., revised and enlarged by Stanley Boyd (London: Henry Renshaw, 1989), p. 56.

20. Ibid., 8th ed., revised and enlarged by H. Montague Murray (London: Henry Renshaw, 1895), pp. 207, 214–242.

21. R. Tanner Hewlett, *Pathology: General and Special* (London: J. and A. Churchill, 1906), p. 276.

22. Green, *Introduction*, 9th ed., revised and enlarged by H. Montague Murray (London: Henry Renshaw, 1890), p. 447.

23. J. Martin Beattie and W. E. Carnegie Dickson, *A Textbook of General Pathology* (London: Rebman, 1908), p. 174.

24. G. Sims Woodhead, *Practical Pathology*, 4th ed., 2 vols. (London: Henry Frowde, 1910), p. 309.

25. Joseph L. Millar and S. A. Mathews, "Effect on the Heart of Experimental Obstruction of the Left Coronary Artery," *Archives of Internal Medicine* 3 (1909):476–484, 483.

26. *Green's Manual of Pathology*, revised and enlarged by H. W. C. Vines (London: Baillière, Tindall and Cox, 1934), pp. 441–443.

27. William Boyd, *A Textbook of Pathology* (London: Henry Kimpton, 1932), pp. 352–357.

28. Thomas D. Savill, *A System of Clinical Medicine* (London: J. and A. Churchill, 1903), pp. 40, 79, 80.

29. Alexander Wheeler and William R. Jack, *Handbook of Medicine and Therapeutics*, 3d ed. (Edinburgh: E. and S. Livingstone, 1908), p. 249.

30. William R. Jack, *Wheeler's Handbook of Medicine*, 4th ed. (Edinburgh: E. and S. Livingstone, 1912), p. 264.

31. John Henderson, *Wheeler and Jack's Handbook of Medicine*, 9th ed. (Edinburgh: E. and S. Livingstone, 1932), p. 334.

32. Savill, *Clinical Medicine*, 8th ed. (London: Edward Arnold, 1930), p. 69.

33. William Broadbent, *Heart Disease*, 4th ed. (London, Baillière Tindall and Cox, 1906), pp. 324, 348.

34. For a general survey of functional disorders see C. Handfield Jones, *Clinical Observations on Functional Nervous Disorders* (London: John Churchill and Sons, 1864).

35. Savill, *System*, 4th ed., 1914, p. 34.

36. Robert H. Babcock, *Diseases of the Heart and Arterial System* (New York: D. Appleton, 1903), pp. 703–704. Babcock preferred the term "cardiac neuroses."

37. Clifford Allbutt, *Diseases of the Arteries, including Angina Pectoris*, 2 vols. (London: Macmillan, 1915).

38. John Lindsay Steven, "Fibroid Degeneration and Allied Lesions of the Heart, and their Association with Disease of the Coronary Arteries," *Lancet* ii (1887):1154.

39. William Osler, "Angina Pectoris," *Lancet* i (1910):698, 699, 840, 974.

40. E. R. LeCount, "Pathology of Angina Pectoris," *Journal of the American Medical Association* 70 (1918):975–977.

41. James Mackenzie, *Angina Pectoris* (London: Henry Frowde, 1923), pp. 122, 152.

42. Ibid., fig. 33.

43. Ibid., p. 154.

44. Contemporaries were well aware that such a transformation was taking place; see Robert L. Levy, "Cardiac pain," *American Heart Journal* 4 (1929):377–389.

45. Bloor, "Durkheim," pp. 274–278.

46. Herrick, "Clinical Features."

47. Joseph T. Wearn, "Thrombosis of the Coronary Arteries with Infarction of the Heart," *American Journal of Medical Science* 165 (1923):250–276, 252.

48. J. W. McNee, "The Clinical Syndrome of Thrombosis of the Coronary Arteries," *Quarterly Journal of Medicine* 19 (1925–1926):44–52, 44.

49. P. D. White, "The Prognosis of Angina Pectoris and of Coronary Thrombosis," *Journal of the American Medical Association* 87 (1926):1525–1530.

50. Herrick, "Clinical Features," p. 2015.

51. Wearn "Thrombosis," p. 250.

52. Herrick, "Thrombosis of the Coronary Arteries," *Journal of the American Medical Association* 72 (1919):387–390, 390.

53. Wearn, "Thrombosis," pp. 252–253.

54. Herrick, "Clinical Features," p. 2017.

55. Henry Christian, "Cardiac Infarction (Coronary Thrombosis); An Easily Diagnosable Condition," *American Heart Journal* 1 (1925):129–137, 129.

56. McNee, "Clinical Syndrome," p. 44.

57. Herrick, "Clinical Features," p. 2018.

58. Wearn, "Thrombosis," p. 255.

59. Parkinson and Bedford, "Cardiac Infarction," p. 6.

60. McNee, "Clinical Syndrome," p. 48.

61. Wearn, "Thrombosis," pp. 254–258.

62. Parkinson and Bedford, "Cardiac Infarction," p. 6.

63. L. Whittington Gorham, "The Significance of Transient Localised Pericardial Friction in Coronary Thrombosis (Pericarditis Epistenocardica)," *Albany Medical Annals* 41 (1920):109–130; Savill, *System*, 8th ed., p. 70.

64. Christian, "Cardiac Infarction," p. 134.

65. Wearn, "Thrombosis," p. 259; NcNee, "Clinical Syndrome," p. 49.

66. Christian, "Cardiac Infarction," p. 134.

67. Parkinson and Bedford, "Cardiac Infarction," p. 7.

68. Christian, "Cardiac Infarction," p. 134.

69. Wearn, "Thrombosis," p. 260.

70. Alexander Gibson, "The Clinical Aspects of Ischaemic Necrosis of the Heart Muscle," *Lancet ii* (1925): 1270–1275, 1271, 1272.

71. Christian, "Cardiac Infarction," p. 135.

72. Henderson, *Handbook*, p. 333.

73. Parkinson and Bedford, "Cardiac Infarction," p. 7.

74. Ibid., p. 6; Wearn, "Thrombosis," p. 259.

75. Wearn, "Thrombosis," p. 261.

76. Christian, "Cardiac Infarction," p. 135.

77. McNee, "Clinical Syndrome," p. 50.

78. Parkinson and Bedford, "Cardiac Infarction," p. 8.

79. M. H. Nathanson, "Diseases of the Coronary Arteries," *American Journal of Medical Science* 170 (1925):240–255.

80. Paul Dudley White, *Heart Disease* (New York: Macmillan, 1931), p. 411.

81. Howell, "Early Perceptions," p. 90.

82. Symposium on Circulation, Association of American Physicians Society Proceedings," *Journal of the American Medical Association* 64 (1915):1939.

83. Howell's judgment that Herrick was told the inverted T wave was of "little importance" does not seem quite correct. See Howell, "Early Perceptions," p. 90, "Symposium on Circulation" (note 82).

84. Fred M. Smith, "The Ligation of Coronary Arteries with Electrocardiographic Study," *Archives of Internal Medicine* 22 (1918):8–27, 19.

85. Howell, "Early Perceptions," p. 89.

86. James B. Herrick, "Thrombosis of the Coronary Arteries," *Journal of the American Medical Association* 72 (1919):386–390, 390. The 1918 paper is "Concerning Thrombosis of the Coronary Arteries," *Transactions of the Association of American Physicians* 33 (1918):408–418, 417. In fact, the papers are virtually identical.

87. Herrick, "Thrombosis," p. 389; "Concerning Thrombosis," p. 416.

88. Harold E. B. Pardee, "An Electrocardiographic Sign of Coronary Artery Obstruction," *Archives of Internal Medicine* 26 (1920):244–257, 253.

89. H. Eppinger and C. J. Rothberger, "Zur Analyse des Elektrokardiogramms," *Wien. Klin. Wochnschr.* 22 (1909):1091–1098; Smith, "Ligation of Coronary Arteries."

90. Smith, "Ligation of Coronary Arteries," pp. 9, 19.

91. Pardee, "Electrocardiographic Sign," p. 253.

92. There is plenty of retrospective evidence. J. B. A. Herrick, in his *Memories of Eighty Years* (Chicago: Chicago University Press, 1949), pp. 196–200, recounted that his electrocardiographic work "woke up" physicians to the diagnosis. A chronicle of the rising incidence of diagnosis of the disease since 1920, by American authors R. L. Levy, H. G. Bruenn, and D. Kurtz ("Facts on Disease of the Coronary Arteries, Based on a Survey of the Clinical and Pathologic Records of 762 Cases," *American Journal of Medical Science* 187 (1934):376–390) noted that it was "perhaps, more than coincidental" that Herrick had published in 1919. Nonetheless, citation of Herrick's work in the early 1920s is not easily come by.

93. Wearn, "Thrombosis," pp. 265–267.

94. McNee, "Clinical Syndrome," p. 50.

95. Christian, "Cardiac Infarction," pp. 130–131, 135.

96. Gibson, "Ischaemic Necrosis," p. 1271.

97. Howell, "Early Perceptions," p. 97.

98. Although not central to my argument, Howell's claim is weakened by the fact that it was Lewis who had used the ECG to map the spread of the excitatory process through the heart's muscle, and also to identify left and right ventricular hypertrophy. That is, Lewis was involved in showing that muscular conformation determined the shape of the curve. Pardee, "Electrocardiographic Sign," p. 249, cited Lewis's work on normal depolarization as theoretical proof of his own conclusions.

99. Joel D. Howell, "Machines' Meanings: British and American Use of Medical Technology, 1890–1930" (Ph.D. thesis, University of Pennsylvania, 1987), ch. 2, pp. 17–82.

100. A. N. Drury, "Arborization Block," *Heart* 8 (1921):23–30.

101. Bernard S. Oppenheimer and Marcus A. Rothschild, "Electrocardiographic Changes Associated with Myocardial Involvement," *Journal of the American Medical Association* 69 (1917):429–431, 430.

102. Pardee "Electrocardiographic Sign," p. 255.

103. Edward Perkins Carter, "Further Observations on the Aberrant Electrocardiogram Associated with Sclerosis of the Atrioventricular Bundle Branches and Their Terminal Arborizations," *Archives of Internal Medicine* 22 (1918):331–353.

104. Drury, "Arborization Block," pp. 25–28.

105. The view that the ECG signs of coronary thrombosis were not agreed on until the late twenties makes possible other readings of Lewis's work. For instance, Howell cites Lewis's 1909–1910 work on coronary ligation as similar to that of Smith in 1918–1919 and suggests that Lewis "failed to note" the inverted T waves obtained when the coronary arteries of dogs are ligated. What Lewis was doing in these experiments, however, was following a well-defined physiological problem previously investigated by Millar and Mathews and by Porter. That is, he was studying the abnormal rhythms produced by ligation and the mode of death that follows. Further, the bulk of the published experiments used not the ECG but the kymograph, in which there is no T wave. In addition, Lewis published three galvanometer traces of these experiments, two showing extrasystoles in which normal T waves seem to be present and the third showing paroxysmal tachycardia (intermittent rapid heart rate) in which no T waves are visible. Howell, "Early Perceptions," p. 91; Thomas Lewis, "The Experimental Production of Paroxysmal Tachycardia and the Affects of Ligation of the Coronary Arteries," *Heart* I (1909–1910):98–137; Miller and Mathews, "Effect on the Heart"; W. T. Porter, "On the Results of Ligation of the Coronary Arteries," *Journal of Physiology* 15 (1893):121–138. Dr. Howell points out, however, that Lewis chose to use the kymograph and to publish those particular curves (personal communication).

106. Lewis, "Experimental Production," p. 122.

107. Thomas Lewis, *Clinical Electrocardiography* (London: Shaw and Sons, 1913), p. 28. Howell in "Early Perceptions," p. 92, as evidence of Lewis's slowness to take up the significance of the ECG in coronary thrombosis, cited a 1928 letter from H. M. Marvin at Yale purporting to ask Lewis why the latest edition of his *Diseases of the Heart* failed to mention the coronary T wave. In fact, Lewis's *Diseases of the Heart* did not appear until 1933. Marvin was referring to the 1924 edition of *Clinical Electrocardiography*, which, as we have seen, contained a comment on inverted T waves and had done so since 1913. The fourth edition, in

1928 (presumably in response to Marvin's letter), described (p. 117) the curves in coronary thrombosis as "sometimes quite characteristic, at others more obscure," but the "most easily recognizable" was the raised ST segment.

108. Christopher Lawrence, "Moderns and Ancients: The 'New Cardiology' in Britain 1880–1930," *Medical History*, suppl. 5, (1985):1–33. For example, Lewis and Mackenzie were attempting to supplant a simple correlation of symptoms with organic changes, an approach that arguably produced coronary thrombosis (see pp. 21, 26–27).

109. Parkinson and Bedford, "Cardiac Infarction," p. 9.

110. Smith, "Ligation," p. 27.

111. A. R. Barnes and M. B. Whitten, "Study of the R–T Interval in Myocardial Infarction," *American Heart Journal* 5 (1929–1930):142–171, 164.

112. Latour and Woolgar, *Laboratory Life*, p. 246.

113. Ibid., pp. 174–183.

114. One is reminded of the 1400-year-old "natural" fact of the human rete mirabile.

115. Lawrence, "Moderns and Ancients."

116. Ibid. In "Moderns and Ancients" I showed how the new cardiology was made and then treated it exactly the way I now suggest the ECG should not be treated. I depicted a cognitive technology being used unproblematically to discover facts of nature (new heart diseases). I am grateful to Steve Sturdy for pointing out that I should have shown how each of these disorders was socially constructed in a step-by-step manner, in which the constructors' interests determined both what was perceived as data and how it was evaluated.

117. Edgar Lea, *Heart Past and Present* (London: Baillière, Tindall and Cox, 1919), pp. 270–273.

118. East and Bain, *Recent Advances*, p. 321.

119. Ibid., p. 24.

120. The Americans seem to have been more often associated with academic medicine in general, the British more with specialized cardiology. For example, Wearn studied at Harvard and Johns Hopkins and became a professor at Western Reserve University. Christian had been a dean at Hopkins, then moved to Harvard. See A. McGehee Harvey, *Science at the Bedside: Clinical Research in American Medicine* (Baltimore: Johns Hopkins University Press, 1981). Another significant American was Samuel Levine, whose *Coronary Thrombosis* (London: Baillière, Tindall and Cox, 1929) was the first monograph on the subject. On some of the other figures see Lawrence, "Moderns and Ancients"; ibid., Joel Howell, "'Soldier's Heart': The Redefinition of Heart Disease and Speciality Formation in Early Twentieth-Century Great Britain," pp. 34–52; Joel Howell, "Hearts and Minds: The Invention and Transformation of American Cardiology," in Russell C. Maulitz and Diana E. Long, eds., *Grand Rounds: One Hundred Years of Internal Medicine* (Philadelphia: University of Pennsylvania Press, 1988), pp. 243–275.

121. Gordon Wolstenholme, ed., *Lives of the Fellows of the Royal College of Physicians of London*, 7 vols. (Oxford: IRL Press, 1984), VII:444.

122. It might be argued that a symbiotic process was also at work, the simultaneous re-evaluation of functional disorders by psychiatrists.

123. One aspect of the realist argument has not been particularly noted here: the view that heart disease was "discovered" because attention was drawn by a massive increase in incidence. See, for example, Finlayson, "Ischaemic Heart Disease." Clearly, this view assumes to be real what I am calling into question. If this view is correct, retrospective epidemiology becomes feasible. This is not to deny the occurrence of epidemiological changes; they just did not have to be conceptualized in the way that they were. Nature's input underdetermined the disease entity.

Mel Bartley has offered an alternative interpretation of this epidemiological explosion: The appearance of coronary thrombosis was part of a more general development that first focused prominently on the vascular system and then on the heart. This was part of an increasing shift of the responsibility for disease from society to the individual. Bartley argues that the rise in cardiovascular disease in the twentieth century is an artifact of a progressively narrower focus, from the cardiopulmonary system to the vascular system to the heart and finally to the coronary arteries. She suggests that the actual incidence of cardiovascular degenerative disease has remained relatively stable. In Britain this perceptual shift can be chronicled, she claims, by examining the changes in the coding system of the Registrar General. The consequence (and causes) of these changes, she argues, was the focus on the heart in a new and growing literature that emphasized the individual's responsibility. Paradoxically, she suggests that

the picture is completed, of a modern disease which, beginning at the time of the economic depression of the late 1920s–1930s, is about to become the commonest cause of death, a picture of a disease which—ironically—can allow it to be attributed to *rising standards of living*.
—Mel Bartley "Coronary Artery Disease and the Public Health," *Sociology of Health and Illness* 7 (1985):289–313

124. Bloor, "Durkheim," p. 267.

DISEASE
AS FRAME

5 THE MEDICALIZATION OF SUICIDE IN ENGLAND: Laymen, Physicians, and Cultural Change, 1500–1870

MICHAEL MACDONALD

Michael MacDonald's study of suicide in early modern England underlines both the antiquity and the negotiated quality of disease definitions; it illustrates as well the possibility of medicalizing behavior and thus changing its moral—and, in the case of suicide, legal—meaning. Perhaps most strikingly, it shows how medical personnel and formal medical thinking constituted only one factor in a diverse social and intellectual context. Increasingly, suicide became retrospective evidence of disease and not a culpable moral offense—as in instance after instance ordinary Englishmen preferred that exculpatory medical option to its legally and traditionally mandated alternative. They chose to label suicides among family members, friends, and neighbors as sick rather than criminally responsible and thus liable to forfeit their property to the state. The study also provides an example of how behavior alone could serve as the crucial element in a "diagnosis" of "sickness."

Certainly the circumstances surrounding suicide are a bit atypical—the need to negotiate concrete determinations at a particular moment, for example, and the gradient defined by the harsh alternatives to a verdict of unsound mind. And so is the chronology; a willingness to expand the boundaries of legitimate illness to include deviance is more typical of the nineteenth century than of the seventeenth and eighteenth.

—C. E. R.

Suicide was regarded as a heinous crime in sixteenth- and early-seventeenth-century England, a kind of murder committed at the instigation of the devil.[1] Suicides were tried posthumously and, if found to have been sane when they took their lives, were severely punished. Their moveable property was forfeited to the crown or to the holder of a royal patent; their bodies were buried profanely, interred in a public highway or at a crossroads, pinioned in the grave with a wooden stake.[2] These savage penalties originated in the early Middle Ages; they were matters of common law and religious custom by the thirteenth century.[3] But they were rigorously enforced for less than 200 years, between about 1500 and 1660. The law of self-murder was seldom used before 1500, in spite of the crown's financial interest in the goods of suicides and the theological condemnation of self killing.[4] After 1660 the law was increasingly evaded, and the scope of the "insanity defense"—if one may use such an anachronistic term—was gradually so broadened that eventually almost all suicides were acquitted as lunatics. The "insanity defense" was as old as the law of self-murder. Indeed, culpable self-murder was distinguished from innocent suicide by the use of two different verdicts. The guilty were designated *felones de se*, felons of themselves; the innocent were *non compos mentis*, lunatics.

In this paper I shall argue that the palliation of the law of suicide and the rise of secular, medical interpretations of it owed little to the leadership of the medical profession. The secularization of suicide was, ironically, almost entirely the work of laymen—philosophers, men of letters, journalists, and, most of all, coroners and their juries. It was a facet of much wider cultural changes that were accelerated by the religious and constitutional conflicts of the midseventeenth century. I shall first briefly describe the age of severity—the era in which suicide was criminalized and diabolized. I shall then pass on to a discussion of the dynamics of decriminalization and secularization. Finally, I shall conclude with some reflections on the causes of change and the implications of the history of suicide for historians. Because space is limited, I shall simply ignore any but the most conspicuous resistance to the trends under discussion. Nor shall I treat suicide in other European nations. Not much is known about the history of suicide elsewhere, but what has been published shows that attitudes and responses varied markedly from region to region. I make no claim that the English experience holds for the whole of Europe.

The Age of Severity

The leniency displayed by medieval coroners' juries was an expression of local solidarity, a display of sympathy for the survivors of suicide. It came to an abrupt halt soon after 1500. The crown reformed the administrative machinery to insure more rigorous enforcement of the law, and the church mounted a campaign to fortify the popular conviction that suicide was a supernaturally evil act. The monarchy was anxious to improve enforcement of the law mainly for financial motives—it profited from the forfeiture of self-murderers' goods.[5] The church condemned suicide for more complex reasons. Confronted with a population whose religious beliefs have been described (more or less accurately) as a blend of paganism, magic, and Catholicism, the reformers of Queen Elizabeth's reign faced a formidable task in trying to convert the people to Protestantism. Like the early Christian missionaries before them, they incorporated some old beliefs and popular customs into their sermons and ceremonies, reinterpreting them theologically rather than rejecting them outright.

Suicide is a case in point. Protestant evangelists, taking up a theme first developed by medieval preachers, stressed that self-murder was directly caused by the devil. They interpreted suicide as the antithesis of the faith that every Christian must have in order to be saved—it was a sort of apostasy, the product of the dreadful sin of despair, the opposite of pious hope. The clergy also tacitly accepted popular beliefs about the spiritual consequences of suicide. The rituals used to desecrate the corpses of self-murderers were ancient demotic customs based on pre-Christian religion. They expressed a profound abhorrence of suicide and a powerful conviction that the act was spiritually polluting: a stake was driven through the bodies of the self-murderers to prevent their ghosts from walking.[6] Ignoring the pagan origins of the rituals of desecrations, the ministers of the Church of England regarded them as essential aspects of the punishment of the sin of self-murder, even though they were nowhere mandated in the canons or the liturgy.[7]

The clergy's hostility to self-murder was also aroused by the contemporary revival of Greek and Roman ideas that excused and even glorified suicide in certain circumstances. Renaissance humanists and skeptics called attention to ancient philosophical justifications for self-killing—notably Epicurean and Stoic doctrines.[8] Scholars, historians, poets, and playwrights celebrated classical suicides—especially Cato, Brutus, and Lucretia—for their heroism.[9] John Donne showed in *Biathanatos* that theological prohibitions against suicide were weak.[10] Humanism also reawakened interest in classical science and brought to the fore medical ideas that tended to palliate suicide.[11] Philip Barrough, whose medical textbook was reissued for three generations after its first publication in 1583, observed that people suffering from the disease of melancholy "desire death, and do verie often behight and determine to kill them selves."[12] Robert Burton declared in his famous *Anatomy of Melancholy*: "In some cases those hard censures of such as offer violence to their own persons . . . are to be mitigated, as in such as are mad, beside themselves for the time, or found to have been long melancholy, and that in extremity."[13]

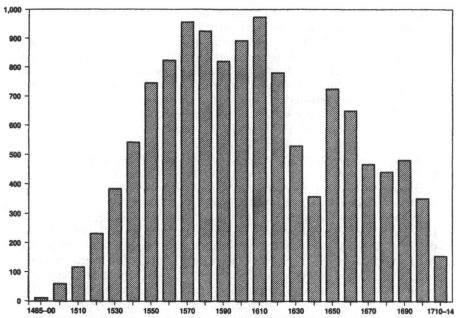

Fig. 5.1. Suicides reported to the King's Bench. *Source:* U.K. Public Record Office. King's Bench 9, 10, 11 (1485–1714)

The campaign to stiffen enforcement of the law and promote popular hostility to suicide was immensely successful in spite of these ambivalent voices. Thanks to the heroic research of Terence R. Murphy, we know that the number of suicides reported to the central government soared in the first half of the sixteenth century. I have statistically analyzed Professor Murphy's notes on suicide inquisitions returned to the court of King's Bench between 1485 and 1714. They show that the level of reporting rose rapidly after 1500, achieved a sort of jagged plateau between about 1560 and 1640, and declined decisively after 1660 (see fig. 5.1).[14] The vast majority of suicides reported by coroners' juries prior to 1660 were judged self-murders and punished. Fewer than two percent were acquitted as persons *non compos mentis*.

Efforts by preachers to increase abhorrence of suicide among the common people and to minimize the influence of classicism and medical science among the elite also were successful. Admiration for classical heroes notwithstanding, only one suicide was publicly justified in the whole of the sixteenth and seventeenth centuries by an appeal to the examples of Cato and Brutus.[15] Donne's *Biathanatos* remained in manuscript until 1647, and the belief that suicide was condemned by divine law was almost universal. The behavior of juries, the best index of popular belief, was resolutely harsh. They convicted suicides as *fel-*

ones de se even when ambiguous circumstances or evidence of mental illness could easily have justified less severe verdicts. Many men and women who experienced suicidal urges or attempted to kill themselves reported that they had actually seen or heard the Tempter himself encouraging them to die.

Finally, the medical arguments for a more merciful attitude toward suicides driven to their deaths by extreme melancholy were simply absorbed into the prevailing supernatural interpretation of the crime. Satan, it was argued, took advantage of the natural gloom of melancholy people and magnified it into suicidal despair. Both Robert Burton and the puritan sage William Perkins agreed that the melancholy humor was the *balneum diaboli* (the devil's bath).[16] The phrase was something of a cliché. John Sym, the author of the first published treatise on suicide, warned that Satan preyed particularly on people plagued by melancholy, "speaking to and persuading a man to kill himself."[17] Evidence of melancholy moods preceding suicide was used by coroners' jurors and royal officials as proof that people had committed the ungodly, satanic act of self-murder.[18]

In Tudor and Stuart England, then, the medical and the supernatural explanations for suicide were not necessarily contradictory. Like other mental and physical afflictions, suicidal impulses could have either natural or supernatural causes or both. This fusion of the natural and supernatural was validated by the "Elizabethan world picture," the cosmology the Renaissance had inherited from the Middle Ages, which described the hierarchy of things and forces in the universe. Susceptibility to disease was a consequence of the Fall, and any illness might be punishment for sin—a retribution or "judgment" either directly from God or indirectly from the Devil acting as God's malevolent instrument. Suicide likewise could come from God or the Devil, and its instrumental cause could be a disease, most often melancholy.[19]

The Secularization of Suicide

The meaning of suicide transformed utterly between about 1660 and 1800. The ruling classes lost faith in the Devil's power to drive people to kill themselves; coroners' juries gradually ceased punishing suicides. The shift from severity to tolerance was a complex phenomenon. It was caused by social, political, and cultural changes, accepted variously by different social and religious groups, and remained incomplete well into the nineteenth century. Discussing all of these causes and effects fully would occupy far more space than a conference paper allows.[20] I want here merely to describe the dynamics of the change and to consider the role the medical profession played—or rather did not play—

in promoting the rise of a more tolerant and secular view of suicide. I shall also concentrate on the behavior of coroners' juries. For although attitudes to suicide in the wider society were crucial and fascinatingly varied, their most forceful expression was the response to actual deaths. How people thought about suicide is important; what they did when troubled men and women killed themselves is more important. That was when they put their ideas into practice. Moreover, since the punishments for self-murder were not finally abolished until the nineteenth century, the palliation of responses to suicide took place on a case-by-case basis. The coroners' jury was the focal point of cultural change.[21]

Coroners' juries mitigated the law of suicide in two ways. First, soon after the Revolution of 1640–1660, they increasingly helped families to evade forfeiting the property of suicides judged *felo de se*. The proportion of inquisitions listing the value of goods said to have been confiscated for forfeiture fell steadily after 1660, from about 35 percent in the 1660s, to about 25 percent in the 1680s, to about 13 percent in the first ten years of the eighteenth century. Second, beginning about the same time, juries returned more and more *non compos mentis* verdicts, from 8.4 percent in the 1660s, to 15.8 percent in the 1680s, to 42.5 percent in the first decade of the eighteenth century. These verdicts continued to increase more or less steadily until they comprised almost 80 percent of suicide verdicts in the 1750s, over 90 percent in the 1760s, and so on up to over 97 percent by the end of the century (see fig. 5.2).

These two different ways of lessening the severity of the law of suicide had different causes and implications. There was a sharp divergence in popular attitudes toward the religious and secular punishments for self-murder. Born of custom, explained by folklore, and validated by religion, the rites of desecration enjoyed widespread support as long as they were practiced. There are few signs of resistance to them, except for occasional attempts by families to bury bodies before inquests or to dig them up after profane interment. A small number of self-murderers were also apparently interred, like Ophelia, with rites that fell short of either desecration or full Christian burial.[22] Forfeiture, by contrast, encountered continuous opposition. Hundreds of delinquent families and juries were prosecuted in the courts of King's Bench and Star Chamber between 1500 and 1640.[23] In 1593, for example, one coroner was presented in Star Chamber for telling his jury that the King's Almoner, the official who collected forfeitures, had no right to the goods of self-murderers: "In the time of popery the goods of felons of themselves were distributed by the Almoner to poor people in hospitals and such like, but in these days . . . the Almoner had nothing to do with the said goods, chattels and debts . . . but the same was to pass by administration to the next of kindred."[24]

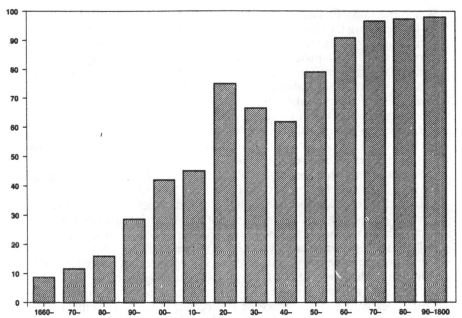

Fig. 5.2. Percentage non compos mentis, King's Bench, selected counties and towns. *Sources:* U.K. Public Record Office. King's Bench 9, 10, 11, 13, 14; PL 26; HCA 1/83; CHES 18/1–6; Cumbria R.O.; Norwich Inquests; London Inquests; Middlesex Inquests; Westminster Inquests; Somerset R.O. D/B/bw 1917; Hunnisett 1981. Very few inquisitions for the period 1720 to 1750 could be discovered. N = 2,360 NCM; 2,144 FDS.

This was, of course, bad history and worse law, but it captured a real resentment. And like many other such resentments, it was voiced openly in the Interregnum. Radical reformers disliked the whole idea of forfeiture for felony. Expressing a sentiment that would often be voiced against forfeiture in the future, John March declared in 1651: "I think there cannot be a more rigid and tyrannical Law in the world, that the children should thus extremely suffere for the crime and wickedness of the Father; the innocent for the nocent."[25] When the parliamentary commission headed by the great jurist Matthew Hale undertook its attempt at law reform in 1653, it included in its proposals the suggestion: "That such as kill themselves shall not forfeit any thing by Reason thereof, unless at the Time of the Fact committed they be under Restraint or Prosecution for some capital Offence."[26]

The Hale Commission report came to nothing, like the other revolutionary proposals for law reform, but the revolution nevertheless contributed to the erosion of forfeiture both directly and indirectly.

When the Long Parliament struck down the Court of Star Chamber in 1641, believing it to have become a menace to religious liberty and private property, it removed the single most effective tool for supervising coroners and their juries. The responsibility for enforcing the law of suicide reverted to King's Bench, a far less efficient tribunal. More broadly, the constitutional disputes that accompanied the midseventeenth-century revolutions weakened respect for the royal prerogative and fostered a veritable cult of private property. This in turn sharpened the longstanding antagonisms between gentlemen who owned the right to forfeitures in particular localities and the officials of the crown. To protect the property rights of these lesser lords, parliament passed a law in 1693 that greatly hampered the ability of King's Bench to exercise the royal prerogative to the chattels of self-murderers.[27]

Ironically, this law—so in tune with the ideology of the Restoration gentry—eventually destroyed the ability of lesser lords to exercise the very rights it was supposed to protect. King's Bench first lost interest in enforcing the suicide law, then began actively to side with heirs who tried to evade forfeiture.[28] After about 1700 juries that concealed a suicide's goods had very little to fear; lords who sought the protection of the law against defiant juries and heirs received no sympathy from the court. Describing the practice of the court, Blackstone remarked in 1766: "The court of king's bench hath generally refused to interfere on behalf of the lord of the franchise, to assist so odious a claim."[29] Blackstone was only one of many Georgian gentlemen who had come to believe that forfeiture was unfair, if not "odious." Throughout the century writers echoed the words of revolutionary law reformers, with whom they otherwise had precious little in common. Defoe remarked as early as 1704 that society was inclined to pity the family of a suicide: "the children should [not] be starv'd because the Father has destroy'd himself."[30] It was a view often repeated, even by traditionalists who deplored the growing tolerance to suicide.[31]

Unlike the decline of forfeiture, the rise of the *non compos mentis* verdict directly challenged older interpretations of the meaning of suicide itself. It labeled a suicide as a psychiatric calamity, the consequence of insanity, rather than a spiritual crime. Moreover, because the rites of desecration could follow only a verdict of *felo de se*, the secular interpretation placed on the death by a *non compos mentis* verdict could not be contradicted by a collective expression of abhorrence. The initial attractiveness of the *non compos mentis* verdict was undoubtedly that it spared the families of suicides the loss of their property and lessened the stigma of a *felo de se* verdict.[32] But as the century progressed juries increasingly broadened the circumstances in which the *non compos mentis* verdict was used until it became the usual judgment in cases of suicide. Contemporary critics had no doubt that juries had embraced a secular, medical interpretation of self-destruction. As early as 1700

John Adams complained: "There is a General Supposition that *every one* who kills himself is *non Compos,* and that nobody wou'd do such an Action unless he were Distracted."[33] Laymen and divines repeated this charge so often that it became an axiom of eighteenth-century discussions of the law of suicide and its enforcement.[34] And although critics exaggerated the speed with which the medical interpretation triumphed, the inquisition evidence shows that they were right about what was happening.[35]

Cultural Change

It is impossible to enter the minds of jurors to discover precisely why they gradually came to believe that suicide was an insane act. But it is possible to chart the main currents of opinion that must have influenced them. Signs of increasing tolerance toward suicide were apparent among the upper classes within a generation after the Restoration. William Ramesey in *The Gentlemans Companion* (1672) observed conventionally (and incorrectly) that suicide is forbidden by scripture but counseled that those who killed themselves ought to be regarded with compassion, for they were frequently the victims of mental illnesses:

> They should rather be objects of our greatest pity than condemnation as murtherers, damn'd Creatures and the like. For, tis possible even for Gods elect, having their Judgments and Reasons depraved by madness, deep melancholly, or [some]how otherwise affected by Diseases of some sorts, to be their own executioners. . . . Wherefore lets be slow to censure in such cases.[36]

During the course of the eighteenth century, philosophical ideas that had once been the property of a small band of skeptics had become the common coin of polite conversation. Radical apologies for suicide were published in England by the deist Charles Gilden, by foreign libertines living in exile, and by the great philosopher David Hume.[37] The works of these local freethinkers were amplified by the arguments of the *philosophes*—Montesquieu, Voltaire, and Rousseau all debated the question, pro and con—and the great law reformer Beccaria called for abolition of laws against suicide.[38]

Few people embraced the views of radical philosophers wholeheartedly, but openness to non-Christian attitudes toward self-killing increased. In the early eighteenth century, the so-called Augustans celebrated Roman examples of noble suicide with far fewer reservations than earlier writers had displayed. Cato was a kind of household god among the fashionable elite. Jonathan Swift's admiration for Cato was unbounded—he even held him up to Stella as a model to follow in matters of honor:

> In Points of Honour to by try'd
> All Passions must be laid aside:
> Ask no Advice, but think alone,
> Suppose the Question not your own:
> How shall I act? is not the Case,
> But how would *Brutus* in my Place?
> In such a Cause would *Cato* bleed?
> And how would *Socrates* proceed?[39]

Cato achieved his apotheosis as a hero in England in Joseph Addison's hugely successful tragedy *Cato* (1713), which depicted Cato's suicide as an act of pathetic, surpassing nobility. The play brought tears to Pope's eyes and captured the imagination of the crowd: "Ministers and Oxford students, Grub Streeters and country rectors, squires and royal physicians, deans and printers—everyone saw something of himself in Cato and a great deal of Cato in himself."[40]

The cult of Cato gradually waned after 1750, to be replaced by a new stereotype of sentimental suicide. In England, as elsewhere, the publication of Goethe's *Sorrows of Young Werther* excited the admiration— and occasionally the emulation—of writers and romantic youths. Robert Merry's "Elegy Written after Having Read the Sorrows of Werter," claimed that there was "a class distinct" of persons whose emotions were so exquisite that their suicides were excused and even pitied by the Lord himself:

> Th'Eternal Pow'r, to whom all thoughts arise,
>   Who ev'ry secret sentiment can view,
> Melts at their flowing tears, their swelling signs,
>   Then gives them force to bid the world adieu.[41]

Werterism found its English hero in Thomas Chatterton, the famous poetical forger, a genius who had poisoned himself in 1770 when he was only 17. Within months of his death, commemorative verses had appeared, but only after the publication of *Werther* did his death became the stuff of legend. One of the book's translators, echoing Robert Merry, declared that Werther's "feelings, like those of our Chatterton, were too fine to support the load of accumulated distress." Artists and writers idealized Chatterton's wretched life, his final despair, and even his suicide itself. An engraving of Chatterton composing in his garret was even transferred onto a souvenir handkerchief in 1782.[42] All of the English Romantics celebrated Chatterton's genius and lamented his tragic death; Keats, the boy wonder among them, dedicated *Endymion* to his memory.[43]

The adulation of Chatterton marked the zenith of sentimental suicide in England. It is impossible to imagine a contemporary suicide achieving such posthumous celebrity two centuries earlier. There were, of course, howls of protest. Deism, classicism, and Romanticism all had their enemies among the clerical establishment and the evangelical middle classes. The Methodists clung steadfastly to the notion that suicide was caused by the devil, and many of the common people also continued to believe that the Tempter offered despairing persons the instruments of their own destruction.[44] But secular and tolerant views of suicide increasingly prevailed among the educated laity. A letter in the *London Journal* in 1724 contrasted Aristides, who manfully endured poverty and disappointment, with Philander, who slew himself in the midst of luxury and (improbably) happiness. "I approve, I applaud *Aristides*," the writer declared, "but at the same Time, I think my self at Liberty to pity *Philander*."[45] Newspapers and periodicals reported thousands of suicides, and most of them were simply noticed or held up as the object of pity, like Philander.[46] Diarists recorded suicides without condemning them.[47]

Perhaps the most conspicuous feature of the upper classes' growing tolerance to suicide was the small role played by physicians and medical writers. Although juries normally excused suicide as lunacy, so far as I am aware only one physician, William Rowley, argued for this view in print: "Everyone who commits suicide is indubitably *non compos mentis*," he wrote in 1788, "and therefore suicide should ever be considered an act of insanity."[48] Rowley's approval of juries' merciful practice was unique; his complete lack of interest in what *kind* of insanity prompted suicide was typical. Physicians made no notable contributions to the understanding of suicide in the late seventeenth and eighteenth centuries. When they mentioned the subject at all, they were content to repeat the Renaissance commonplace that melancholy (sometimes rechristened the vapors or the spleen) often led to self-destruction.[49] Richard Blackmore, for instance, remarked in his best-selling *Treatise of the Spleen and Vapours* that when melancholy patients "through great Despondency and Inquietude, discover Marks of a Design upon their own Lives, their Distemper exceeds its proper Nature and Extent, and has contracted a Degree of Lunacy."[50] A few years later Dr. George Cheyne in his famous book, *The English Malady*, blamed the prevalence of spleen for the alleged epidemic of suicides in England: "The late Frequency and daily Encrease of wanton and uncommon self-murders, produc'd mostly by this *Distemper,* and their [the deists'] *blasphemous* and *frantick Apologies* grafted on the Principles of the *infidels*, and propagated by their *Disciples.*"[51]

The most conspicuous champions of the notion that mental illness was the cause of suicide and excused its victims from the charge of self-murder were medical laymen. As early as the 1670s, William Ramesey and Thomas Philipot urged their readers to forgive suicides by people afflicted with melancholy.[52] In the eighteenth century, even church-men and moralists joined in the call for merciful treatment on medical grounds. The prominent cleric John Jortin was inclined to believe that "in our country, where spleen and melancholy, and lunacy, abound" it was wise of coroners' juries to be lenient, for "it is surely safer and better to judge too favourably than too severely the deceased."[53] Adam Smith, agreeing with Jortin, added a passage to his refutation of philo-sophical arguments in favor of suicide:

> There is, indeed, a species of melancholy . . . which seems to be accompanied with, what one may call, an irresistible appetite for self-destruction. . . . The unfortunate persons who perish in this miserable manner, are the proper objects, not of censure, but of commiseration. To attempt to punish them, when they are beyond the reach of human punishment, is not more absurd than it is unjust.[54]

Medical opinion therefore provided the main rationale for suspending the old penalties for suicide, but no thanks to statements made by physicians.

Nor do medical men seem greatly to have hastened the palliation of suicide by their involvement in actual coroners' inquests. Few coroners were medically qualified before the end of the eighteenth century.[55] Also, the records of the few who were seem to suggest that they were actually likely to enforce the law more narrowly than their lay col-leagues. This was certainly the case in Norwich in the middle of the eighteenth century and in Wiltshire during the reign of George III, the only two jurisdictions for which data are available.[56] Medical witnesses did testify at suicide inquests more frequently as the eighteenth cen-tury progressed, but they seldom testified about the mental state of the deceased, describing instead the suicide's wounds or efforts to save his life. In London physicians and mad-doctors occasionally avouched that they had treated suicides for mental maladies, and the rich normally produced batteries of medical witnesses to prove that their suicide rela-tives had been delirious or melancholy. But by this time, juries were normally excusing all kinds of people for trivial mental disturbances and, in London at least, had already begun to declare that unidentified men and women, about whose mental state nothing was known, had died innocent lunatics.[57]

The Causes of Change

My argument has been that attitudes and responses to suicide were secularized in the late seventeenth and eighteenth centuries. Coroners' juries slowly adopted the medical explanation and excused suicides as innocent lunatics. Fashionable society embraced a more tolerant and even sentimental view of suicide that was expressed in tracts, periodicals of all kinds, and imaginative literature. There was, in other words, a complete switch in the alignment of opinion about suicide. During the age of severity, hostility had prevailed, and tolerance had been the less influential view; after about 1660, tolerance increasingly predominated, and hostility became the weaker opinion. This reversal in attitudes and responses cannot be explained entirely in terms of the positive attraction of new ideas. The eighteenth century's taste for scientism, Enlightenment humanitarianism, neoclassicism, and, finally, Romanticism all contributed to more tolerant attitudes. But none of these intellectual movements was decisive, either by itself or in combination.

Physicians did little to advance the medical interpretation of suicide; Enlightenment philosophy had only a limited impact on the governing elite, who ignored the *philosophes'* calls to repeal the laws against self-murder; and neoclassical and Romantic views of suicide were roundly denounced by conservatives. It is notable that the philosophical and medical arguments for excusing suicide actually contradicted one another. The philosophers justified suicide as a rational choice; the advocates of medical approaches regarded it as an act of unreasoning insanity. This contradiction did not escape the notice of contemporaries.[58] To understand the causes of change, we must therefore also recognize that political and religious strife had charged the supernatural with a negative force.

All over western Europe in the eighteenth century, governing elites were sick of religious strife and eager to find new and less controversial grounds for intellectual and social discourse. This was particularly true in England, where the events of the Puritan Revolution left a lasting impression on the governing classes. For over a century after 1660 establishment propagandists denounced Protestant "enthusiasts" and Catholics as subversives whose bogus claims to divine inspiration and miraculous powers endangered the English church and civil society. They therefore strove to discredit the presumption that good and evil spirits intervened directly and frequently in human affairs. Modern claims to inspiration were dismissed as the symptoms of mental or physical illnesses; natural causes were adduced for the pious emotions of the enthusiasts and the demonic afflictions that they and the papists were supposed to alleviate.[59] The connection between religious and political strife and the growing tolerance toward suicide was made by the greatest eighteenth-century authority on the subject, Charles Moore. The "affectation of piety and bigotry of puritanism in Cromwell's days," Moore argued, had led to "the opposite extreme of licentious and atheistical principles," which eventually culminated in defenses of suicide like David Hume's notorious essay.[60] Moore's analysis of the cultural changes that fostered new attitudes to suicide was perceptive in spite of its obvious tendentiousness. The ruling classes' horror of religious fanaticism after the Restoration coincided with new developments in philosophy and science and encouraged among the educated elite a "hankering after the bare Mechanical causes of things," in spite of foot-dragging by clerics and conservatives.[61]

The hankering for natural explanations for suicide was easily satisfied because of the longstanding recognition that mental illnesses caused some people to take their own lives. Even during the period of greatest severity to suicide, when almost every self-killer had been condemned as a self-murderer, educated laymen and coroners' juries had recognized that Satan was not the sole cause of suicide. Melancholy, lunacy, and delirium played their parts as well, even if they were usually regarded as secondary causes. The eighteenth-century loss of confidence in diabolical powers did not, therefore, require the invention of a new psychology of suicide. It meant merely that lay opinion makers stressed one rather than the other of the traditional explanations for it, and that juries chose to use the *non compos mentis* verdict more frequently. In other words, attitudes and responses to suicide were medicalized by default. The ancient eclectic model of psychological causation that united the natural and supernatural was demystified, and only the medical possibilities remained.

By pointing out that the physicians were riding in the caboose of change, not driving its engine, I am not seeking to discredit doctors, then or now. My aim instead is to restore a realistic perspective on the role that the medical profession played in the intellectual and social life of the nation. Medicine in the eighteenth century lacked the authority it would gradually gain in the Victorian age and after. After about 1650 medical ideas did play an increasingly important role in justifying society's responses to its discontents, but this was as much or more the work of laymen as it was the consequence of changes in medical thought and organization. A faction of physicians who advocated the notion that suicide was the consequence of mental illness did emerge in the mid-nineteenth century. It included the progressive coroners Edwin Lankester and Thomas Wakely. Even at this time, however, medical ideas that excused suicides from legal punishment were fiercely opposed by other physicians, including the famous Henry Maudsley. Wakely himself came to favor a compromise verdict of "state of mind unknown" in cases of suicide.[62] When the old penalties for self-murder were finally removed by stages in statutes passed by parliament in 1823, 1870, and 1882, the doctors' voices were entirely silent—the cause of reform belonged to the lawyers and politicians.[63]

The history of suicide in England is a cautionary tale with three morals. First, it demonstrates the perils of attempting to study mental illness from the perspective of physicians and medical writers. Because the medical interpretation of suicide and suicidal moods changed little over the century, one would conclude from such an approach that nothing of interest to medical historians happened during the early modern period. But in fact, medical ideas gained immensely in prestige and became for the first time the basis for the usual societal response to suicidal deaths. Second, it also suggests that historians prematurely depicting the medical profession as the agents of social control have singled out the wrong group of villains. For although physicians were certainly men of their own class, the medical profession lacked the authority and organizational strength it would gain in the nineteenth and twentieth centuries. Medical knowledge was wielded for social purposes—whether to justify the exculpation of suicides or the incarceration of lunatics—mainly by laymen. Finally, it shows that the study of even a phenomenon whose transmigration from the realm of the supernatural into the domain of the secular was as apparently straightforward and linear as that of suicide must be seen in its entire historical context to be fully understood. Attitudes and responses to mental disorders were shaped not only by the ideas and intentions of influential men but also—and even more—by the political, religious, social, and cultural environment.

NOTES

1. Much of this article is based on Michael MacDonald and Terence R. Murphy, *Sleepless Souls: Suicide in Early Modern England* (Oxford: Clarendon Press, 1990), and its arguments are more fully documented there.

2. Michael Dalton, *The Countrey Justice*, 3d ed. (London, 1626), pp. 234–235; Matthew Hale, *Historia Placitorum Coronae: The History of Pleas of the Crown*, ed. S. Emlyn, G. Wilson, and T. Dogherty, new ed., 2 vols. (London: F. & J. Rivington, 1800), vol. 1, pp. 411–418; J. Stephen, *A History of the Criminal Law of England*, 3 vols. (London: Macmillan, 1883), 3:104–105.

3. Charles Moore, *A Full Inquiry into the Subject of Suicide*, 2 vols. (London, 1790), vol. 1, pp. 286–305; Henry de Bracton, *Bracton on the Laws and Customs of England*, ed. Samuel E. Thorne, 5 vols. (Cambridge: Harvard University Press, 1968), 2:423–424

4. Barbara A. Hanawalt, *Crime and Conflict in English Communities, 1300–1348* (Cambridge: Harvard University Press, 1979), pp. 101–104, and *Crime in East Anglia in the Fourteenth Century: Norfolk Gaol Delivery Rolls, 1307–1316*, Norfolk Record Society, 44 (Norwich, 1976).

5. R. F. Hunnisett, *Calendar of Nottinghamshire Coroners' Inquests*, Thoroton Society Record Series, 25 (Nottingham, 1969), pp. *xviii–xix*, and *Sussex Coroners'*

Inquests, 1485–1558, Sussex Record Society, 74 (Lewes, 1985), pp. *xiii–xiv;* R. H. Wellington, *The King's Coroner* (London: William Clowes & Sons, 1905), pp. 66–70.

6. Rowland Wymer, *Suicide and Despair in the Jacobean Drama* (New York: St. Martin's Press, 1986), pp. 118–119.

7. Michael MacDonald, "The Inner Side of Wisdom: Suicide in Early Modern England," *Psychological Medicine* 7 (1977):574–578; Richard Greaves, *Society and Religion in Elizabethan England* (Minneapolis: University of Minnesota Press, 1981), pp. 531–537.

8. Montaigne, *The Essayes of Michael Lord of Montaigne*, tr. J. Florio, 2 vols. (1603; repr. London: Dent, 1965), 2:26–41; P. Charron, *Of Wisdome Three Bookes*, tr. S. Lennard (London, 1608).

9. Wymer, *Suicide and Despair*, chs. 5, 7.

10. John Donne, *Biathanatos*, ed. M. Rudick and M. P. Battin (New York: Garland, 1982).

11. L. Babb, *The Elizabethan Malady: A Study of Melancholia in English Literature from 1580 to 1642* (East Lansing: Michigan State College Press, 1951).

12. Philip Barrough, *The Method of Physick, Conteyning the Causes, Signs, and Cures of Inward Diseases in Mans Body*, 3d ed. (London, 1596), p. 46.

13. Robert Burton, *The Anatomy of Melancholy*, ed. H. Jackson, 3 vols. (London: J. M. Dent, 1972), 3:439.

14. For further data and discussion, see MacDonald and Murphy, *Sleepless Souls*, ch. 1; S. J. Stevenson, "The Rise of Suicide Verdicts in South-East England, 1530–1590: The Legal Process," and "Social and Economic Contributions to the Pattern of 'Suicide' in South-East England, 1530–1590," *Continuity and Change* 2 (1987):37–75, 225–62; Michael Zell, "Suicide in Pre-Industrial England," *Social History* 11 (1986):303–317.

15. W. Allen, "Killing No Murder," in *The Harleian Miscellany*, vol. 9 (London: Robert Dutton, 1810), p. 306.

16. Burton, *Anatomy of Melancholy*, 1:429, 3:395; William Perkins, *The Workes of that Famous and Worthy Minister of Christ in the University of Cambridge, Mr William Perkins*, 3 vols. (London, 1626–1631), 2:46–47, 3:381.

17. John Sym, *Lifes Preservative Against Self-Killing*, ed. Michael MacDonald (1637; repr. London: Routledge, 1988), pp. 246–247.

18. Michael MacDonald, "The Secularization of Suicide in England, 1660–1800," *Past and Present*, no. 111 (1986):66.

19. E. M. W. Tillyard, *The Elizabethan World Picture* (Harmondsworth, Middlesex: Penguin Books, 1979); Stuart Clark, "The Scientific Status of Demonology," in B. Vickers, ed., *Occult and Scientific Mentalities in the Renaissance* (Cambridge: Cambridge University Press, 1984), pp. 351–374; Michael MacDonald, *Mystical Bedlam: Madness, Anxiety and Healing in Seventeenth-Century England* (Cambridge: Cambridge University Press, 1981), pp. 198–206; Keith Thomas, *Religion and the Decline of Magic* (New York: Scribner, 1971), pp. 469–477.

20. For fuller treatments, see MacDonald, "Secularization of Suicide"; MacDonald and Murphy, *Sleepless Souls*, chs. 4–6.

21. MacDonald, "Secularization of Suicide," pp. 64–68; R. F. Hunnisett, *Wiltshire Coroners' Bills, 1752–1796*, Wiltshire Record Society, 36 (Devizes, 1981); idem. "The Importance of Eighteenth-Century Coroners' Bills," in

E. W. Ives and A. H. Manchester, eds., *Law, Litigants, and the Legal Profession* (London: Royal Historical Society, 1983), pp. 126–139.

22. Michael MacDonald, "Ophelia's Maimèd Rites," *Shakespeare Quarterly* 37 (1986):309–317.

23. Public Record Office (London), STAC 2–5, 7–8.

24. P.R.O., STAC 5/A1/21.

25. John March, *Amicus Reipublicae. The Common-Wealths Friend* (London, 1651), p. 109.

26. The Hale Commission's report is reprinted as "Several Draughts of Acts," in *A Collection of Scarce and Valuable Tracts*, 1st ed. (also known as *Somers' Tracts*), vol. 1 (London, 1748), pp. 497–592; the quotation is on p. 584.

27. 4 & 5 William & Mary, *cap.* 22; Narcissus Luttrell, *The Parliamentary Diary of Narcissus Luttrell, 1691–1693*, ed. Henry Horwitz (Oxford: Clarendon Press, 1972), p. 348.

28. Public Record Office (London), KB 33/25/2.

29. William Blackstone, *Commentaries on the Laws of England*, 2d ed., 4 vols. (Oxford, 1766), 1:302. The language of this passage varies slightly from edition to edition of this famous work. For its source, see M. Foster, *A Report of Some Proceedings on the Commission of Oyer and Terminer and Gaol Delivery for the Trials of the Rebels in the Year 1746* (Oxford, 1762), p. 266.

30. Daniel Defoe, *Defoe's Review*, ed. A. W. Secord (New York: Columbia University Press, 1938), p. 255.

31. See, e.g., *Gentleman's Magazine* 24 (1754):507; Caleb Fleming, *A Dissertation Upon the Unnatural Crime of Self-Murder* (London, 1773), p. 17; Moore, *Full Inquiry into Suicide*, pp. 336–337, 339.

32. Coroners' Inquisitions, Liberties of Cockermouth and Egremont, Cumbria Record Office (Carlisle), D/Lec/CR I, 14/2.

33. John Adams, *An Essay Concerning Self-Murther* (London, 1700), pp. 120–121.

34. *Defoe's Review*, p. 255; William Fleetwood, *The Relative Duties of Parents and Children, Husbands and Wives, Masters and Servants, Consider'd in Sixteen Sermons: With Three More Upon the Case of Self-Murther* (London, 1705), p. 482; Isaac Watts, *A Defense Against the Temptation to Self-Murther* (London, 1726), pp. 48–49; *Self-Murther and Duelling the Effects of Cowardice and Atheism* (London, 1728), p. 4; *Gentleman's Magazine* 19 (1749):341–342; *A Discourse upon Self-Murder: or, the Cause, the Nature, and the Immediate Consequences of Self-Murder, Fully Examined and Truly Stated*, 2d ed. (London, 1754), p. 14; F. Ayscough, *A Discourse Against Self-Murder* (London, 1755), p. 13; *Duelling and Suicide Repugnant to Revelation, Reason, and Common Sense* (London, 1774), p. 11.

35. MacDonald, "Secularization of Suicide," p. 91.

36. W. Ramesey (or Ramesay), *The Gentlemans Companion: or, A Character of True Nobility and Gentility* (London, 1672), pp. 240–241.

37. Charles Blount, *The Miscellaneous Works of Charles Blount*, ed. "Lindamour" [Charles Gildon] (London, 1695), sigs. A6r–A12r; A.-F. Boureau Deslandes, *A Philological Essay: Or, Reflections on the Death of Free-Thinkers* (London, 1713), pp. 119–121; idem, *Dying Merrily: Or, Historical and Critical Reflexions on the Conduct of Great Men Who, In Their Last Moments, Mock'd Death and Died Facetiously*, tr. T. W. (London, 1745), pp. 115–119; A. Radicati [Count of Passerano], *A Philosophical Dissertation upon Death* (London, 1732), esp. p. 86;

David Hume, "On Suicide," in his *Of the Standard of Taste and Other Essays*, ed. John W. Lenz (Indianapolis: Bobbs-Merrill, 1965), pp. 151–60.

38. L. G. Crocker, "The Discussion of Suicide in the Eighteenth Century," *Journal of the History of Ideas* 13 (1952):47–72; S. E. Sprott, *The English Debate on Suicide; From Donne to Hume* (LaSalle, Il.: Open Court, 1961), ch. 4.

39. Jonathan Swift, *Collected Poems of Jonathan Swift*, ed. J. Horrell, 2 vols. (London: Routledge & Kegan Paul, 1958), 2:724.

40. J. W. Johnson, *The Formation of English Neoclassical Thought* (Princeton: Princeton University Press, 1967), p. 100.

41. S. P. Atkins, *The Testament of Werther* (Cambridge: Harvard University Press, 1949), p. 37.

42. E. H. W. Meyerstein, *A Life of Thomas Chatterton* (London: Ingpen & Grant, 1930), pp. 475–476. The most influential eighteenth-century sentimentalization of Chatterton's life and death was H. Croft, *Love and Madness. A Story Too True*, 4th ed. (London, 1780), pp. 125–244.

43. Meyerstein, *Chatterton*, chs. 18–20; Linda Kelly, *The Marvellous Boy: The Life and Myth of Thomas Chatterton* (London: Weidenfield & & Nicolson, 1971), chs. 8–12.

44. MacDonald, "Secularization of Suicide," pp. 88–91.

45. Sprott, *English Debate on Suicide*, pp. 109–110.

46. Michael MacDonald, "Suicide and the Rise of the Popular Press in England," *Representations* 22 (1988):36–55.

47. Horace Walpole, *The Letters of Horace Walpole*, ed. H. Paget Toynbee, 19 vols. (Oxford: Clarendon Press, 1903–1925), 14:52; J. Woodforde, *The Diary of a Country Parson*, ed. John Beresford, 5 vols. (Oxford: Oxford University Press, 1981), 1:195, 338; 2:324; 3:291; 5:375.

48. W. Rowley, *A Treatise on Female, Nervous, Hysterical, Hypochondriacal, Bilious, Convulsive Diseases . . . With Thoughts on Madness, Suicide, &c* (London, 1788), p. 343.

49. A. Moore, *Backgrounds of English Literature, 1700–1760* (Minneapolis: University of Minnesota Press, 1953), ch. 5; J. F. Sena, "The English Malady: The Idea of Melancholy from 1700 to 1760" (Ph.D. thesis, Princeton University, 1967).

50. R. Blackmore, *A Treatise of the Spleen and Vapours* (London, 1725), p. 163.

51. George Cheyne, *The English Malady: Or, A Treatise of Nervous Diseases of All Kinds*, 3d ed. (London, 1734), p. *iii*.

52. Ramesey, *Gentlemans Companion*, pp. 240–241; Thomas Philipot, *Self-Homicide Murther* (London, 1674), p. 8.

53. John Jortin, *Sermons on Different Subjects*, 3d ed., 7 vols. (London, 1787), 5:147–148.

54. Adam Smith, *The Theory of Moral Sentiments*, ed. D. D. Raphael and A. L. Macfie (Oxford: Clarendon Press, 1976), p. 287.

55. Hunnisett, "Eighteenth-Century Coroners' Bills."

56. Hunnisett, *Wiltshire Coroners' Bills*; Coroners' Inquisitions, 1670–1800, Norfolk and Norwich Record Office (Norwich).

57. Westminster Coroners' Inquisitions, 1760–1800, Westminster Abbey Library and Muniments Room (London); Coroners' Inquisitions for London and Southwark, 1788–1800, Corporation of London Record Office (London); Middlesex Coroners' Inquisitions, 1753–1800, Greater London Record Office (London).

58. *The Case and Memoirs of the Late Rev. Mr. James Hackman . . . And Also Some Thoughts on Lunacy and Suicide* (London, 1779), pp. 25–26.

59. Michael MacDonald, "Religion, Social Change, and Psychological Healing in England," in W. J. Sheils, ed., *The Church and Healing*, Studies in Church History, 19 (Oxford: Blackwell, 1982), pp. 101–125.

60. Moore, *Full Inquiry into Suicide*, 2:68–70.

61. Henry Halliwell, *Melampronoea: Or a Discourse of the Polity and Kingdom of Darkness* (London, 1681), pp. 77–78; J. Glanvill, *Saducismus Triumphatus: Or, Full and Plain Evidence Concerning Witches and Apparitions*, ed. C. O. Parsons (1689; repr. Gainesville, FL: Scholars' Facsimiles and Reprints, 1966); Thomas, *Religion and Magic*, chs. 18–22; Michael Hunter, *Science and Society in Restoration England*. (Cambridge: Cambridge University Press, 1981), ch. 7; B. J. Shapiro, *Probability and Certainty in Seventeenth-Century England* (Princeton: Princeton University Press, 1983), chs. 1–3, 6; Charles Webster, *From Paracelsus to Newton; The Magic and Making of Modern Science* (Cambridge: Cambridge University Press, 1982).

62. Olive Anderson, *Suicide in Victorian and Edwardian England*. (Oxford: Clarendon Press, 1987), pp. 222–231.

63. Ibid., pp. 263–282.

6   AMERICAN PHYSICIANS' "DISCOVERY" OF HOMOSEXUALS, 1880–1900: A New Diagnosis in a Changing Society

BERT HANSEN

One of the most striking aspects of the interaction between Western medicine and society in the late nineteenth century was the tendency to medicalize deviance, to frame a variety of stigmatized or problematic behaviors in medical terms—to see guiltless disease where earlier generations might have diagnosed sin. It was true of drug and alcohol addiction, compulsions, anxieties, and depression. It was certainly the case with strong physical attraction to one's own sex.

Bert Hansen seeks to place the "discovery" and clinical construction of "homosexuality" within this general framework. But his argument has a second fundamental component, linked chronologically and structurally with these events in medical history. He associates the clinical perception of this new disease with urbanization and the gradual development of self-conscious homosexual communities in large cities whose density of single men and women facilitated the creation of single-sex subcultures.

Chronologically and spatially associated as well was the development of another urban phenomenon: medical specialization. The discovery of "sexual inversion" was in good measure the work of neurologists, a newly prominent outpatient specialty whose sphere of expertise naturally encompassed that diversity of symptoms traditionally designated as "nervous." Such specialists were an obvious choice for self-referring patients uncomfortable with particular manifestations of their sexuality. In the physician's ministrations and diagnoses, troubled patients could develop a framework to explain their otherwise unfathomable, and often guilt-inducing, desires. Presumably individuals free of such ambivalence did not find their way into the neurologist's consulting room.

Once added to the accepted categories of medical nosology, homosexuality became, as Hansen contends, culturally available as an option for individuals seeking a rational understanding of their needs and actions. In the late nineteenth century, of course, socially unacceptable behaviors were often construed as the consequence of an underlying somatic, constitutional state. For at least some men and women, such speculative mechanisms must in their very determinism have provided a sort of comfort. The beginnings of public discussion of these sensitive matters also demonstrated that such "sufferers" were not alone in the world. What might seem in retrospect an oppressive exercise in the hegemonic use of medical ideas and status to label deviance as—literally—pathological, might also have been a source of guilt-reducing resignation, Hansen suggests, and an element in the complex shaping of a subculture's collective identity.

—C. E. R.

I. Medical Novelties

ONLY IN THE LAST THIRD of the nineteenth century did European and American physicians begin to publish case reports of people who participated in same-sex sexual activity. In 1869 the *Archiv für Psychiatrie und Nervenkrankheiten* in Berlin published the very first medical case report of a homosexual, with six more cases following quickly in the German psychiatric literature.[1] A French report appeared in 1876 and an Italian one in 1878. The first American case report appeared in 1879, the second in 1881, and in the latter year the British journal *Brain* offered an entry, though that report's subject was German and the reporting physician Viennese. (British physicians seem to have been rather reticent on this subject through the 1890s.) The years 1882 and 1883 saw several major American additions to the literature on homosexuals, by which time nearly twenty European cases had been reported.[2] These persons were variously described as exhibiting "sexual inversion," "contrary sexual instinct," or "sexual perversion," though this last term was not common in the narrow sense and usually referred to a wider range of sexual possibilities.

By the century's end, the medical literature on the subject was large and growing fast, with dozens of American contributions and considerably more in Europe. For the most part, these reports treated the problem as a new phenomenon, unprecedented and previously unanalyzed. The case histories read like naturalists' enthusiastic reports of a new biological species. American physicians and their European counterparts not only described their subjects' lives, but also elaborated the novel conception that homosexual behavior was the manifestation of an underlying morbid condition, a view that remained official psychiatric doctrine into the 1970s. A majority of the early reports were of males, but many concerned female homosexuals, and although the social histories of lesbians and gay men have differed in certain periods, this first stage of medicalization treated them simultaneously—and similarly.

In several ways, the new diagnosis of homosexuality—or sexual inversion, as it was commonly labeled until well into the twentieth century—resembled other "new" diagnoses of that era, categorizing as medical problems behaviors not previously interpreted as such, for example, masturbatory insanity, neurasthenia, inebriety, anorexia nervosa, sexual psychopathy among women, multiple personality, and kleptomania. "Deviance was increasingly—if by no means universally—being defined as the consequence of disease process and, thus appropriately, the physicians' responsibility."[3] Yet unlike the nineteenth-century masturbators, kleptomaniacs, and neurasthenics, many of the sexual inverts were securing public space and recognition in a complex process of social and sexual change that was not limited to the medical sphere, however crucial medical diagnosis was to the new group's self-consciousness and the public conception of inversion. The doctors' discovery and framing of this new diagnosis were prompted by changing social phenomena: New kinds of people were in fact gaining public visibility—and consulted physicians who then discussed cases among themselves and in their journals. The medical formulation of sexual inversion as a disease, in turn, helped shape the behavior and the self-understanding of many inverts. In this interactive process, the physicians' reports of cases came to be normative, as well as descriptive.

The present study aims not only to describe the American medical profession's earliest encounters with homosexuals, but also to set the physicians' case reports and commentaries into the context of broad changes in the social structures of sexuality. It also calls attention to the vivid documents of social history preserved in the literature of clinical medicine; the first generation of medical case reports—in contrast to succeeding materials—naively recorded patients' own accounts of their lives, which now give rare glimpses of individuals in the process of creating new social forms.[4]

II. Social History: Action and (Medicalized) Reaction

Earlier historians saw this episode in the history of medicine as an evolution only of attitudes and labels. For them, an unchanging homosexuality was no longer regarded as a sin but simply became a sickness, relabeled, "medicalized," or "morbidified" by Krafft-Ebing, Havelock Ellis, and their less famous European and American colleagues. Such a formulation, however, masks the depth of the change and mistakes its fundamental character. More recent historiography has challenged this picture and reframes the historical process as a metamorphosis of the social and psychological reality of homosexuality itself—due in part to the activities of physicians—from a form of sexual behavior (a pattern of actions) to a condition (a way of being). The condition resides deeply within certain people who eventually came to be known as homosexuals, a group marked off from the rest who (slightly later) came to be known as heterosexuals. The new interpretation is often designated the "historical construction of homosexuality" or the "making of the modern homosexual" and, less formally, the "emergence thesis" or "social constructionism."[5]

An appreciation of this shift will be facilitated by reflecting a little on the modern ways of thinking about types of persons.[6] People's thoughts today commonly proceed, for example, from the observation of a theft to the recognition of a thief, from a crime to a criminal, and from a homosexual act to a homosexual; but our ancestors did not think that way about those acts. To them theft and sodomy were sins, which anyone might commit. A sinner might subsequently be socially labeled a "thief" or a "sodomite," but this label was only shorthand for "person who perpetrated this sinful act." It was *not* an indication of one's being a fundamentally different kind of person from one's peers. Furthermore, in this traditional viewpoint, a person could not be a thief or a sodomite without having committed the relevant act.

Several features of sexuality's "old regime," which was replaced in America only gradually over the last few decades of the nineteenth century,[7] are exemplified in a brief newspaper report of 1867 from the *California Police Gazette*, a San Francisco scandal sheet.

Miss Mary Walker, a fast young lady of Richmond, Va., took it into her head, a few months ago, to don male attire, and engaged as a barman. Somehow, the breeches seem to have put bad notions into her head, for she went making love to the pretty girls who came after the family beer. One of them corresponded with her, and was engaged to be married to her. As making presents costs money, the feminine barman borrowed from the till, was detected, pleaded guilty, and is now waiting for her sentence. Her enamored fiance [*sic*] visited her in jail. They are not to be married at present, as women's rights have not attained to that degree of development.[8]

This account acknowledges (homo)sexual interest only casually and emphasizes instead the anomalous gender role; it does not treat either of the women as a sexually odd sort of person.

When one turns to some of the medical case reports written only fifteen years later, one encounters the "modern" way of thinking about persons and their sexuality, in which actions came to be seen as less critical than essence or being. Consider, for example, this excerpt from one of the earliest cases published in America, a long report on "Mr. X" by Dr. G. Alder Blumer, an Assistant Physician at the lunatic asylum in Utica, New York.

Mr. X. is about twenty-seven years of age, and of high social status. Height, considerably below the average; muscular development, good; hair, dark, profuse in growth, parted in the middle and brushed well back; eyes, dark brown, large, brilliant and swimming; pupils dilated; long eyelashes; mouth small; teeth, regular and sound; chin, somewhat pointed; general contour of face, oval; expression, womanly. Lower limbs, short in proportion to trunk, but well developed. Sexual apparatus believed to be normal. Gait, precise; strides, quick and short. Voice and intonation, like a woman's; lisps in pronouncing certain words. Capillary circulation, poor; extremities, frequently cold. Head often flushed and hot, and complains of headache. . . . Probably practiced onanism when younger. Nocturnal emissions of unusual frequency.

. . . As a child, showed no disposition to mingle with boys of his own age and adopt their pastimes [sic]. Very precocious, and developed an early interest in literature, himself writing prose and poetry when quite young. Passionately fond of music, a brilliant pianist and composer of weird-like impromptus. Occupations and tastes essentially womanly. Fond of discussing women's dress, in which he is always *au courant*. His own dress is always precise and natty, showing more especially in pattern, style and arrangement of necktie a taste and deftliness rarely found in men. Conscious of his youthful, unmanlike appearance, he is very sensitive on the subject, and resents imputations of womanliness. Admires manly men, frequently speaks of them, and extols all that is noble in his own sex. Seldom speaks of women, is indifferent to their charms, and expresses a horror of matrimony, the very idea of marital relations being repugnant to him. Admits that he has on several occasions been approached by men of unnatural desire, and declares his unspeakable horror of paederastia. Has in conversation said that these latter individuals are able to recognize each other. Began the study of law, but finding its dry details distasteful, soon abandoned it for literature in which he achieved early success. . . . Is good to the sick and poor, and takes pleasure in charitable work. . . . Punctilious in regard to table etiquette; has a fastidious and capricious appetite, eating in a nibbling, mincing manner like many women.[9]

Like the California newspaper writer quoted a little above, Dr. Blumer attended carefully to X's cross-gender mannerisms and behavior. But Blumer then proceeded to characterize Mr. X as exhibiting "contrary sexual instinct" even while recording that X had never had a sexual experience with another man. Clearly, sexuality was in the process of coming to stand for something other than a person's sexual behavior. Blumer's 1882 report also illustrates how patient and physician might collaborate in framing (homo)sexuality as a condition, rather than a set of behaviors.

What was the historical process through which previously isolated, individual sexual actions came to be associated with a "type of person" embodying an abnormal "instinct" or "condition"? In my view, this transition can best be described and analyzed as the evolution of four interacting structures: social roles (patterned behaviors and the expectations held by others of these patterns), a self-conscious identity (the personal internalization that one *is* that kind of person, not just acting out a role), social institutions (recognized settings for particular activities), and community (a shared sense of belonging to a group with others who have the same role or identity).[10]

In the decades after 1800 in fast-growing American cities, especially in the Northeast, more and more men were working and living outside of the productive farm family typical of earlier generations. While we have no reason to assume that more of these men manifested homosexual desire than those in earlier times, their new social circumstances did allow those with such desires more opportunity to find others who shared their tastes. Living outside of the traditional family also allowed such men to spend more time in the pursuit of sex, with the result that specialized meeting places gradually came into existence (certain bars, parks, transportation depots, military installations, etc.). Such institutions fostered a vague sense of community—at first, however, this was only a community of taste, not of identity or style of life. But the growth of such institutions and the more numerous social and sexual encounters they facilitated led to confrontations with the police, with moral reformers, and eventually with doctors. These clashes led to labeling and, over several decades, to some self-consciousness. A tentative sense of identity facilitated further interaction with other members of the community, which then facilitated the formation of a homosexual identity for more individuals. The chain-reaction continued, yielding at each step more labeling, further community growth, higher visibility, and the proliferation of institutions.[11]

This general process took different forms and moved at different rates in different countries, varying also by region, class, level of urbanism, and, of course, sex.[12] It seems that many of these changes occurred about a generation or two later among lesbians in America than among male homosexuals. That the development of homosexual and lesbian identities depended on being able to live and support oneself outside of the family seems partly confirmed by the fact that the first noticeable group of women to do this, the graduates of the new women's colleges, had both a low rate of marriage and a large proportion of lesbian couples.[13]

With the evidence so far available, it seems useful to see the first two thirds of the century as a period characterized by individual explorations of new life-styles and the next stage as a consolidation of identities and the creation of a new public awareness of sexually deviant social roles. Data for the earlier period are as yet rather limited but nonetheless suggestive. Some records tell of homosexual affairs, meeting places for men, and arrests for sodomy, and hint at a growing awareness that other people shared the same tastes and interests—especially in the eastern seaboard cities and in post–Gold Rush San Francisco. One indication of some level of popular awareness is in the glimpses we get here and there of new stereotypes, for example, of the dry-goods clerk, or "counter-jumper," of the new urban emporia, regarded as an unmanly occupation. Soldiers near military depots and bellboys in grand hotels came to have some notoriety for their sexual accessibility.[14]

Three things made the sexual encounters of this first stage historically different from the couplings of previous centuries. First, they now sometimes occurred in recognized locales. Second, they more frequently involved men on their own, living outside of the productive, family units that dominated living space before the rise of wage labor in cities. Third, previously private and personal activities were taking on a social dimension. Men could pursue sexual adventures not as short-term diversions from family responsibilities but as a full-time life-style; and by the process of interacting, some of these men came to see themselves not only as having different tastes from the majority, but as being a fundamentally different type of person. In the same era, however, the homosexual experiences of many people did not include even a hint of the newer consciousness of being a different sort of person.[15] Physicians' case reports, cited below, illustrate the simultaneous presence of both old and new ways of thinking during this transitional period.

One remarkable first-person account can flesh out this rather abstract portrayal of how people in those years articulated a new identity and fashioned a sense of community; it also portrays one of the social institutions that facilitated these developments. Under the auspices of Dr. Alfred W. Herzog, editor of the *Medico-Legal Journal*, a man named Earl Lind, who called himself Jennie June, published his memoir, *Autobiography of an Androgyne*, in the hope that

every medical man, every lawyer, and every other friend of science who reads this autobiography will thereby be moved to say a kind word for any of the despised and oppressed step-children of Nature—the sexually abnormal by birth—who may happen to be within his field of activity.[16]

The book also solicited reports of similar cases, offering a two-page questionnaire that could be completed by "every friend of science," whether a physician or one of "the homosexualists" himself. (See fig. 6.1.) In his sequel, *The Female Impersonators*, edited by Dr. Herzog four years later, Lind reported an experience he had had nearly thirty years before in a New York City bar.

> On one of my earliest visits to Paresis Hall—about January, 1895—I seated myself alone at one of the tables. I had only recently learned that it was the androgyne headquarters—or "fairie" as it was called at the time. Since Nature had consigned me to that class, I was anxious to meet as many examples as possible. . . .
>
> In a few minutes, three short, smooth-faced young men approached and introduced themselves as Roland Reeves, Manon Lescaut, and Prince Pansy—aliases, because few refined androgynes would be so rash as to betray their legal names in the Underworld. Not only from their names, but also from their loud apparel, the timbre of their voices, their frail physique, and their feminesque mannerisms, I discerned they were androgynes. . . .
>
> Roland was chief speaker. The essence of his remarks was something like the following: "Mr. Werther—or Jennie June, as doubtless you prefer to be addressed—I have seen you at the Hotel Comfort, but you were always engaged. A score of us have formed a little club, the Cercle Hermaphroditos. For we need to unite for defense against the world's bitter persecution of bisexuals. We care to admit only extreme types—such as like to doll themselves up in feminine finery. We sympathize with, but do not care to be intimate with, the mild types, some of whom you see here to-night wearing a disgusting beard! . . .
>
> "We ourselves are in the detested trousers because having only just arrived. We keep our feminine wardrobe in lockers upstairs so that our everyday circles can not suspect us of female-impersonation. For they have such an irrational horror if it!"[17]

This brief extract illuminates how—in one small group at one point in time—community, identity, and a variety of homosexual social roles were being formed and reinforced.

Except for the medical auspices of its eventual publication, this quotation does not, however, make explicit the single most important novelty in the decades after 1870: a new public discourse, primarily medical, about the homosexual as a type of person. Although the popularizing of this medical discourse was slow and uneven, I believe it played a major role in reinforcing gay people's consciousness of being different and also of just how they were different. This process took place directly in the doctors' consulting rooms and indirectly through medical publications and the nonmedical writings they influenced. Even though some of the medical writings were not supposed to be distributed to the public, they were eagerly sought out—and found—by individuals searching for clues to their own nature.[18]

QUESTIONNAIRE ON HOMOSEXUALITY

[The governments of all cultured lands take from time to time censuses of the blind, the deaf, and other defective classes. None has ever taken a census of homosexualists, although the latter are fully as numerous as the two definite classes previously named, and their effect on the social body is even more marked. The Medico-Legal Journal, on the basis of the following questionnaire, makes the first essay, in the history of culture, in lining up the defective class in question so that science may have a broader knowledge of them than that afforded by the comparatively few detached biographical and analytical notes at present extant. The reader is therefore requested to fill out the following questionnaire — or have the intelligent homosexualist do so — and mail it to the Medico-Legal Journal, New York. If unable to answer all queries, kindly give as much information as possible. Additional schedules will be furnished on request. Unpublished textual descriptions of cases would be welcome, and will be returned on request. The results of this questionnaire will be published, and the addresses of respondents will be filed for due notification.]

(1) Physical sex of homosexualist...... (2) Age at date....years (or....years at death)
(3) No. of brothers.... Sisters.... (4) Approximate age of father at subject's birth.... Of mother....
(5) Underline applicable physical type: Brunette. Blonde. Red-haired. Not definitely any of these.
(6) Principal occupation as adult....................
(7) Lineage (i.e., from what foreign countries did forebears emigrate) ....
....
(8) Environment in which life principally passed (Indicate by x's):

	Rural or village under 2,500	Municipality 2,500 to 25,000	Municipality 25,000 to 100,000	Municipality 100,000 to 500,000	Municipality over 500,000
Up to 10 years old	.....	.....	.....	.....	.....
11 to 20 years old	.....	.....	.....	.....	.....
21 to 50 years old	.....	.....	.....	.....	.....
After 50 years old	.....	.....	.....	.....	.....

(9) Ever legally married..... Children, how many.....

(10) Plays any musical instrument...
(11) Underline interest in sport: Practically none. Slight. Extensive.
(12) Underline interest in music: Practically none. Slight. Extensive.
(13) Underline interest in other art (designate): Practically none. Slight. Extensive.
(14) Underline interest in religion: Practically none. Slight. Extensive.
(15) Underline applicable schooling: Less than 8 years. High school. Liberal-arts college. Postgraduate. Professional school.
(16) If liberal arts course, favorite subjects in order of preference.........................
(17) Number — if any — of foreign languages ever spoken with considerable ability..... Number studied for translation.....
(18) What — if any — mental diseases suffered..... ....
(19) A dipsomaniac..... Other drug addiction.....
(20) What rather serious (excluding the practically universal) bodily diseases suffered....................
.... Particularly underline applicable:
Venereal warts. Syphilis. Gonorrhea. Locality (initial) of last three..........
(21) If dead, cause of death..
(22) What — if any — disease has run in the family of either parent..........................
(23) What other blood relatives have shown sexual abnormality:
Definite relationship   Nature of abnormality
....
....
....
(24) Are sexual organs normal..... Describe any abnormalities ....
....
....
(25) Underline applicable fundamental or original instinct:
Fellatio (active buccal). Passive buccal. Masturbation of other. Mutual onanism. Paedicatio (anal), indicating whether subject is active or passive, or both. Cunnilingus (corresponding to active fellatio in the male). Tribadism (corresponding to masturbation of other in the male).
(26) Approximate age when instinct first manifested itself in the feelings.....years. In actions.....years.

Fig. 6.1. Questionnaire on homosexuality, from Earl Lind, *Autobiography of an Androgyne* (New York, 1918).

(27) Secondary or acquired methods of coitus..
....
(28) Is subject a psychical hermaphrodite (attracted toward both sexes)..........
(29) Does coitus stimulate any erogenous center (i.e., afford a pleasurable titillation of any portion of the body) or is the satisfaction entirely or almost entirely mental ....
(30) State of health day following coitus..
(31) If a male, does coitus induce an emission..... If so, is sensation pleasurable or horrifying.............
....
(32) Is there love or adoration for the associate, or is the latter used merely to secure the stimulation of the subject's erogenous center..
(33) Upper and lower age limits of individuals that attract, with indication of subject's own age at the different periods ....
(34) Any quality or apparel (such as plumpness, military uniform) that constitutes a special attraction....
....
(35) If a physical male, is he undersized..... If female, unusually large..... Muscles vigorous or feeble..........
(36) Is there a striking contrast between the real and apparent age.....
(37) Any peculiarity about the hair system, particularly the facial...................................
....
(38) If a physical male, any tendency to wear the hair several inches long................................
(39) Describe any anatomical approach toward the psychical sex (e.g., milk glands in a male)...............
....
(40) Ever desired to wear apparel of psychical sex..... Ever worn such apparel since childhood.........
(41) If a physical male, fondness for loud or fancy apparel..... If female, for plain apparel..............
....
(42) Ever arrested or imprisoned for following instincts of psychical sex..... If so, aggregate number of months or years imprisoned...
(43) Please note any other significant data as to particular homosexualist on separate sheets.

GENERAL QUERIES:

(44) Approximate number of homosexualists — positively known as such — have you encountered in your life-time: Passive inverts.... Active pederasts.... Male psychical hermaphrodites.... Physically female homosexualists....
(45) How many additional individuals have been under your suspicion: Passive inverts.... Active pederasts.... Male psychical hermaphrodites.... Physically female homosexualists....
(46) The author of "Autobiography of an Androgyne" estimates, roughly, passive inverts as 1 out of 300 physical males. Please give your estimate: Passive inverts, 1 out of....males. Active pederasts, 1 out of....males. Male psychical hermaphrodites, 1 out of....males. Female psychical hermaphrodites, 1 out of....females. Other physically female homosexualists, 1 out of....females. (In query 46, only adults are to be considered.)

NAME AND ADDRESS OF PHYSICIAN SUBMITTING SCHEDULE. THIS INFORMATION WILL BE CONSIDERED CONFIDENTIAL AND IS ESSENTIAL.

Name....................
Street and number....................
Post-office....................
Date..........   State....................

Fig. 6.1. continued

III. Discovering a New Disease

In the medical reports about sexual inverts, the clinicians were self-conscious about opening up new territory, and they were proud of being the first generation to study the subject in a scientific manner.[19] The authors of these pioneer case studies agreed as to the history of their collective project. "Casper was the first to call attention to the condition known as sexual perversion," explained one typical account.

"But not until several works had been published by a Hanoverian lawyer, Ulrichs, himself a sufferer from the disease, did the matter become the subject of scientific study on the part of physicians. Westphal was the first to discuss it."[20]

In the 1870s and 1880s, when sexual inversion still seemed a rarity, individual cases were carefully described, numbered, and added to the stock of specimens. In 1883 Dr. J. C. Shaw, a Brooklyn practitioner, and Dr. G. N. Ferris, First Assistant Physician at the Kings County Asylum, announced that theirs was the first case in the United States and the nineteenth case worldwide.[21] Their report, though shorter than most, is sufficiently typical to be worth examining in full.

> Our case . . . came under observation twice in the spring of 1880. The patient has not been seen since; in fact, was so reserved that it has been impossible to find him again.
>
> Case 19.—A German, aged thirty-five years. Height, about five feet four inches. Weight, about one hundred and forty pounds. Black hair. Dark complexion. Well built. Physiognomy of an intelligent expression. Very reticent on some points of his history. Says he is engaged in mercantile business and occupies a good position. For some time past has had an almost uncontrollable desire to embrace men. Fears that some time this horrible morbid desire may overcome him and he will really embrace some of his fellow-clerks. When in the presence of men, he is tormented by constant erections of his penis and a desire to embrace the men. He regards his desire as abnormal, and laments his condition. Remarks that he is ashamed to tell the doctor of his condition, as he must consider him a horrible creature, and look upon him with disgust. Has never given way to his desires, but is afraid that they will overcome him some day and make known his unnatural condition. He has tried to overcome this morbid state by having connection with women, but intercourse with them gave him no pleasure, and he was obliged to force himself to it, and has only tried it on three occasions.
>
> Examination of genital organs shows them to be well developed. At the time of examination penis is in full erection, and this the patient says is the condition of the organ whenever he is near men. He denies nervous disease in his family. Patient is an intelligent man, and perfectly natural in his appearance and manner, except that he is distressed by his abnormal state, and wishes medicine to overcome it. Would disclose neither his name, residence, or family history.[22]

In 1884 Dr. James G. Kiernan of Chicago brought the total, he claimed, to twenty-seven (with five from America) and observed that, with only four females and twenty-three males, men are clearly "predisposed to the affection."[23] Over time as the cases grew more numerous, it was possible for physicians to write about the condition without retelling individual life histories.

The most complete reports of homosexuals and lesbians ran twenty pages or more, but some were as brief as two paragraphs. Extensive case reports included information about age, physique, physiognomy, occupational history, medical history, family history (especially insanity or nervous disorders among relatives), gender attributes, emotional life, and sexual experiences.

Both women and men were described as inverts, although clinicians agreed that men predominated. Chicago neurologist James G. Kiernan suggested that the condition might be noticed more frequently among women if it were not so difficult, in general, to elicit a full sexual history from females.[24] Cases included a full range of occupations and social levels, though clerks and small-business proprietors appear frequently. While a number of the cases were first discovered among asylum patients, most of these exhibited no mental symptoms beyond their odd sexuality, and few of those outside the asylum system had mental problems beyond discontent with their peculiar sexual urges, shame, or fear of exposure and disgrace.

Among those men and women who felt attraction to members of their own sex, a number revealed cross-gender feelings, actions, and even physique, ranging from casual tomboy activities and a somewhat cross-gendered style of attire to the feeling of having a soul mismatched to one's body. Others were "perfectly natural in appearance."[25] While the medical literature of this era also recorded some men having elaborate erotic involvement with female clothing and masquerade, it reported them being sexually oriented toward women rather than other men; they used their feminine finery as a masturbatory stimulus.[26] Sometimes the adoption of an outwardly feminine style served among homosexual men as a means of mutual recognition. No men were recorded as having "passed" consistently in public living as women, though there are numerous cases in which two females lived as husband and wife for years or even decades without detection, sometimes even being legally married.[27] For example, Murray Hall, a prominent Tammany Hall leader for many years, voted regularly as a man and was married twice to other women before her successful masquerade was discovered and revealed by her physician when Hall died of cancer in 1901.[28]

As with gender variation, the actual sexual activity of this first generation of medically observed inverts varied widely. The described sexual relations ranged from the purely genital to feelings and forms of affection far removed from the physical. The unfortunate Mr. X, described above, pursued the affection of his beloved for months without seeking a physical connection, having an "unspeakable horror" of it. An immigrant businessman was tormented by desire "to embrace men," but had "never given way to his desires."[29] One young woman, in the words of Dr. Kiernan,

feels at times sexually attracted by some of her female friends, with whom she has indulged in mutual masturbation. These feelings come at regular intervals, and are then powerfully excited by the sight of female genitals. . . . She is aware of the fact that while her lascivious dreams and thoughts are excited by females, those of her female friends are excited by males. She regards her feeling as morbid.[30]

In contrast, Lucy Ann Slater, alias "Rev. Joseph Lobdell," and an un-named cigar dealer were both sexually experienced and demanding. When committed to an asylum, Slater "embraced the female attendant in a lewd manner and came near overpowering her. . . . Her conduct on the ward was characterized by the same lascivious conduct, and she made efforts at various times to have sexual intercourse with her associates."[31] The cigar dealer went to considerable lengths to secure sexual satisfaction on a regular basis, even in a less than ideal relationship.[32]

A number of the inverts described by American physicians acknowledged heterosexual experience, sometimes with husbands or wives, sometimes with prostitutes. While some of these experiences were forced (often at the urging of friends or doctors) and others were reported as unsatisfying, the cases taken together do not actually confirm Krafft-Ebing's characterization, repeated frequently by American doctors,[33] that inverts have a congenital absence of sexual feeling toward the opposite sex.

That the cities might harbor numerous inverts was recognized early by some New York physicians, such as George M. Beard.[34] In 1884 Dr. George F. Shrady, editor of *The Medical Record*, printed without comment the claim of a German homosexual then living in America that an estimate of the rate of sexual inversion at one in five hundred was too low, since he was personally acquainted with twelve in his native city of 13,000 and also knew at least eighty in a city of 60,000.[35] When Dr. Blumer's Mr. X rejected the approaches from "men of unnatural desire" out of his distaste for the sexual behavior it implied to him, he noticed that they were "able to recognize each other," and he was aware that these men shared some common nature, even if he thought he was not one of them.

Relatively few lesbians and homosexuals described in the medical literature, however, seem to have been aware of others like themselves. Many felt no possibility of finding like-minded individuals, yet some, especially in larger cities, did locate gathering places and make contacts within tentatively forming communities.[36] It seems likely that they were seeking not only social and sexual relations, but also a confirmation of their odd feelings, for numerous cases refer to persistent efforts toward self-understanding. By voluntarily approaching physicians, many seem to have sought knowledge as much as therapy; case accounts regularly refer to patients' concern for self-justification. Blumer, for example, described a young man's letters and essays which sought to "explain, justify or extenuate his strange feeling."[37] Two decades later, Dr. William Lee Howard of Baltimore observed of homosexuals, "they are well read in literature appertaining to their condition; they search for everything written relating to sexual perversion; and many of them have devoted a life of silent study and struggle to overcome their terrible affliction."[38]

Despite their scientific aspirations, these pioneering clinicians were quick to generalize from a very small number of cases. At first, while the discussion was closely tied to the cases being reported, the diversity of the described experiences slowed efforts to fashion a well-defined syndrome under the rubric of sexual inversion, but after some publications outlined general features, other observers were drawn into elaborating a consistent invert type. Even though George M. Beard had earlier pointed out that most inverts had "no occasion to go to a physician; they enjoy their abnormal life . . . or are too ashamed of it to attempt any treatment,"[39] an increasingly uniform medical image came to stand for the whole group. That the doctors saw sexual inversion as a unitary condition (despite extreme variations), rather than as an accidental collection of disparate phenomena, probably resulted in part from the self-consciousness exhibited by at least some of the inverts themselves.

In searching for the characteristic feature of their sexually inverted patients, the physicians were unselfconsciously formulating the modern notion of a person's "sexuality" as something distinct from that individual's sexual behavior. After the initial group of cases, the reports came to focus less on particular sexual actions than on consistent impulses—today one might say "orientation" or "preference"—and by implication, personality. In moving to a level of characterization that departed from gross sexual behavior, the doctors were not inventing a scheme *ex nihilo;* they were following the lead of their patients, most of whom felt that there was some interior quality that made them "different," whether it appeared in their behavior or not. As the reporting clinicians tried to draw this condition into view for examination–and possibly therapy—they juggled and juxtaposed old and new concepts. For example, when they observed homosexual behavior in persons who seemed to lack this interior condition, they termed it "vice," and condemned it with a vigor that very few of them applied to the behavior of those they considered true inverts.[40] The distinction between being and behavior could even operate in such a way that the sexual partners of the lesbians and homosexuals were in many cases not regarded as inverts by either the doctors or the patients.

IV. Professional Reputation
and Patient Self-referral

Some reports provide considerable evidence of an early pattern of self-referral to appropriate physicians. Looking back from the perspective of 1904, Dr. Howard described it this way:[41]

They have but little faith in the general practitioner; in fact, in our profession, and their past treatment justifies their lack of confidence. Hence it is that when they do find a physician who has taken up a conscientious study of their distressing condition, they open their hearts and minds to him.

And later in the same article, Howard remarked of a patient who was a Princeton student and a music lover:

He was well informed as to the attitude of the family physician in such cases as his, hence had studied up the subject for himself, having quite a library dealing with sexual perversions.

How such people found sympathetic physicians is not indicated in the case records, but since many were aware of the growing medical literature (both European and American) on their condition, it seems probable that some physicians gained a reputation for trust and tolerance. The fact, as noted below, that practitioners of the emerging specialization of outpatient neurology were overrepresented among authors on inversion is another indication that some homosexuals at least were aware of specialty groupings within the profession—and of the neurologists' reputation as healers sympathetic to sufferers from "nervous" ills.

Because, with rare exceptions, the evidence on these men and women's motivations for self-referral comes only through the physicians' reports, it is impossible to know precisely why they approached physicians for assistance at this particular juncture in time. Yet three general transformations of that era probably helped shape individuals' response to a sense of personal difference. First, science slowly and broadly came to hold a position of social authority by the end of the nineteenth century in the United States, also enhancing the status of medicine. One consequence for some Americans was that physicians replaced the clergy as authoritative personal consultants in the realm of sex. Second, for contemporaries who observed the behaviors being medicalized in these decades—kleptomania and the judicial defense of "innocent by reason of insanity" were prominently discussed in the press, for example—the shift in status from crime to illness might well have appeared humanitarian and progressive. The consequent image of a sympathetic, forward-looking profession probably encouraged individuals to risk sharing their secrets with physicians active in these developments. A third context of self-referral was the contemporary emergence of a public awareness of homosexuals. That novelty showed some people with "odd feelings" that they were not unique in their experience, and while such an awareness would not necessarily lead a troubled person to choose to consult a physician, it promoted the idea that other people might assist in understanding or changing one's feelings. Whether a troubled person turned for help to a physician or to fellow sufferers clearly depended on such factors as the person's occupation, place of residence, social standing, awareness of communities of inverts, and individual psychology, as well as familiarity with the new types of medical practitioners.

The force of these three factors is clear in Earl Lind's *Autobiography of an Androgyne*. When at age seventeen Lind first shared concern about his sinful impulses with a minister, he was counseled to see the family doctor. This physician "advised me to enter into courtship with some girl acquaintance, and said that this would render me normal. Like most physicians in 1890, he did not understand the deepseated character of my perversion."[42] Lind later sought specialized help first from Prince A. Morrow, the eminent venereologist, and then from an alienist, Robert S. Newton. Thereafter, Lind read all the medical literature on inversion he could find at the library of the New York Academy of Medicine. For a while he abandoned physicians, who mostly administered anaphrodisiacs. But in time he found a doctor (unnamed) who advised him to follow his sexual desires as a less harmful alternative to the risk of nervous problems from frustrating an exceedingly amorous nature. Through many meetings this physician supported Lind's sense that his peculiar character was natural to him and should be accepted and permitted expression. Eventually Lind published his memoir so that other physicians—and their patients—would suffer less from ignorance of natures like his.

While the pioneer specialists consistently acknowledged social disapproval of the sexual behaviors involved, most tempered moralism with a tone of scientific detachment, exhibiting a tolerance based, at least rhetorically, on a medical materialism powerfully reinforced by contemporary achievements. Reports often included some sort of evaluation, though not always so ferocious as Dr. Monroe's litany: abominable, disgusting, filthy, worse than beastly.[43] Most physician authors, however, distanced themselves from such scornful attitudes—without openly rejecting them—by coupling their disapproval with some justification for the legitimacy of their interest. George F. Shrady's 1884 editorial opened with a typical example:

Sir Thomas Brown once wrote . . . that the act of procreation was "the foolishest act a wise man commits in all his life. Nor is there anything that will more deject his cooled imagination." The physician learns, however, and finds . . . far down beneath the surface of ordinary social life, currents of human passion and action that would shock and sicken the mind not accustomed to think everything pertaining to living creatures worthy of study. Science has indeed discovered that, amid the lowest forms of bestiality and sensuousness exhibited by debased men, there are phenomena which are truly pathological and which deserve the considerate attention and help of the physician.[44]

Kiernan opened one of his articles in a similar fashion: "The present subject may seem to trench on the prurient, which in medicine does not exist, since 'science like fire, purifies everything.'"[45] In William A. Hammond's monograph on impotence, the section on sexual inversion began with a disclaimer different in tone, but similar in function:

> Several cases of sexual inversion in which the subjects were disposed to form amatory attachments to other men have been under my observation. They are even more distressing and disgusting than cases . . . I have just given; but it is necessary for the elucidation of the subject to bring their details before the practitioner. So long as human nature exists such instances will occur and physicians must be prepared to treat them.[46]

Despite his declaration of disgust and distress, Hammond's reports are quite sympathetic to the two inverts he described; and while such sympathy from Dr. Hammond and his colleagues may not have been unconditional, the humane impulse in many of these pre-1900 cases is prominent. This feature of the new viewpoint may well have arisen from its formation within the clinical context, where categories are not simply abstractions and where "problems" are embodied in persons.

When clinicians asserted that homosexuality should no longer be regarded as criminal and forbidden, but tolerated in private as pathology that might be treated, some of them may at times have been aware that this would expand the medical profession's power.[47] For example, George F. Shrady in *The Medical Record*, which he edited, declared on behalf of the profession:

> we believe it to be demonstrated that conditions once considered criminal are really pathological, and come within the province of the physician. . . . The profession can be trusted to sift the degrading and vicious from what is truly morbid.[48]

But such a motivation seems subordinate to and entirely consistent with their interest in according science and nature a higher status than traditional morality and religion. After the Chicago physician G. Frank Lydston began his 1889 article[49] by quoting Kiernan's statement that "science, like fire, purifies everything," he announced that the subject of sexual perversion was being taken from the "moralist"; it was far better "to attribute the degradation of these poor unfortunates to a physical cause, than to a wilful viciousness" and to think of them as "physically abnormal rather than morally leprous."

Other physicians, nonetheless, saw moral dangers in the profession's accepting the invert as natural, even as a *lusus naturae*. For example, Dr. J. A. De Armand, of Davenport, Iowa, actively opposed the trend toward medicalizing—and thus offering inappropriate sympathy for—sin:

Sexual perversion is the direct outgrowth of sexual abuse. It is the sensual Alexander seeking new worlds to conquer. It is the legitimate heritage of vicious associations and acquired weakness. The complimentary offering of "mental derangement" which excuses the man who seeks sexual gratification in a manner degrading and inhuman, is seldom more than a cloak within whose folds there is rottenness of the most depraved sort. . . . It surely is unnecessary to complicate medico-legal nomenclature by attributing such conduct to morbid mentality, when it clearly is deviltry. . . . The individual who starts out with jackass proclivities and billy-goat capabilities will soon tire of the normal sexual act, and the fact that he enters into an era of sexual abandon . . . is no proof that his mentality is at fault. . . . I have no patience with the ready excuse which the medical profession volunteers.[50]

However common De Armand's sentiments may have been among the profession in general, they were not frequently expressed in print.[51] In the published literature on sexual inversion the medical model was becoming dominant by the turn of the century.

If the materialism of a scientific approach within medicine was one of the intellectual supports for reclassifying homosexuality from sin to natural anomaly, the shift was further advanced by a contemporary realignment of power and prominence among specialists in mental disease. An old guard, primarily asylum superintendents who were conservatives on political, religious, moral, and scientific issues, lost ground to neurologists, characterized as a group by private practice, appointments to general hospitals rather than asylums, European training in science, and a tendency toward agnosticism and materialism.[52] The neurologists, including the eminent figures William A. Hammond, Charles H. Hughes, James G. Kiernan, and Edward C. Spitzka, predominated among the mental disease specialists who published cases of sexual inversion. With their highly visible outpatient practice, they were a natural first recourse for self-referral by troubled men and women.[53]

V. Persistence of Older Conceptions

While some physicians were establishing the invert as a type of person, and others were resisting this reformulation, still others were continuing to deal with sexuality in traditional terms, as simply an aspect of behavior rather than a fundamental aspect of being. For example, when Dr. Randolph Winslow of Philadelphia reported in 1886 on an "epidemic of gonorrhoea contracted from rectal coition" in a boys' reformatory, he described the extent and manner of "buggery" with precision, but without acknowledging any awareness that a burgeoning medical literature was endeavoring to describe the type of person engaged in such practices and to determine whether and when constitutional factors were more significant than depravity in explaining their occurrence.[54]

As late as 1899 George J. Monroe, a Louisville proctologist devoting an entire article to sodomy and pederasty, considered these sexual acts as "habits" ("abominable," "disgusting," "filthy," and "worse than beastly" to be sure) with no attention to what kinds of persons might engage in them and with no psychological etiology. His assumption was that such acts occur either situationally, "where there is enforced abstinence from natural sexual intercourse," such as among "soldiers, sailors, miners, loggers," etc., or in those "satiated with normal intercourse." Without apparent irony, he declared, "There must be something extremely fascinating and satisfactory about this habit; for when once begun it is seldom ever given up."[55]

Noticing these differences in thinking, however, should not lead one to conclude that Winslow, Monroe, and others like them were consciously rejecting the new observations and the style of thinking that defined the homosexual as a particular, if pathological, personality type. Since they neither addressed nor challenged this novel conception, it seems more likely that they were simply unaware of it. In their traditional view, someone need only "tire of the normal sexual act" to engage in homosexual practices.

VI. Medical Discovery,
the Discovered, and the Wider World

Reviewing the ways late-nineteenth-century clinicians discovered homosexuals and then shaped homosexuality into a disease entity brings two points to the fore. First, this initial stage in the morbidification of homosexuality does not fit into a unidirectional model of stigmatization and social control. Patients were often active conspirators with the physicians, not passive victims of the new diagnostic labeling.

Although in the twentieth century the homosexuality diagnosis became a central feature in the social oppression of homosexuals, to the benefit of some members of the medical profession offering "cures,"[56] it is wrong to attribute such later baneful developments to the first generation of observers or their self-referring homosexual patients. The actions of those doctors and patients must be appreciated for the ambiguous character they had at the time and not dismissed out of a twentieth-century distress over people victimized by medicalization. Carroll Smith-Rosenberg's interpretation of female hysteria is an apt guide here for its emphasis on the collaboration between patients and their physicians, the advantages the diagnosis gave women (at least some of them some of the time), the way both parties' thoughts and actions were shaped by cultural expectations concerning properly gendered norms of behavior, and the way "physicians, especially newly established neurologists with urban practices, were besieged by patients."[57]

Physicians played another and more public role as well. Since few homosexuals before 1900 described their lives publicly from a nonmedical point of view, doctors were, by default, the leaders in accumulating and organizing knowledge of homosexuals for society as a whole. In time, this "knowledge" (even distorted as it was in parts) entered the public domain, where it offered thousands of uncertain people an identity, a way to think of themselves as fundamentally different but neither immoral nor vicious. Charles Rosenberg has described the process this way: "The physician, not the priest or judge, made the most appropriate guardian for the rights of both society and the individual. The sufferer from phobias and anxieties, the victim of sexual incapacity, the man or woman consumed by desire for a socially unacceptable love object could be seen as the product of his or her material condition rather than as an outcast."[58]

As noted by several of the physicians quoted above, some inverts were often quite active in seeking out writings, including medical writings, on their condition. It cannot be proven conclusively that the activities and publications of American physicians and related nonmedical writings significantly affected the self-understanding of perhaps thousands of lesbians and gay men who were not patients, but convincing confirmation comes from numerous individual accounts documenting some of the ways *public discourse* directly and substantially changed the *private experience* of individuals.[59] These accounts, many of which portray both a personal discovery of an identity and an enthusiasm for a new-found sense of community, also indicate the crucial role played by medical models.

From the many personal stories I have collected, two illustrate most graphically the powerful role of public discourse and medical conceptions in the dialectic of gradual change in people's experience of identity, community, social role, and institutions.[60] One of the stories is from the famous autobiography by political activist Emma Goldman and concerns an experience on her lecture tour of 1913:

Men . . . and women . . . used to come to see me after my lectures on homosexuality, and . . . [they] confided to me their anguish and their isolation. . . . Most of them had reached an adequate understanding of their differentiation only after years of struggle to stifle what they had considered a disease and shameful affliction. One young woman . . . had never met anyone, she told me, who suffered from a similar affliction, nor had she ever read books dealing with the subject. My lecture had set her free; I had given . . . back her self-respect.[61]

The second is by Elsa Gitlow, a lesbian who was born just before the turn of this century. In 1914, at the age of sixteen, she moved with her family to Montreal. Her recollection in the mid-1970s of her youthful pursuit of other poets includes a remarkable vignette of the process we've been examining.

When I was about seventeen, I'd been working for a year and a half in tiresome jobs, and I just couldn't see myself going through life that way. So in search of my people—of where, perhaps, I might belong—I had an inspiration one day. Under an assumed name I wrote a letter to the editor's column in the Montreal *Daily Star*, asking, "Is there in Montreal any kind of an organization to which writers, or prospective writers, might belong?" I forget the exact wording, but that was the gist of it. I wrote [that inquiry] under an assumed name; [and] then, under my own name, I sent an answer, saying, "The undersigned knows of no such organization, but one is in the process of being formed and anyone interested should write to . . . ," and I gave my name and address. I received all kinds of letters, and I called a meeting at my parents' house for the formation of the group. One of the men who came was a young man who these days would be called gay. . . . Roswell George Mills— the most extraordinary being I'd ever seen in my life. He was beautiful. About nineteen, exquisitely made up, slightly perfumed, dressed in ordinary men's clothing but a little on the chi-chi side. And he swayed about. . . . We became friends almost instantly because we were both interested in poetry and the arts.
Roswell apparently recognized immediately my temperament. He said, "Do you know about Sappho?" I don't remember if I'd heard anything about her, but I went to the library, found writings about her and translations of her fragments, and immediately became interested. Through Roswell I started to hear about some literature that would lead me to some knowledge about myself and other people like me. Other than the literary, I think the first books I read were Edward Carpenter's *The Intermediate Sex*, and Kraft-Ebbing [*sic*], and Lombroso—and all of these were revelatory to me because I could have no doubt, having read them, of where my orientation lay. Though they wrote on the level of morbid psychology, and I couldn't accept the morbidity side of it, it was very interesting to read all this and to find out there had been other people like me in the world.[62]

Elsa Gitlow's finding confirmation of her identity in the new medical discourse on sexually deviant persons may well have been that of thousands of other men and women for several decades after 1880.

Stepping back from the experiences of the individual physicians and the individual homosexuals described here will allow for a more general consideration of how this new disease was framed. While the shaping of the diagnosis stemmed from several factors intrinsic to medicine, the timing was due largely to extrinsic factors, primarily the historical emergence of the social roles and personal identities of homosexual men and women in Europe and North America over the course of the nineteenth century. Their growing presence might eventually have provoked a spontaneous response from American physicians, but the medical response was unlikely to have gained momentum so rapidly or to have occurred when it did without the particular impetus created by the personal essays of the homosexual jurist Karl Heinrich Ulrichs. Ulrichs's writings engendered much discussion in the medical and psychiatric journals of western Europe and, shortly later, in those of the United States.

As sexual inversion quickly advanced from the subject of almost pure description of odd cases to an articulated disease concept, the form it took was shaped by currents present in American medicine,[63] which was then aspiring to scientific status without relinquishing its humanitarian commitment to patients. American medical thinking of the 1880s and 1890s was dominated by a somatic bias (which would decline after the turn of the century), and this led to linking sexual behavior to defective heredity and its putative physical signs. It was probably the materialism and naturalism of their scientific outlook—so characteristic of the late nineteenth century—that helped physicians challenge older views of inverts as sinners or criminals. "A growing secularism paralleled and lent emotional plausibility to this framing in medical terms of matters that had been previously construed as essentially moral. Science not theology should be the arbiter of such questions."[64] But medicine's new outlook and new public status came to frame and legitimate more than a diagnosis and an interpretation of behavior; by helping to give large numbers of people an identity and a name, medicine also helped to shape these people's experience and change their behavior, creating not just a new disease, but a new species of person, "the modern homosexual."

NOTES

1. Westphal, "Die conträre Sexualempfindung: Symptom eines neuropathischen (psychopathischen) Zustandes," *Archiv für Psychiatrie und Nervenkrankheiten* 2 (1869):73–108.

2. References to all these cases can be found in J. C. Shaw and G. N. Ferris, "Perverted Sexual Instinct (Contrare [*sic*] Sexualempfindung: Westphal; Inversione dell' instincto sessuale: Arrigo Tamassia; Inversion du sens genital: Charcot et Magnan)," *Journal of Nervous and Mental Disease* (New York) 10:2 (= n.s. 8:2) (April 1883):185–204.

3. Charles E. Rosenberg, "Disease and Social Order in America: Perceptions and Expectations," *Milbank Quarterly* 64, suppl. 1 (1986):34–55, p. 44.

4. The present article is an expansion of "American Physicians' Earliest Writings about Homosexuals, 1880–1900," *Milbank Quarterly* 67, suppl. 1 (1989): 92–108, which limited itself largely to "what the doctors said" and interpreted their writings only in the context of the profession without attending to American society more broadly. Sections III through V below draw closely on the *Milbank Quarterly* article.

5. Mary McIntosh, in "The Homosexual Role," *Social Problems* 16 (1968): 182–192, was the first to develop this point of view, which has now been taken up by many sociologists and historians. A collection of influential essays, edited by Kenneth Plummer, appeared in 1981 under the title, *The Making of the Modern Homosexual* (London: Hutchinson); it reprints McIntosh's original article with a new postscript by her (pp. 30–49). Two sociologists have recently addressed these issues in a critical fashion: Stephen O. Murray, *Social Theory, Homosexual Realities* (New York: Gay Academic Union, 1984) and Steven Epstein, "Gay Politics, Ethnic Identity: The Limits of Social Constructionism," *Socialist Review* 93/94 (1987):9–54. The most recent contribution to this school of interpretation is David M. Halperin, *One Hundred Years of Homosexuality and Other Essays on Greek Love* (New York/London: Routledge, 1990), especially the title essay, pp. 15–40, which includes extensive references to the literature.

Some historians of homosexuality ignore or oppose this interpretation, most notably John Boswell; see his "Revolutions, Universals and Sexual Categories," *Salmagundi* 58–59 (1982–1983):89–113, now reprinted in *Hidden from History: Reclaiming the Gay and Lesbian Past*, ed. Martin Bauml Duberman, Martha Vicinus, and George Chauncey, Jr. (New York: New American Library, 1989).

Aspects of the American medical literature on homosexuality have been analyzed in John C. Burnham, "Early References to Homosexual Communities in American Medical Writings," *Human Sexuality* 7:8 (1973):34–49; Vern L. Bullough, "Homosexuality and the Medical Model," *Journal of Homosexuality* 1 (1974):99–110; Lillian Faderman, "The Morbidification of Love between Women by 19th-Century Sexologists," *Journal of Homosexuality* 4 (1978):73–90; George Chauncey, Jr., "From Sexual Inversion to Homosexuality: Medicine and the Changing Conceptualization of Female Deviance," *Salmagundi* 58–59 (1982–1983):114–146; and Vern L. Bullough, "The First Clinicians," in *Male and Female Homosexuality: Psychological Approaches*, ed. Louis Diamont (Washington, DC: Hemisphere Publishing Corporation, 1987), pp. 21–30. Much of the medical literature was first reprinted in Jonathan Ned Katz, *Gay American History: Lesbians and Gay Men in the U.S.A.* (New York: Thomas Y. Crowell, 1976).

6. Along with sociologists and historians, some philosophers have taken up the social constructionist position; see, for example, Ian Hacking, "Making Up People," in *Reconstructing Individualism: Autonomy, Individuality, and the Self in Western Thought*, ed. Thomas C. Heller, Morton Sosna, and David E. Wellbery (Stanford: Stanford University Press, 1986), pp. 222–236; and Arnold I. Davidson, "Sex and the Emergence of Sexuality," *Critical Inquiry* 14 (Autumn 1987):16–48.

7. Regarding the absence of a concept of homosexual-as-a-type-of-person in America before the nineteenth century, one may consider the evidence of sodomy trials in the North American colonies, which furnish important records of homosexual and lesbian behavior under the rubrics of "sodomy," "the crime against nature," "lewdness," etc.

When Edmund Morgan wrote about these cases in the 1940s, he dealt with them using twentieth-century sexual categories unselfconsciously; see "The Puritans and Sex," *New England Quarterly* 15:4 (1942):591–607. But three more recent studies by Robert F. Oaks—"Perceptions of Homosexuality by Justices of the Peace in Colonial Virginia," *Sexualaw Reporter* 4(2)(April/June 1978):35–36 (reprinted in *Journal of Homosexuality* 5 (1979–1980):35–41); "'Things Fearful to Name': Sodomy and Buggery in Seventeenth-Century New England," *Journal of Social History* 12 (1978):268–281; and "Defining Sodomy in Seventeenth-Century Massachusetts," *Journal of Homosexuality* 6 (1980–1981):79–83—and Jonathan Ned Katz, *Gay/Lesbian Almanac: A New Documentary* (New York: Harper & Row, 1983) show that the wording of the original documents reveals no conception of the sodomite as a type of person, even when an offender was implicated more than once and was regarded as persisting in the vile behavior.

8. This article was discovered by Allan Bérubé and is quoted with his permission from an unpublished essay, "Lesbians and Gay Men in Early San Francisco: Notes Toward a Social History of Lesbians and Gay Men in America" (1979).

9. G. Alder Blumer, "A Case of Perverted Sexual Instinct (*Conträre Sexualempfindung*)," *American Journal of Insanity* (Utica, NY) 39 (1882):22–35, pp. 23–25.

10. McIntosh, "The Homosexual Role"; Lawrence Stone, *The Family, Sex, and Marriage in England, 1500–1800* (New York: Harper & Row, 1977); and Jeffrey Weeks, "'Sins and Diseases': Some Notes on Homosexuality in the Nineteenth Century," *History Workshop* 1 (1976):211–219 were the most important publications in stimulating my own work on the social construction of homosexuality. Conversations with Natalie Zemon Davis, John D'Emilio, Michael Lynch, and Robert Padgug have also played a critical role in helping to shape my interpretation.

11. Evidence on this period's gay history has not yet been collected systematically, and little of it has been published. John D'Emilio, "Capitalism and Gay Identity," in *Powers of Desire: The Politics of Sexuality*, ed. Ann Snitow, Christine Stansell, and Sharon Thompson (New York: Monthly Review Press, 1983), pp. 100–113, though brief, offers the best published narrative of these developments in American gay history, and the present summary depends on his account. John D'Emilio and Estelle B. Freedman have recently set homosexual experiences into a broader context in *Intimate Matters: A History of Sexuality in America* (New York: Harper & Row, 1988).

An unpublished paper by Michael Lynch, "The Age of Adhesiveness: Male-Male Intimacy in New York City, 1830–1880" (presented at American Historical Association Meeting, New York, December 1985), is the richest account of specific developments in New York City at midcentury. George Chauncey has examined the later period in an unpublished Yale dissertation (Ph.D., 1989), forthcoming as *Gay New York: Urban Culture and the Making of a Gay Male World, 1890–1970*.

In addition to my own research, my knowledge of specific details draws on many conversations with friends, who have generously shared information of mutual interest. These include Allan Bérubé, George Chauncey, Lisa Duggan, Joseph Interrante, Jonathan Ned Katz, Gary Kinsman, Michael Lynch, Allan

V. Miller, and the late Gregory Sprague. Though not historians, Carole Vance and Ian Hacking have also made critical contributions through our conversations about sex and about "making up people." (On the latter, see Hacking's article cited above.)

12. For aspects of the social history of lesbians in American, see Lillian Faderman, *Surpassing the Love of Men: Romantic Friendships and Love Between Women From the Renaissance to the Present* (New York: William Morrow, 1981); Katz, *Gay American History*; Katz, *Gay/Lesbian Almanac*; Carroll Smith-Rosenberg, *Disorderly Conduct: Visions of Gender in Victorian America* (New York: Alfred A. Knopf, 1985); and *Hidden from History*.

National studies are essential for understanding these developments in a comparative perspective. For Canada, see Gary Kinsman, *The Regulation of Desire: Sexuality in Canada* (Montreal: Black Rose Books, 1987). For England, see R. F. Claus, "Confronting Homosexuality: A Letter from Francis Wilder," *Signs: Journal of Women in Culture and Society* 2 (1977):928–933, and Jeffrey Weeks, *Coming Out: Homosexual Politics in Britain, from the Nineteenth Century to the Present* (London: Quartet Books, 1977). For France, see Pierre Hahn, ed., *Nos ancêtres les pervers: La vie des homosexuels sous le Second Empire* (Paris: Olivier Orban, 1979), and Robert A. Nye, "Sex Difference and Male Homosexuality in French Medical Discourse, 1830–1930," *Bulletin of the History of Medicine* 63 (1989):32–51. For Germany, see Klaus Pacharzina and Karin Albrecht-Désirat, "Die Last der Ärzte: Homosexualität als klinisches Bild von den Anfängen bis heute," in *Der unterdruckte Sexus*, ed. J. S. Hohmann (Lollar: Achenbach, 1979), pp. 97–112. For Russia, see Laura Engelstein, "Lesbian Vignettes: A Russian Triptych from the 1890s," *Signs* 15 (1990):813–831, and her forthcoming book on sexuality in late-nineteenth-century Russia.

For a transnational perspective, Michel Foucault, *The History of Sexuality. Volume 1: An Introduction*, trans. Robert Hurley (New York: Pantheon, 1979), is stimulating, though my interpretation diverges from his on the exact role doctors' labeling played in the emergence of "the modern homosexual." See also Peter Conrad and Joseph Schneider, *Deviance and Medicalization: From Badness to Sickness* (St. Louis: Mosby, 1980), and David F. Greenberg, *The Construction of Homosexuality* (Chicago: University of Chicago Press, 1988).

13. D'Emilio, "Capitalism and Gay Identity," p. 106.

14. Dr. Irving C. Rosse mentioned messenger boys, as well as teenagers employed by military personnel; see his "Sexual Hypochondriasis and Perversion of the Genesic Instinct," *Journal of Nervous and Mental Disease* (New York) 17:11 (Nov. 1892):795–811, pp. 803–804; which was also published in *Virginia Medical Monthly* 19, n.s. 17, (1892):633–649. Dr. William A. Hammond described a bellboy's ready acceptance of his patient's sexual solicitation; see his *Sexual Impotence in the Male* (New York: Bermingham, 1883), pp. 66–67.

Walt Whitman's expansive "Song of Myself" in *Leaves of Grass* was parodied in *Vanity Fair* on March 17, 1860 as follows:

I am the Counter-jumper, weak and effeminate.
I love to loaf and lie about dry-goods.

. . .

For I am the creature of weak depravities;
I am the Counter-jumper;
I sound my feeble yelp over the woofs of the World.

Reprinted in Henry S. Saunders, *Parodies of Walt Whitman* (New York: American Library Service, 1923), p. 18.

15. See, for example, the sexual experiences documented in Martin Bauml Duberman, "'Writhing Bedfellows': 1826–Two Young Men from Antebellum South Carolina's Ruling Elite Share 'Extravagant Delight,'" *Journal of Homosexuality* 6 (1980–1981):85–101, now reprinted in *Hidden From History*.

16. Earl Lind ("Ralph Werther"—"Jennie June"), *Autobiography of an Androgyne*, edited and with an introduction by Alfred W. Herzog (New York: The Medico-Legal Journal, 1918; reprinted New York: Arno Press, 1975), p. 3. Though not published until 1918, this memoir was first written in 1899.

17. Ralph Werther—Jennie June ("Earl Lind"), *The Female Impersonators: A Sequel to the Autobiography of an Androgyne and an Account of Some of the Author's Experiences During His Six Years' Career as Instinctive Female-Impersonator in New York's Underworld; Together with the Life Stories of Androgyne Associates and An Outline of His Subsequently Acquired Knowledge of Kindred Phenomena of Human Character and Psychology*, edited and with an introduction by Alfred W. Herzog (New York: The Medico-Legal Journal, 1922; reprinted New York: Arno Press, 1975), pp. 150–152; this section is also reprinted in Katz, *Gay American History*, pp. 367–368.

18. In addition to the medical writings, the extensive international publicity in the 1890s surrounding Oscar Wilde's trial in England advanced the process of self-discovery and identity formation for isolated lesbians and gay men, as did the more localized news coverage of raids on homosexual bars and on houses of ill repute that offered "the two prostitutions, feminine and pederastic." When Dr. Irving C. Rosse, in "Sexual Hypochondriasis and Perversion" (p. 803), reported on the police breakup of such a place, called The Slide, in New York City, he remarked particularly on the publicity this exposure received in the *New York Herald*. For more on The Slide, see Katz, *Gay/Lesbian Almanac*, pp. 219 and 233.

19. The following observations are based on my review of what I believe are all the cases published by physicians in the United States through the end of the nineteenth century. Bullough, "Homosexuality and the Medical Model"; Faderman, "The Morbidification of Love between Women"; Chauncey, "From Sexual Inversion to Homosexuality"; and Bullough, "The First Clinicians," have analyzed aspects of this medical literature, and much of it was first reprinted in Katz, *Gay American History*. Chauncey's interpretation and mine diverge on the importance of the medical literature in gay people's lives.

20. James G. Kiernan, "Perverted Sexual Instinct [Report of paper read 7 January 1884]," *Chicago Medical Journal and Examiner* 48 (March 1884):263–265, p. 263.

The role played by the actions and writings of Karl Heinrich Ulrichs in prompting intense medical study of homosexuality is explained by Vern L. Bullough in "The Physician and Research into Human Sexual Behavior in Nineteenth-Century Germany," *Bulletin of the History of Medicine* 63 (1989): 247–267, pp. 254–257.

21. Although my own search of the literature revealed that Shaw and Ferris's was in fact the seventh American case report (preceded by the American cases published by Hagenbach, "Dr. H.," Blumer, Hammond, and Wise), their sense of novelty and their enthusiasm for discovery seem justified. These two doctors significantly advanced Americans' familiarity with this new phenomenon by publishing their English versions of eighteen prior case reports from medical publications in French, German, and Italian. For the six earlier reports see: (1) Allen W. Hagenbach, "Masturbation as a Cause of Insanity," *The Journal of Nervous and Mental Disease* (Chicago) 6 (n.s. 4) (1879):603–612; (2) "Dr. H," "'Gynomania'—A Curious Case of Masturbation," *The Medical Record* (New York) 19:12 (19 March 1881):336; (3) Blumer, "Case of Perverted Sexual Instinct"; (4 and 5) Hammond, *Sexual Impotence in the Male*, 55 ff.; and (6) P. M. Wise, "Case of Sexual Perversion," *The Alienist and Neurologist: A Quarterly Journal of Scientific, Clinical and Forensic Psychiatry and Neurology, Intended Especially to Subserve the Wants of the General Practitioner of Medicine* (St. Louis) 4:1 (January 1883):87–91.

22. Shaw and Ferris, "Perverted Sexual Instinct," pp. 202–203.

23. Kiernan, "Perverted Sexual Instinct," p. 263.

24. Kiernan, "Perverted Sexual Instinct," p. 263.

25. Shaw and Ferris, "Perverted Sexual Instinct."

26. "Dr. H," "Gynomania."

27. In the nineteenth century, apparently thousands of women passed as men in public at one time or another. The advantage was significant, in terms of safety, social privilege, and the ability to vote, travel, and work freely, and this clearly motivated many of them. Linda Grant DePauw, "Women in Combat: The Revolutionary War Experience," *Armed Forces and Society* 7 (1981):209–226, reports that during the Civil War as many as four hundred Northern women adopted a masculine identity and passed successfully as soldiers: "Some served more than two years before being detected. Of course, those who got away with the masquerade and were never caught are beyond our ability to count" (p. 218). Most of these were probably seeking contact with husbands and boy friends, not with female lovers, but their ability to pass successfully confirms the feasibility of this option for many women. For a number of detailed, primary source accounts of women passing as men, see Katz, *Gay American History*, pp. 209–279, and San Francisco Lesbian and Gay History Project, "'She Even Chewed Tobacco': A Pictorial Narrative of Passing Women in America," in *Hidden From History*, pp. 183–194.

28. Katz, *Gay American History*, pp. 232–238.

29. Shaw and Ferris's case, quoted above at note 22.

30. Kiernan, "Perverted Sexual Instinct," p. 264.

31. Wise, "Case of Sexual Perversion," pp. 87–88; also reprinted in Katz, *Gay American History*, pp. 221–225.

32. Hammond, *Sexual Impotence in the Male*, pp. 55–64. These pages on the cigar dealer are of special interest because the man was one of the few American cases to appear in a book instead of a journal. The account by Dr. Hammond includes interesting information on the subject's social life and on the doctor-patient encounter, which often went unrecorded. Additionally, it is striking evidence of the rapid rise of interest in the subject that, although this case of sexual inversion and a second one together occupy only sixteen pages in a 274-page book, the book review by William J. Morton singled out for specific mention "those remarkable perversions of the sexual appetite recently studied by Westphal, Charcot and Magnan, Tamassia, and others." See his "Review of *Sexual Impotence in the Male* by William A. Hammond," *Journal of Nervous and Mental Disease* (New York) 10 (1883):649–651, p. 650.

33. For example, Shaw and Ferris, "Perverted Sexual Instinct," p. 203.

34. George M. Beard, *Sexual Neurasthenia (Nervous Exhaustion): Its Hygiene,*

Causes, Symptoms, and Treatment, with a Chapter on Diet for the Nervous, ed. A. D. Rockwell (New York: Treat, 1884), p. 102.

35. George F. Shrady, "Editorial: Perverted Sexual Instinct," *Medical Record* (New York) 26 (19 July 1884):70–71.

36. We have no evidence that the reticent German immigrant (reported by Shaw and Ferris) regarded himself as a member of a group rather than as a uniquely peculiar individual; he seemed to understand his "abnormal state" as an individual ailment that he simply wished medicine "to overcome." In contrast, when Mr. X rejected approaches from "men of unnatural desire" out of distaste for the implied sexual behavior, he noticed that the men were "able to recognize each other" and shared some common nature, even if he thought he was not one of them. Dr. Hammond gave no evidence that his cigar-dealer patient was aware of a community of like-minded individuals, but this patient's offering the standard formulation of having a woman's soul in a man's body, his ability to find partners, and his use of the name "Lida" with some of his friends all suggest at least a slight involvement in an informal community of inverts.

37. Blumer, "Case of Perverted Sexual Instinct," p. 23.

38. William Lee Howard, "Sexual Perversion in America," *American Journal of Dermatology and Genito-Urinary Diseases* (St. Louis) 8 (1904):9–14, p. 11.

39. Beard, *Sexual Neurasthenia*, pp. 101–102.

40. The next section will consider how the physicians distinguished between true inverts and vicious persons: by regarding the former as having a congenital, i.e. inherent if not precisely hereditary, condition and the latter as having an acquired taste.

41. Howard, "Sexual Perversion in America," pp. 11 and 13.

42. Lind, *Autobiography*, p. 47.

43. George J. Monroe, "Sodomy—Pederasty," *Saint Louis Medical Era* 9 (1899–1900):431–434.

44. Shrady, "Editorial," p. 70.

45. James G. Kiernan, "Sexual Perversion and the Whitechapel Murders," *Medical Standard* (Chicago) 4:5 (November 1888):129–130, and 4:6 (December 1888):170–172, p. 129.

46. Hammond, *Sexual Impotence in the Male*, p. 55.

47. See Foucault, *History of Sexuality*.

48. Shrady, "Editorial," p. 71.

49. G. Frank Lydston, "Sexual Perversion, Satyriasis, and Nymphomania," *Medical and Surgical Reporter* (Philadelphia) 61:10 (7 September 1889):253–258, and 61:11 (14 September 1889):281–285, pp. 253–254.

50. J. A. Armand, "Sexual Perversion in Its Relation to Domestic Infelicity," *American Journal of Dermatology and Genito-Urinary Diseases* (St. Louis) 3 (1899):24–26, pp. 24–25.

51. De Armand's views closely resemble the moral stance expressed some years earlier by John Gray and contrast with those of the neurologists, for which see Charles E. Rosenberg, *The Trial of the Assassin Guiteau: Psychiatry and Law in the Gilded Age* (Chicago: University of Chicago Press, 1968), p. 70.

52. On the character and membership of these two groups, see Bonnie Ellen Blustein, "New York Neurologists and the Specialization of American Medicine," *Bulletin of the History of Medicine* 53 (1979):170–183; Jeanne L. Brand, "Neurology and Psychiatry," in *The Education of American Physicians*, ed. R. L. Numbers (Berkeley: University of California Press, 1980), pp. 226–249; F. G.

Gosling, *Before Freud: Neurasthenia and the American Medical Community, 1870–1910* (Urbana: University of Illinois Press, 1987); Rosenberg, *Trial*; and Barbara Sicherman, "The Uses of a Diagnosis: Doctors, Patients, and Neurasthenia," *Journal of the History of Medicine and Allied Sciences* 32 (1977):33–54.

53. John Gray, perhaps the most famous representative of the older alienist school, published no articles on inverts, as far as I know. Although three of the authors cited above were employed in asylums (Blumer, Ferris, and Wise), they were all (significantly?) in the post of assistant physician and not superintendent. The conservative professional association of alienists admitted as members only superintendents, not other asylum doctors (Rosenberg, *Trial*, p. 62).

54. Randolph Winslow, "Report of an Epidemic of Gonorrhoea Contracted from Rectal Coition," *Medical News* (Philadelphia) 49 (14 August 1886):180–182.

55. Monroe, "Sodomy—Pederasty," pp. 432–433.

56. Ronald Bayer, *Homosexuality and American Psychiatry: The Politics of Diagnosis* (New York: Basic Books, 1981).

57. Carroll Smith-Rosenberg, "The Hysterical Woman: Sex Roles and Role Conflict in Nineteenth-Century America," *Social Research* 39:4 (1972):652–678; reprinted in Smith-Rosenberg, *Disorderly Conduct*, pp. 197–216, at 204.

58. Rosenberg, "Disease and Social Order," p. 45. This passage continues: "By no means all contemporaries accepted such views, of course. But these hypothetical diagnoses may well have been palatable to the stigmatized themselves; given the choice, an individual might well prefer to think of his or her deviant behavior as the product of hereditary endowment or disease process. It might well have offered more comfort than the traditional option of seeing oneself as a reprehensible and culpable actor. The secular rationalism so prevalent in the late nineteenth century freed many Americans from a measure of personal guilt at the cost of being labelled as sick. Only in the second half of the twentieth century has this come to seem a problematic bargain."

Rosenberg's hypothesis of how these changes "might well have been" received by late-nineteenth-century individuals seems strongly confirmed by many sexual inverts' enthusiasm for the medical literature and their active collaboration with physicians as documented here.

59. See, for example, the unsigned report edited by Charles H. Hughes, "The Gentleman Degenerate: A Homosexualist's Self-Description and Self-Applied Title: Pudic Nerve Section Fails Therapeutically," *Alienist and Neurologist* 25:1 (February 1904):62–70. A section is reprinted in Katz, *Gay American History*, pp. 145–146, where Katz suggests Dr. Hughes himself may be the unnamed doctor reporting this case.

60. See also an early homosexual memoir by Claude Hartland (b. 1871), *The Story of A Life: For the Consideration of the Medical Fraternity* (St. Louis, 1901); reprinted with a Foreword by C. A. Tripp (San Francisco: Grey Fox Press, 1985). In 1904, when New York neurologist Charles H. Hughes edited and published an unnamed invert's self-description in a medical journal, he mentioned this book and identified Hartland as one of his patients ("The Gentleman Degenerate," p. 68).

61. Quoted in Katz, *Gay American History*, p. 377.

62. *Word Is Out: Stories of Some of Our Lives*, ed. Nancy Adair and Casey Adair (San Francisco: New Glide Publications, 1978), pp. 15–17. Another installment of Elsa Gitlow's recollections, also dealing with Roswell Mills, appeared in *The Body Politic* (Toronto), May 1982. She later published a book-length memoir, *Elsa: I Come With My Songs* (San Francisco: Bootlegger Press, 1986).

63. For a concise review of American medicine's changing features in this era, see Rosenberg, "Disease and Social Order," especially pp. 44 and 45.

64. Rosenberg, "Disease and Social Order," p. 45.

7 FROM PSYCHIATRIC SYNDROME TO "COMMUNICABLE" DISEASE: The Case of Anorexia Nervosa

JOAN JACOBS BRUMBERG

Until the appearance of AIDS in the United States, many Americans would have named anorexia as the characteristic disease of the past generation. Eating disorders in general seem very much a contemporary phenomenon, and anorexia seems particularly a context-dependent ailment, incorporating intrapsychic, cultural, and ultimately—in terms of disease process—biological elements in a pervasive, ambiguous, yet dangerous pathological reality.

Brumberg centers her argument on the epidemiology, and especially the "contagiosity," of anorexia. At the same time it is necessarily a study of the ailment's etiology as well. The spread of anorexia is to some extent a reflection of its cultural availability, its visibility. If a particular disease concept becomes widely diffused and discussed it must in some sense be seen as communicable. In the case of anorexia we have seen a disease entity articulated and rapidly disseminated through the mass media; it is an episode in the communication of medical ideas presumably facilitated by the incorporation of widely shared cultural values in the anorexia concept. (Diseases can, of course, also become visible as a result of changes in medicine's technical capacity or conceptual assumptions). Once visible, a disease entity becomes an emotional option, available to predisposed individuals. Thus a particular disease can become an actor in countless family dramas and doctor-patient interactions.

If anorexia is a choice, it is to a degree self-defeating. Although it reflects the cultural values of self-denial as moral stature and—relatedly—of slimness as moral worth, it is a diagnosis that confers problematic status. The ambiguity and punitiveness that surround other "functional" ills are associated with anorexia as well. Unlike most unambiguously somatic diagnoses, it does not automatically confer the secondary gains that often come with patienthood. Like many such functional ills, eating disorders are too clearly tied to volition.

The career of the individual anorectic illustrates the slippery slope separating mind from body and body from mind. The course of the illness illustrates the way incremental patterns of willed behavior can transform themselves into physical form—in the case of anorexia, malnutrition and perhaps ultimate starvation, in the case of cigarette smoking, lung cancer and other pulmonary and circulatory ills. The interrelations between body and mind are also interrelations between cultural values and individual personality; it is a fundamental reality that the social history of eating disorders demonstrates with instructive clarity.

—C. E. R.

[T]he neuroses in particular have a contemporary style—they flourish in certain situations and are almost invisible in others.

—KARL JASPERS[1]

ALTHOUGH ANOREXIA NERVOSA was named and identified in the 1870s, it remained a relatively rare disorder for nearly a century. Despite uncertainties in its epidemiology, most experts agree that its incidence began to rise after World War II and then accelerated during the past thirty years. Today, it is hard not to know about anorexia nervosa, a disorder that may afflict as many as one million young women in the United States alone.[2]

In our increasingly medicalized society, we are besieged with references to anorexia nervosa: on late night television comedy programs and on day and prime time sitcoms, series, and movies; on lurid front pages of national tabloids; in popular fiction about adolescent and women's issues; and in the self-help sections of most bookstores. In the bestseller *Cultural Literacy: What Every American Needs to Know*, E. D. Hirsch puts anorexia nervosa on "the list" of names, concepts, and events educated Americans need to know; and in the popular guidebook *Teenagers: When to Worry and What to Do*, Douglas H. Powell, Chief of Psychology Service and Director of Training for the Harvard University Health Service, called eating disorders the "main event" of the

1980s.[3] By the summer of 1987, concern about this emerging public health problem was sufficient to justify a special U.S. House Subcommittee report on "Eating Disorders: The Impact on Children and Families."[4] Anorexia nervosa has become the characteristic psychopathology of middle-class adolescent women.

How did anorexia nervosa move from relative obscurity to its preeminent place in the repertoire of contemporary psychiatric disorders? This is a complex problem in psychiatric epidemiology warranting the attention of behavioral scientists, educators, and physicians as well as social and cultural historians. What we witnessed in the past two decades was a transformation peculiar to mass culture: the shift of a predominantly psychiatric disorder into the category of a "communicable" disease. (By psychiatric disorder I mean a syndrome generated by unique internal and intrafamilial conflicts; "communicable" is used metaphorically because, as far as we know, no microorganism is involved in the spread of anorexia nervosa.)

This chapter is organized around a set of questions that bear on our understanding of communicability: In the absence of a specific pathogen, what accounts for the transmission of anorexia nervosa? What is actually transmitted in the eating-disorders "epidemic"?[5] And, is it possible that group education and excitation have changed the physical as well as the emotional experience of anorexia nervosa?

Social psychology sheds some light on the nature of communicability. For example, the psychological phenomena of contagion and convergence provide a useful model for understanding the current social trajectory in eating disorders.[6] Contagion here refers to the spread of affect or behavior from group member to group member, one member in effect stimulating an imitative act by another. In eating disorders, this dynamic is frequently associated with young women in dormitories at boarding schools or residential colleges. After one woman in the dormitory develops anorexia nervosa or bulimia, others follow. A 1986 study in the *American Psychologist* confirmed this scenario, also demonstrating that young women who know the most about the disorders are most at risk to develop them. This means that young women with anorexic and bulimic friends are more susceptible than those who have no experience with these disorders.[7] The peer group and the social environment make eating disorders "catchy."

Convergence describes situations in which group members develop common affects and behavioral patterns independently but then express them (or act them out) simultaneously. For example, the incidence of anorexia nervosa appeared to accelerate spontaneously in the general population of middle-class adolescent and young adult women beginning in the 1960s and, in much the same way, bulimia emerged as a widespread problem in the 1980s. Contagion and convergence need not be mutually exclusive; in fact, they can be mutually reinforcing.

Both seem to be operative in the case of eating disorders, although there has been no systematic social psychological study of how the "epidemic" actually progressed among particular groups of people in particular places.

Although most clinicians attribute anorexia nervosa and bulimia to individual and familial causes,[8] they also acknowledge that symptoms such as food refusal and other deregulating food behaviors (such as binging and purging) have become increasingly prevalent in the contemporary symptom repertoire. This does not mean that psychopathology is merely learned behavior, but the existence of a learning curve in eating disorders does point to the salience of pathoplastic factors. That is, causal factors that are not strictly psychogenic but nevertheless contribute to the structure of an illness, exert a modeling role, or even act as a triggering agent.[9] Certainly, the recent increase of anorexia nervosa and bulimia is related to the pathoplastic factors to be described in this essay.

But more than a rising incidence needs explanation. My work on the history of anorexia nervosa suggests that symptomatology also has changed so that the disorder has a different character today than it did in the Victorian Era, or even fifty years ago. In fact, the social and cultural environment that frames contemporary anorexia nervosa has molded and shaped the illness into a particular form.

The Changing Face of Anorexia

If one approaches anorexia nervosa from a purely biological perspective, it is hard to explain changes in symptomatology over time. But when we take into account the heightened cultural pressure to be thin, as well as changes in the role of women, the structure of the middle-class household, and the state of technology—all elements that are not ordinarily part of the biological disease model—we can see how the social and cultural climate prepared the ground for a new experience of the disease. A comparative look at clinical case records from the nineteenth and twentieth centuries suggests how social possibilities shape symptomatology.[10]

Food refusal, of course, represents a basic continuity in the reports: Both Victorian and contemporary anorectics refuse to eat. However, they articulate different reasons. Although Victorian anorectics valued slimness for its spiritual and social meaning, they tended to give somatic reasons ("I can not eat because it hurts"). Today's anorexic patient almost always displays a morbid preoccupation with her weight and a dread of fatness ("I don't need to eat; I am too fat"). Much has been written about the obesophobia of contemporary society, the cult of diet and exercise, and the intensity of the Western cultural imperative to be thin.[11] Fear of becoming fat can obviously become a morbid preoccupation in this particular social and cultural climate. Clinicians confirm that a dread of fatness is the most frequent and predictable psychological symptom of contemporary anorectics.[12]

Two new, or at least now common, symptoms—compulsive or ritualistic physical activity and bulimia (bingeing and purging)—were not central in the disease when it emerged in the nineteenth century, nor were they prevalent before 1970. This is not to say that anorectics had never before exhibited these symptoms, only that anorexia nervosa acquired a distinctly formulaic quality in the 1980s. Clinicians as sophisticated and experienced as Hilde Bruch, the foremost international authority on disorders of the human appetite, felt that something had happened by the 1970s to lessen the individualistic aspects of anorexic experience: "The illness used to be the accomplishment of an isolated girl who felt she had found her own way to salvation. Now it is a group reaction."[13] The copycat phenomenon, what Bruch called "me too" anorexia, has become a significant part of the larger nationwide clinical picture. Nearly all contemporary anorectics are obsessively active; a growing number combine bingeing and purging with restrictive eating.

Consider the first aspect of the new formula: compulsive activity. Contemporary anorectics are often observed to be constantly in motion. While "hyperactivity" in an emaciated state may have a biological basis, cultural forms clearly determine its expression. Some case reports from the Victorian Era describe anorexic girls who insisted upon long solitary walks, playing at shuttlecocks all day, or doing somersaults in bed until late at night. More to the point is Clifford Albutt's 1905 portrait of anorexic patients who were extraordinarily frenetic about their volunteer and charity work: "A young woman thus afflicted [with anorexia nervosa], her clothes scarcely hanging together on her anatomy . . . this wan creature, whose daily food might lie on a crown piece, will be busy with mother's meetings, with little sister's frocks, with university extension—with what else of unselfish effort, yet on what funds God only knows."[14] In the past, compulsive activity among women seems to have taken a different—and culturally sanctioned—form: "doing good."

Until the 1970s, when the "fitness and exercise cult" began to accelerate, compulsive physical activity was not a predictable part of the anorexic experience. Today, however, we see a pattern of excessive and compulsive running, swimming, and other aerobic exercise that persistently accompanies not eating and diminishes with successful treatment.[15] Run more and eat less is the prevailing moral calculus, and both components—parsimonious eating and habitual exercise—are culturally sanctioned among middle- and upper-class followers of the religion of health. Thus, the new symptomatology echoes our contemporary *mentalité* and, as we shall see, has some devastating consequences.

Since the 1980s, an increasing number of patients appear to "mix" bulimia with anorexia nervosa, a condition known as bulimia nervosa, or bulimarexia.[16] Nineteenth-century clinical case records give no evidence of bulimia in anorexia nervosa.[17] Victorian doctors did identify and write about "hysterical vomiting," but that was a different phenomenon altogether and involved a different patient group. Although bulimia may have been covert and kept secret from parents and physicians, this seems unlikely on the basis of what we know about Victorian girls, their families, and their homes. A reasonable argument is that Victorian domestic life simply did not provide the opportunities to be bulimic that young women have today.

First, the necessary degree of privacy was not then generally available. If a home had an indoor toilet with plumbing facilities, it was shared with parents, siblings, and guests, so that chronic regurgitation and its telltale signs and smells would be hard to hide on a regular basis. Also, a flush toilet of some kind is required to eliminate all traces of regurgitation, and in the late nineteenth century many homes still used chamber pots that would have collected "evidence."[18]

Second, for the nineteenth-century middle class (in which anorexia nervosa emerged) eating was a highly visible and interactive activity, arranged for certain times of day and structured around a fixed center of sociability: the family dinner table. The kitchen, in contrast to the dining room, was the territory of domestic servants, who were the ones to eat any leftovers from the family meals.[19] It is difficult to conceive how a Victorian girl could have secured large amounts of food on her own and then had the chance to eat it undetected. Where could she get it? Did she raid the larder when everyone was in bed? Could she buy it, quickly, whenever she wanted it? Where could she binge? At the family table? At teatime while visiting an aunt or neighbor? Historical data (and common sense) imply that these opportunities were simply not available to Victorian young women, and that bulimia nervosa could not exist as we know it.

Bulimia in anorexia nervosa seems to be a recent development suited to the pace and psychology of contemporary society. To a large extent, it depends on women's personal freedom, a desocialized eating environment, a lack of supervision, ample disposable income, and the availability—at almost any time—of food for purchase.[20] The autobiographies of bulimic women include regular mention of picking up fast foods in supermarkets or convenience stores and eating alone in cars or other solitary settings. But university cafeterias are also a conducive environment for bulimic behavior, particularly if a student eats alone and is allowed unlimited quantities or passes through the food line. Reports of "group regurgitation rites" in sororities or among groups of girlfriends add to the accumulating evidence for a learned symptomatology.

The emergence of bulimia as a symptom may also be related to two other important but essentially contradictory aspects of our culture: the extraordinary emphasis on dieting (a form of denial) and the equally powerful urge for instant gratification. Many current psychological theories of bulimia nervosa propose that the increasingly thin beauty ideal plays a major role in promoting binging and purging.[21] And data on the effects of starvation on mood and behavior make a compelling argument for a link between dieting and binge eating. In other words, emphasis on dieting to attain beauty makes bingeing and purging more likely. Astute clinicians increasingly report that chronic and self-conscious denial of food sets the stage for bulimic behaviors.

Despite its pervasive preoccupation with dieting, contemporary American society also promotes instant gratification and the notion that "you can have it all." Traditionally, Americans admired those who attain success through hard work. More recently, we admire and envy those who are happy and successful without observable effort. We are told "to work smarter, not harder."[22] At the same time, self-indulgence—in exotic foods, elegant perfumes, a massage, or a sensual vacation—once the object of moral indignation, if not vilification, is now regarded as a sign of mental health.

The symptomatology in bulimia nervosa parallels these contemporary cultural characteristics. Bingeing—almost always on easily digested foods high in carbohydrates, sugar, and fat—is based on instant gratification. No appetizers prepare the palate, and no delays separate courses. A lot is consumed as quickly as possible. Regurgitation then achieves a temporarily efficacious compromise: evasion of the caloric disaster of indulgence without the hard work of chronic dieting, but a dangerous way to beat the metabolic system.

Taken together, these new symptoms—compulsive physical activity and bulimia—have physiological consequences. Anorexia nervosa may be more hazardous in the 1980s than in earlier times for two reasons. First, today's anorectic is thinner. An index of body mass for anorexia nervosa hospital admissions since the 1930s shows a marked decline over time, suggesting that more patients with more severe weight loss are being admitted today than ever before.[23] The greater severity may possibly be due to delayed treatment or longer outpatient therapy before hospitalization, but this seems unlikely given popular awareness of the disease and the many local and regional organizations available to help with eating disorders. The severity of current cases probably reflects patients' zealous commitment to both exercise and diet as well as a generalized acceptance of thinner bodies by peers, parents, and medical personnel. Second, bulimia in anorexia nervosa complicates the therapeutic picture. Clinical data suggest a hardening of the symptomatology once a young woman starts inducing regular vomiting; those who combine restrictive anorexia with bulimia are the most difficult to cure and most at risk for life-threatening complications.[24] Clearly, both the physical and emotional experience of anorexia nervosa has been transformed by the contemporary young woman's social context.

Popularizing Problematic Behaviors

In an enlightened and liberal society, information is supposed to lead to improvement; in the case of anorexia nervosa it seems also to enrich the medium in which the disease can flourish and spread. I use the term popularization for the informal but widespread educational process by which medical and psychological information about anorexia nervosa, its causes, and its symptoms, was disseminated outside the clinical world, often energetically to the people most at risk. Over the past two decades, we inadvertently created an information environment that fostered the circulation of news about eating disorders throughout a reservoir of susceptible young women. Popularization transformed anorexia nervosa from an enigmatic and rare condition into a recognizable and accessible disease.

In the 1970s and 1980s, the spread of information about eating disorders took a surprising number of forms—some serious and some trivial.[25] The formats were eclectic, ranging from the authoritative *New York Times* to tabloids like *The Weekly World News* and *The Star* to made-for-television movies such as psychiatrist Steve Levenkron's "The Best Little Girl in The World" (1981) and programs such as "Fame." A rash of references to the disease appeared in first-run films (e.g., *Down and Out in Beverly Hills*), on "Saturday Night Live," and in feminist fiction (e.g., Marge Piercy's 1984 novel, *Fly Away Home*). Not surprisingly, the three magazines that gave anorexia nervosa its greatest national coverage—*People*, *Mademoiselle*, and *Seventeen*—all cater to the primary constituency for the disease: adolescent and young adult women.

Hilde Bruch was a key actor in the popularization process. At the time of her death in 1984 from Parkinson's disease, Bruch was emeritus professor of psychiatry at the Baylor University Medical School in Houston, Texas. Described in an obituary as "Lady Anorexia," Bruch had spent nearly forty years studying both overeating and noneating. In addition to her record of scholarly publications in the fields of pediatric endocrinology and psychiatry, she wrote an important text for the mental health professions, *Eating Disorders: Obesity, Anorexia Nervosa and the Person Within* (1974). Because of her expertise, she was consulted by popular women's magazines to answer questions about dieting and weight and, in 1972, first discussed anorexia nervosa in the popular press in *Family Circle*. In 1977 she contributed an article on anorexia nervosa to *Ann Landers Encyclopedia of Answers A to Z* and in 1981 wrote a commentary for *TV Guide* as part of the promotion for a television movie about the disease.[26]

Bruch's translations of the anorexic experience into nonmedical language were effective and generated great interest. Letters to her after the publication of the *Family Circle* piece reveal how difficult it was in the early 1970s to find information about anorexia nervosa: "This condition is rarely discussed and I am at wit's end how to overcome it," wrote a twenty year old. A number of young women were pleased to learn from her column that they were not alone: "You don't know how relieved I was to read the article and discover there are others like me. And you admitted that there is something wrong."[27] Adolescents in particular derived considerable emotional solace from the newly found camaraderie of the disease. In subtle but powerful ways, anorexia nervosa was becoming a generational issue and, in that respect, played to both the desire for group identity and the capacity for mimetic behavior that is a hallmark of adolescence. No popular text has had as wide a circulation (over 150,000 copies in English) or as profound an impact as Bruch's *The Golden Cage: The Enigma of Anorexia Nervosa* (1978), which was deliberately intended for a popular audience.[28] The book was especially influential because it played on the feeling of generational identity and made anorexia nervosa "live." In much the same way that *A Mind That Found Itself, I'll Cry Tomorrow*, and *The Days of Wine and Roses* humanized mental illness and alcoholism, Bruch used her clinical experience to provide an intimate account of the interior life of the anorectic.[29] Based on seventy case histories drawn from thirty years of psychotherapeutic practice, *The Golden Cage* was the first text to clearly and sympathetically relate anorexia nervosa to the developmental crisis of adolescence. In this way, Bruch transmitted an emotional understanding of the disease: she made food refusal comprehensible, discerning meaning in a seemingly inexplicable, even mysterious symptom.

The anorexic patients described by Bruch seemed like the girlfriends of her youthful and affluent readers, who empathized with the "paralyzing sense of ineffectiveness," parental tension, and quest for perfection that pervaded an anorectic's life. In this context, the anorectic's rigid insistence on bodily control through dieting began to make sense. The classic anorectic only acted as if she had no appetite; in fact, her life was a constant struggle to deny her natural hunger and constantly reduce her body weight. Bruch presented food avoidance techniques in the patients' own words. For readers doubtful about the existence of a parent-child dialogue in anorexia nervosa, *The Golden Cage* presented a nearly complete script: case histories laying out the kind of difficulties between girls and their parents provide the modern familial context for anorexia nervosa. A decade after the book's first publication, adolescent and young adult women continue to be touched by it, and many rank it as one of the most moving and influential books they have ever read.

While Bruch represents the powerful role of medical authority in articulating and legitimating new disease concepts, obviously other sources also were influential. Two popular nonmedical sources are important for an understanding of how the symptomatology of anorexia nervosa was transmitted through print: fiction designed specifically for the adolescent market, and autobiographical accounts by sufferers.[30] As a genre, fictive anorexia stories provide adolescent girls with both a dramatic warning and substantive information. Although their anorexic protagonists never die, the books are notable for graphic descriptions of symptomatology—particularly ritual food preoccupations, such as eating no more than three curds of cottage cheese at one sitting and never eating before a certain hour—and for their endorsement of medical and psychiatric intervention. In fact, they seem intended primarily as a device for persuading adolescents that such intervention is imperative. However, they do not adequately depict either the classic battle for control that absorbs so much time and energy in psychotherapeutic treatment or the physical realities of forced feeding.

The anorexia stories are nearly formulaic in emphasizing family tensions and the adolescent girl's confused desire for autonomy.[31] Plots almost always involve an attractive, intelligent high school girl, usually 5 feet 5 inches tall, from a successful two-career family. Naturally enough, the protagonist becomes interested in reducing her weight. Like virtually all American girls, she embraces the current social ideal of slimness. (For example, Francesca, in *The Best Little Girl in the World*, cuts out pictures of models and sorts them according to thinness. Her goal is to be thinner than the thinnest.) In each of the anorexia stories, for various reasons all having to do with the difficulties of adolescence, ordinary dieting becomes transformed into a dominating pattern of ritualistic food and eating behavior. Some girls eat only one food, such as celery, yogurt, or dry crackers; others steal from the refrigerator at night to avoid being seen eating. In one novel parents are persuaded by their daughter to allow her to take supper alone in her room, supposedly so that she can do homework at the same time. The mother later discovers that for over a year her daughter has been throwing her dinner out of the window of their Central Park West apartment. Details of food behavior in the novel have an authentic and convincing ring, resembling material in actual case records of anorectics.

Personal testimonials provided another compelling perspective on "theory and practice" in anorexia nervosa. Between 1980 and 1985, several autobiographical accounts achieved wide readership, particularly Shelia MacLeod's on her anorexic girlhood in Great Britain and Cherry Boone O'Neill's on her anorexia, marriage, and dedication to the evangelical Christian faith.[32] Unlike the fiction, which protects its youthful readers from the harsh realities of recovery, these are testimonials to extraordinary and protracted personal suffering. For example, Boone's eating behavior became so bizarre that she stole slimy scraps from a dog dish. These texts, catering to the American penchant for intimate disclosures, give a complete and graphic catalog of a psychiatric symptomatology.

Celebrity biographies and a tragic death also heightened the public awareness of anorexia nervosa and of women's difficulties with food in our society. In January 1983, the death of thirty-two-year-old popular singer Karen Carpenter due to heart failure associated with low serum potassium—a consequence of prolonged starvation—fueled interest in the disease, as well as a belief in its intractability.[33] In the same year, Jane Fonda revealed that she had binged and purged throughout her years at Vassar College and during her early film career,[34] and other autobiographical narratives by bulimic women have since surfaced. These are among the most disturbing and unhappy documents written by women in our time. They describe obsessive thinking about food and its acquisition; stealing food; secret, ritualistic eating; and compulsive vomiting, often with orgiastic overtones. Compared to the restrictive anorectic subsisting on 200 to 400 calories a day, the bulimic may ingest up to 8,000 calories at one sitting.[35] The public discovery of bulimia in the 1980s meant that anorexia nervosa was no longer the solitary example of aberrant female eating and appetite. Increasingly, medical services and patients began to talk about the larger category of eating disorders.

U.S. colleges and universities are particularly fertile grounds for the spread of such behaviors. This should not be surprising: According to social psychologists, "high density" group situations intensify people's reactions to one another.[36] In fact, eating disorders have become so prevalent on American campuses that the *Chronicle of Higher Education* took up the subject, and in the past few years an impressive number of colleges and universities have developed an institutional response to the problem.[37] A brief look at the culture of women students suggests why a response is necessary and also how much contagion depends on one woman observing and being affected by what another woman is doing.

The anorectic is not the only woman on campus anxious about food and her body: dieting, and talk about dieting, are central features in the culture of women students.[38] Diet-conscious female students report that fasting, weight control, and binge eating are common. Self-esteem and stress levels are tied to body image. Young women not only share a vision of the ideal body type but also learn from one another how to attain it. Circulation of information about dieting, purging, and other deregulating behaviors is an essential part of their subculture. Techniques are culled from the mass-market weight-control industry, women's magazines, popular stories about anorexia nervosa, and the experience of friends. Such a milieu enhances the vulnerability of young women already at risk for a preoccupation with appetite control.

Talk about eating disorders, whether serious or trivial, is actually encouraged by responsive, sympathetic college and university personnel in the mental health professions. Most sophisticated colleges and universities now offer support groups as well as individualized psychotherapy for students (predominantly female, of course) who either have eating disorders or have friends who do. Cornell University, for example, has approached the problem from an appropriately broad perspective: the Office of the Dean of Students sponsors a continuing workshop entitled "Women, Food, and Self Esteem" that has attracted capacity enrollment every year since 1985. In February 1986, representatives from many of the nation's most prestigious colleges and universities met at Radcliffe College for an Intercollegiate Eating Disorders Conference, a two-day program designed to map strategies for effective peer counseling on the nation's campuses. Professional directors of dormitories routinely inform resident advisers about resources on campus for handling eating disorders and how to broach the subject with a student who seems intent on denying a noticeable and severe weight loss.

But college women's constant talk about food and dieting is ultimately a trope for other issues. That is why anorexia nervosa and bulimia have particular power and meaning for the current generation. For a variety of reasons—material, psychological, ideological, and symbolic—food and diet are the arenas in which affluent young women in the postindustrial world work out struggles for autonomy and identity, connection and control. In the world of privileged adolescent women, slimness is equated with perfection. This is why "diet talk" is so pervasive and sustained; it is, in effect, a new religion. As I have argued elsewhere, the current cult of diet and exercise is the closest thing our secular society offers young women in terms of a coherent philosophy of the self.[39]

As a result, it is not uncommon for an undergraduate woman to tell a faculty adviser that a roommate or friend is eating little, regurgitating regularly, or behaving bizarrely around food. She questions: "Should I intervene in the behavior of a friend? Are my friend's choices about food and eating different from her choices about academic and sexual behavior? Should I report her peculiar eating to college authorities?" Some students, caught up in the "therapeutic mentality," become overly zealous in judging and reporting who is anorexic, or use the term loosely (and/or proudly) as a self-description. For those young women already imbued with the ethic of care—a characteristic described by psychologist Carol Gilligan[40]—attention to one another's diet and weight may well be a new component in contemporary female morality. Today, food and eating issues have supplanted religious and political questions as the vehicles for female moral development and reflective thinking, both part of the process of coming of age. For the current college generation of women, learning how to respond to difficulties in handling food constitutes a serious moral dilemma, and the dilemma generates a degree of group excitation about anorexia nervosa and bulimia that is powerful and reinforcing.

The Social Experience of Recovery and Treatment

The popularization of anorexia nervosa is mediated by social experience within the peer group as well as treatment settings designed to forge a sense of solidarity with other sufferers. Treatment, as well as symptomatology, has been collectivized. In the 1970s and 1980s, new styles of treatment and care put an emphasis on group experience and, even, social analysis. Although the anorectic still receives individual biomedical and psychotherapeutic attention as she did fifty years ago, she is more and more likely to attend meetings or lectures about eating disorders, to join a group of other young women undergoing similar struggles, or to be hospitalized in a specialized unit or clinic with other patients suffering eating disorders. These experiences have a collective dimension and underscore the highly social aspects of contemporary eating disorders.

The strategies of national, regional, and local eating-disorders groups provide a key to understanding the current experience of the disease. The program of the most active of such advocacy groups, the American Anorexia and Bulimia Association (AA/BA), founded in suburban Teaneck, New Jersey, is a case in point. In 1985, the AA/BA had almost twenty-five different affiliates across the country. The organization publishes a newsletter that notes conferences on eating disorders, publishes excerpts from speeches given at professional meetings, reviews new books on anorexia and bulimia, reports on research done at the "doctoral level or higher," and lists successful publicity efforts by name and date of publication. "The world is learning about anorexia nervosa and bulimia," the newsletter states, proud of its organizational capacity to keep eating disorders in the public eye.[41]

However, the AA/BA does more than publicize; it is an important national model for community-based eating-disorders groups. One of the most influential features of its program is an emphasis on participation in support groups. Sufferers and family members are encouraged to explore the experience of the disease through frank conversation—the expression of emotions and recounting of shared experiences. One goal of this kind of group therapy is to eliminate feelings of isolation and to promote a sense of connectedness to others—"You are not alone in your suffering." Faith in the therapeutic value of shared experience is reflected in the multiplicity of support groups and the maintenance of a twenty-four-hour telephone "hot-line" on which a caller can speak to a recovered anorectic or bulimic. Such accessibility to and interaction with other sufferers underscore the role contemporary anorexic and bulimic women play in defining their own symptoms.

A strategy of providing support for sufferers through their commonality is not unique to the AA/BA. Alcoholics Anonymous and feminist consciousness-raising groups are obvious precursors. Throughout the 1960s and 1970s, feminist groups stressed networking, affective relationships, and the shared experience of oppression, and eating-disorders groups also have an identifiable feminist thrust. Much of the literature they circulate is actually cultural criticism that highlights the ways women are used and/or manipulated by powerful social forces with a stake in keeping women decorative, thin, and helpless.[42] Like consciousness-raising groups, eating-disorders organizations seek to discredit the notion of individual or familial culpability in favor of "social construction" or systemic aspects of the problem. In group sessions, eating-disordered women spend a great deal of time talking about the broad cultural messages that affect their views of their own bodies. The effect of support groups is, ultimately, to make anorexia nervosa and bulimia "social" in the sense of both experience and ideology.

The Question of Psychic Epidemics

Today, as in the past, psychopathologies exist in reciprocal relationship to the prevailing culture. Despite some basic continuities, psychiatric disorders mirror the deep preoccupations of a society. That is why psychogenic illness in one culture may be characterized by nausea, in another by headaches and dizziness, and in still others by dancing, trance, hallucinations, suicide, noneating, or regurgitation. Despite our best efforts at nosological certainty, symptoms (and symptom constellations) are fueled or dissipated by material, social, and cultural conditions and in the process take on a different life in different historical cohorts and in different class, age, and gender groups. In the case of anorexia nervosa, the current disease frame (e.g., increased incidence, popularization, group excitation, and the social experience of treatment) has generated a "new" disease.

Although communicability has social rather than biomedical meaning in the case of anorexia nervosa, it is still not altogether clear how the phenomenon of contemporary eating disorders relates to other mass psychogenic illnesses, epidemic hysterias, or psychic epidemics.[43] In fact, the whole area of mass psychogenic illness has been likened to a vast "no man's land" between psychiatry and the social sciences.[44] Much of the literature in medical history deals with what George Rosen called "psychic epidemics," outbreaks of bizarre behavior in specific geographical settings (e.g., a convent, village, school, or factory) for specific durations (e.g., a few days, a few months, a year).[45] Obviously, the problem of anorexia nervosa in recent times is different because of a changed temporal and geographical context. If anorexia nervosa is a "psychic epidemic," à la Rosen, it is notable because it has entered its second decade and has spread outside the United States to Western Europe and most recently, Japan.[46]

One striking continuity in the literature on mass psychogenic illness is particularly germane to the case of modern eating disorders, although it remains essentially uninvestigated. Across the centuries, "psychic epidemics" (e.g., dancing manias, demonopathy, ecstatic phenomena, and prophetic trance) appear to have involved disproportionate numbers of young women. According to Francois Sirois, in a retrospective study of epidemic hysterias: "The classical outbreak has the following epidemiological characteristics. It involves a small group of segregated young females. It appears, spreads, and subsides rapidly. . . ."[47] The work of both Rosen and Sirois suggests that for the affected group these epidemics express either cultural alienation or an attempt at rejuvenation or revitalization.[48] Little effort is made, however, to explain why adolescent and young adult women are almost always central to these periodic eruptions of social and cultural maladjustment. The answer to this question is beyond the scope of this chapter and would require a close reading of the history of various psychogenic illnesses and associated cultural circumstances—as well as a better understanding of the pattern and function of mimetic behavior in adolescence.[49]

Yet, the recent history of eating disorders does suggest that anorexia nervosa in adolescent and young adult women probably has an "epidemic" form, much like that of some culture-bound psychopathologies associated with tribal societies.[50] But in the "global village" the spread of socially shared psychopathologies is accelerated, and both excitation and sustaining power may increase. For this reason, knowledge of the information environment surrounding young women at risk is crucial to an understanding of the present and future course of the eating-disorders "epidemic."

The cohort of young women coming of age in the 1970s and 1980s encountered a unique information environment that provided both vicarious and direct experience with anorexia nervosa. They learned from a variety of popular sources about the "practice" of anorexia nervosa—that is, its symptoms and how to have them—and thus made anorexia nervosa their generation's disease. Because of these recent shifts in both incidence and symptomatology, we must conclude that anorexia nervosa is not a stable pattern, to be found in all societies and epochs. Rather, eating disorders appear ultimately to be cultural productions, no matter what biological mechanisms they provoke. For those modern mental health professionals, teachers, and parents who see health education as entirely benign and psychiatric disorders as strictly psychogenic, this history should be unsettling, precisely because it suggests how familiarity with eating disorders may well be implicated in their increase.

NOTES

1. Karl Jaspers, *General Psychopathology*, translated from the 7th ed. by J. Hoenig and Marion W. Hamilton (Chicago: University of Chicago Press, 1963), pp. 732–743.

2. For the story of the identification of anorexia nervosa in England, France, and the United States in the nineteen century, see Joan Jacobs Brumberg, *Fasting Girls: The Emergence of Anorexia Nervosa as a Modern Disease* (Cambridge: Harvard University Press, 1988), chaps. 4–6. I use the term "may" because the extent of the disorder has never been definitively established. Because of diagnostic confusions and ambiguities of definition, it is difficult to know how many people have anorexia nervosa, and the number is a hotly contested issue. As with many diseases, some people have a vested interest in inflating or lowering the numbers. See R. E. Kendall et al., "The Epidemiology of Anorexia Nervosa," *Psychological Medicine* 3 (1973):200–203; D. J. Jones et al., "Epidemiology of Anorexia Nervosa in Monroe County, New York: 1960–76," *Psychosomatic Medicine* 42 (1980):551–558; Alexander Lucas et al., "Anorexia Nervosa in Rochester, Minnesota: A 45-year Study," *Mayo Clinic Proceedings* 63 (May 1988):433–442; and Brumberg, chap. 1. The Lucas study is interesting because it highlights age differences. The incidence of anorexia nervosa remained unchanged from 1935 to 1979 among female residents of Rochester 20 years of age or older, but increased in two periods, 1935–1949 and 1965–1979, among residents aged 10 to 19.

3. On anorexia nervosa in American popular culture, see Brumberg, chap. 1. Douglas H. Powell, *Teenagers: When to Worry and What to Do* (New York: Doubleday, 1987), p. 203. E. D. Hirsch, *Cultural Literacy: What Every American Needs to Know* (Boston: Houghton Mifflin, 1987).

4. "Eating Disorders: The Impact on Children and Families," hearing before the Select Committee on Children, Youth, and Families, House of Representatives, One Hundredth Congress. U.S. Government Printing Office, 1988.

5. Anorexia nervosa does not meet the epidemiological standard for an epidemic since it is relatively infrequent in the general population. Its annual incidence has never been estimated at more than 1.6 per 100,000 population. See R. E. Kendall et al., "The Epidemiology of Anorexia Nervosa," *Psychological Medicine* 3 (1973):200–203.

6. Stanley Milgram and Hans Toch, "Collective Behavior: Crowds and Social Movements," in Gardner Lindzey and Elliot Aronson, eds., *The Handbook of Social Psychology*, 2d ed. (Reading, MA: Addison-Wesley, 1968), pp. 550–553.

7. Ruth Striegel-Moore, Lisa R. Silberstein, and Judith Rodin, "Toward an Understanding of Risk Factors in Bulimia," *American Psychologist* 41 (March 1986). See also D. M. Schwartz, M. G. Thompson, and C. L. Johnson, "Anorexia Nervosa and Bulimia: The Socio-Cultural Context," *International Journal of Eating Disorders* 1 (Spring 1982):20–36.

8. On the etiology of anorexia nervosa, see Brumberg, pp. 24–40.

9. See Gerald F. M. Russell, "The Changing Nature of Anorexia Nervosa: An Introduction to the Conference," *Journal of Psychiatric Research* 19, no. 23 (1985):101–109. Russell introduced me to the term pathoplastic, although it was introduced by Karl Birnbaum in *Der Aufbau der Psychose* (Berlin, 1923).

10. The argument and narrative here are based on a comparative study that I did with a contemporary clinician: Joan Jacobs Brumberg and Ruth Striegel-Moore, "Continuity and Change in the Symptom Choice: Anorexia Nervosa in Historical and Psychological Perspective," forthcoming in Glen Elder, Jr., John Modell, and Ross Parke, eds., *Children in Time and Place* (Chicago: University of Chicago Press, 1991).

11. See for example Brumberg, chaps. 1, 9, Afterword; Kim Chernin, *The Obsession: Reflections on the Tyranny of Slenderness* (New York: Harper & Row, 1981); Hillel Schwartz, *Never Satisfied: A Cultural History of Diets, Fantasies, and Fat* (New York: The Free Press, 1986); Albert J. Stunkard, ed., *Obesity* (Philadelphia: Saunders, 1980); Judith Rodin, Lisa Silberstein, and Ruth Striegel-Moore, "Women and Weight: A Normative Discontent," in *1984 Nebraska Symposium on Motivation*, ed. Theodore Sondereggen (Lincoln: University of Nebraska Press, 1985).

12. Russell, pp. 103–104.

13. Bruch, *The Golden Cage: The Enigma of Anorexia Nervosa* (Cambridge: Harvard University Press, 1978), p. *xii.* This book was translated into German, French, Italian, Swedish, and Japanese.

14. See Samuel Gee, *Medical Lectures and Aphorisms* (London: Frowd, 1908), pp. 43–44; Silas Weir Mitchell, *Lectures on Diseases of the Nervous System, Especially in Women* (Philadelphia: Lea Brothers, 1885), pp. 229–230, 243–244; John K. Mitchell, *Self-Help for Nervous Women* (Philadelphia: Lippincott, 1909), p. 103; W. J. Collins, "Anorexia Nervosa," *Lancet* (January 27, 1894):203; Pierre Janet, *The Major Symptoms of Hysteria* (New York: Macmillan, 1907), p. 228; Clifford Allbutt, *A System of Medicine III* (New York: Macmillan, 1905), p. 474.

15. According to psychologist Rita Freedman, the contemporary emphasis on fitness and exercise is a double-edged sword. See Rita Freedman, *Beauty Bound* (Lexington, MA: Lexington Books, 1986), pp. 166–167.

16. Marlene Boskind-White and William C. White, *Bulimarexia: The Binge/Purge Syndrome* (New York: Norton, 1983); American Psychiatric Association, *Diagnostic and Statistical Manual of Mental Disorders* (Washington, DC, 3d ed., 1980; 3d ed. revised, 1987); R. C. Casper et al., "Bulimia: Its Incidence and Clinical Importance with Patients with Anorexia Nervosa," *Archives of General*

Psychiatry 37 (1980):1030–1035; Gerald F. M. Russell, "Bulimia Nervosa: An Ominous Variant of Anorexia Nervosa," *Psychological Medicine* 9 (August 1979):429–448.

17. I strongly disagree with the interpretation presented by David M. Stein and William Laakso in "Bulimia: A Historical Perspective," *International Journal of Eating Disorders* 7, no. 2 (1988):201–210. Stein and Laasko base their argument for bulimia in nineteenth-century anorexia nervosa on a single published case report by William Gull (1874), which they read in a problematic manner. Their study never looked at unpublished case material. Moreover, they failed to distinguish between bulimia as a symptom and bulimia nervosa, a disease, which involves binging and purging along with concerns about weighing too much. In "The Psychiatric History of Anorexia Nervosa and Bulimia," *International Journal of Eating Disorders* 8 (March 1989):259–272, Tilmann Habermas argues that bulimia nervosa (as defined in the DSM-III-R) emerged only in the twentieth century and particularly after 1940.

18. According to Susan Strasser, *Never Done: A History of American Housework* (New York: Pantheon Books, 1982), indoor plumbing really became a fact of American life only in the 1920s; in Britain, this happened even later.

19. See Faye Dudden, *Serving Women: Household Service in Nineteenth Century America* (Middletown, CT: Wesleyan University Press, 1983).

20. The argument here is from Brumberg, pp. 259–262.

21. Janet Polivy and C. Peter Herman, "Dieting and Binging. A Casual Analysis," *American Psychologist* 40 (1985):193–201.

22. Marilyn Machlowitz, *Workaholics: Living with Them, Working with Them* (New York: New American Library, 1981).

23. W. Stewart Agras and Helena Kraemer, "The Treatment of Anorexia Nervosa: Do Different Treatments Have Different Outcomes?" *Psychiatric Annals* 13 (December 1983):929.

24. Paul E. Garfinkel and David M. Garner, *Anorexia Nervosa: A Multidimensional Perspective* (New York: Brunner-Mazel, 1982).

25. The discussion that follows is drawn largely from Brumberg, chap. 1 and my unpublished paper, "From Fat Boys to Skinny Girls: Hilde Bruch and the Popularization of Anorexia Nervosa in the 1970s," prepared for American Association for the History of Medicine Meetings, New Orleans, 1988.

26. There is no English language biography of Bruch. For the outlines of her personal and professional life I relied on the introduction by Theodore Lidz and the biography by Randy J. Sparks in Randy J. Sparks, *The Papers of Hilde Bruch. A Manuscript Collection in the Harris County Medical Archive* (Houston, TX, 1985); Theodore Lidz, "In Memoriam: Hilde Bruch, M.D. (1904–1984)," *American Journal of Psychiatry* 142 (July 1985):869–870; and a transcribed interview with Hilde Bruch conducted November 1974–January 1975 by Jane Preston and Hannah Decker for the American Psychiatric Association; Reinhard Heit Kamp, "Hilde Bruch: Leben und Werk," unpublished Ph.D. dissertation, Universität zu Koln, Honen Medizinischen Fakultät (1987); Hilde Bruch, "The Constructive Use of Ignorance," in *Explorations in Child Psychiatry*, ed. E. James Anthony (New York: Plenum Press, 1975), pp. 247–264. See Harriet LaBarre, "Ideas for Living: An Interview with Dr. Hilde Bruch, Authority on Obesity," *Family Circle* 81 (September 1972):74, 166–168, 180; Ann Landers, *The Ann Landers Encyclopedia, A to Z* (Garden City, NY: Doubleday, 1978); Hilde Bruch, "Background of Anorexia Nervosa," *TV Guide* (May 9, 1981):28–30.

27. Hilde Bruch Papers, Series VII, Folders 1 and 2, August 22, 1972; January 1, 1972 (Texas Medical Center Archives).

28. See note 13 above.

29. Clifford Beers, *A Mind That Found Itself*, 4th ed. (New York: Longmans, Green, 1910); Lillian Roth, *I'll Cry Tomorrow* (New York: F. Fell, 1954); "Days of Wine and Roses," 1962 film.

30. The material that follows is from Brumberg, pp. 16–18.

31. Some novels about anorexia nervosa are: Deborah Hautzig, *Second Star to the Right* (New York: Knopf, 1988); Steven Levenkron, *The Best Little Girl in The World* (New York: Warner Books, 1978); Rebecca Joseph, *Early Disorder* (New York: Farrar, Straus, Giroux, 1980); Ivy Ruckman, *The Hunger Scream* (New York: Walker, 1983); Margaret Willey, *The Bigger Book of Lydia* (New York: Harper & Row, 1983); John Sours, *Starving to Death in a Sea of Objects* (New York: Aronson, 1980); Emily Hudlow, *Alabaster Chambers* (New York: St. Martin's Press, 1979); Isaacsen-Bright, *Mirrors Never Lie* (Worthington, OH: Willowisp Press, 1982).

32. For personal testimonials see Shelia MacLeod, *The Art of Starvation: A Story of Anorexia and Survival* (New York: Schocken Books, 1981); Cherry Boone O'Neill, *Starving for Attention* (New York: Dell, 1983); Aimee Liu, *Solitaire* (New York: Harper & Row, 1979); Sandra Heater, *Am I Still Visible? A Woman's Triumph over Anorexia Nervosa* (Whitehall, VA: Whitehall Books, 1983); Camie Ford and Sunny Hale, *Two Too Thin* (Orleans, MA: Paraclete Press, 1983).

33. See the coverage of Karen Carpenter's death in *People Weekly* (February 21 and November 21, 1983, May 31, 1985). In the earliest accounts, low serum potassium was reported to have caused an irregularity in Carpenter's heartbeat. By 1985, the reports were that Carpenter had died of "cardiotoxicity" brought on by the chemical emetine. The suggestion is that Carpenter was abusing a powerful over-the-counter drug, ipecac, used to induce vomiting in cases of poison ingestion.

34. See Thomas Kiernan, *Jane: An Intimate Biography of Jane Fonda* (New York: Putnam, 1973), p. 67 and "Private Lives," *Ladies Home Journal* 102 (October 1985):202.

35. See for example Lisa Messinger, *Biting the Hand That Feeds Me: Days of Binging, Purging, and Recovery* (Novato, CA: Arena Press, 1986); Jackie Barrile, *Confessions of a Closet Eater* (Wheaton, IL: Tyndale House, 1983); K. B., "A First Anniversary—A Recovering Bulimic's Story," AA/BA Newsletter 8 (Sept.–Nov. 1985):8; Greg Foster and Susan Howerin, "The Quest for Perfection: An Interview with a Former Bulimic," *Iris, A Journal about Women* 12 (1986):18–22. See also Geneen Roth, *Feeding the Hungry Heart* (Indianapolis: Bobbs-Merrill 1982) and Paulette Maisner and Jenny Pulling, *Feasting and Fasting* (London: Fontana, 1985).

36. See Jonathan L. Freedman, "Theories of Contagion as They Relate to Mass Psychogenic Illness," in Michael Colligan, James Pennebaker, and Lawrence Murphy, eds., *Mass Psychogenic Illness: A Social Psychological Analysis* (Hillsdale, NJ: L. Erlbaum Associates, 1982), pp. 180–181.

37. Elizabeth Greene, "Support Groups Forming for Students with Eating Disorders," *Chronicle of Higher Education* (March 5, 1986), 1,30.8.

38. Much of what follows is drawn from Brumberg, pp. 31–32 and 262–271.

39. Brumberg, p. 269.

40. Carol Gilligan, *In a Different Voice: Psychological Theory and Women's Development* (Cambridge: Harvard University Press, 1982).

41. *AA/BA Newsletter* 8 (November 1985–February 1986):5. On the AA/BA, see Brumberg, pp. 19–20.

42. The best known examples of this genre are Susie Ohrbach, *Fat Is a Feminist Issue* (New York: Berkeley Books, 1978) and *Hunger Strike* (New York: Norton, 1986); Marcia Millman, *Such a Pretty Face: Being Fat in America* (New York: Norton, 1980); and Kim Chernin (note 11 above). For a review of this literature see Carole Counihan, "What Does It Mean To Be Fat, Thin, and Female in the United States: A Review Essay," *Food and Foodways* 1 (1985):77–94.

43. Social psychologists use the term "mass psychogenic illness" for spontaneous collective outbreaks of illness that have no somatic basis. Implied in the definition is the notion that such illness is induced by anxiety or stress. Some instances have also been labeled "epidemic" or "mass hysteria." See the collection of essays in Colligan, Pennebaker, and Murphy, note 36 above. In *Madness in Society: Chapters in the Historical Sociology of Mental Illness* (Chicago: University of Chicago Press, 1968), George Rosen uses the term "psychic epidemic" for many of the same historic incidents described by the social psychologists.

44. Francois Sirois, "Perspectives on Epidemic Hysteria," in Colligan, Pennebaker, and Murphy, p. 217.

45. Rosen, pp. 195–228.

46. On anorexia nervosa as a culture-bound syndrome, see Brumberg, pp. 12 and 280 (footnote 15); Raymond Prince, "The Concept of Culture Bound Syndromes: Anorexia Nervosa and Brain-Fag," *Social Science Medicine* 21, no. 2 (1985):197–202; Leslie Swartz, "Anorexia Nervosa as a Culture Bound Syndrome," *Transcultural Psychiatric Research Review* 22, no. 3 (1985):205–207; R. A. Hahn, "Culture Bound Syndromes Unbound," *Social Science Medicine* 21, no. 1 (1985):165–171.

47. Sirois, p. 233. See also A. Martin, "History of Dancing Mania: A Contribution to the Study of Psychic Mass Infection," *American Journal of Clinical Medicine* 30 (1923):265–271; P. O. Moss and C. P. McEvedy, "An Epidemic of Overbreathing among School Girls," *British Medical Journal* 2 (1966):1295–1300; E. A. Shuler and V. J. Parenton, "A Recent Epidemic of Hysteria in a Louisiana High School," *Journal of Social Psychology* 17 (1943):221–235. For a discussion of sex differences in contagious psychogenic illness, see Joseph E. McGrath, "Complexities, Cautions, and Concepts in Research on Mass Psychogenic Illness," in Colligan, Pennebaker, and Murphy, pp. 71–73 and 80–81. Sex differences are generally explained on the basis of different sex role training; that is, females are socialized more than males to express anxiety and stress through somaticization.

48. Rosen, pp. 223–225. Essentially, Rosen sought to prove that psychic epidemics are not necessarily psychopathological or psychotic.

49. Studies that suggest the imitative quality of adolescent behavior in mass society include: D. P. Phillips and M. S. Cantensen, "Clustering of Teenage Suicides after Television News Stories about Suicide," *New England Journal of Medicine* 315 (1986):685–689; K. A. Bollen and D. P. Phillips, "Imitative Suicides: A National Study of the Effects of Television News," *American Sociological Review* 47 (1982):802–809. See also G. Stanley Hall, *Adolescence* (New York: D. Appleton, 1908), vol. 1, pp. 316–317.

50. I refer here to *amok*, *latah*, and *koro*. Amok is a manic and homicidal condition that occurs in men in the Malay-speaking countries; latah is a behavior found in women living in Malaysia, Indonesia, and the Phillipines that involves extreme suggestability, compulsive imitation, sexual delusions, and vulgar language; in koro, a delusional syndrome found in Malaya and South China, a man believes that his penis is shrinking into his abdomen. For definitions see Leland E. Hinsie and Robert J. Campbell, *Psychiatric Dictionary* (New York: Oxford University Press, 1970) and Charles Winick *Dictionary of Anthropology* (New York: Philosophical Library, 1956); the Human Relations Area Files are also useful. P. W. Ngui, "The Koro Epidemic in Singapore," *Australian and New Zealand Journal of Psychiatry* 3 (1969):263–266; Karl Schmidt, "Running Amok," *International Journal of Social Psychology* 23 (Winter 1977):264–274; H.B.M. Murphy, "History and Evolution of Syndromes: The Striking Case of Latah and Amok," in M. Hammer, K. Salzinger, and S. Sutton, eds., *Psychopathology; Contributions from the Social, Behavioral and Biological Sciences* (New York: Wiley-Interscience, 1973), pp. 33–66.

8 FROM MYALGIC ENCEPHALITIS TO YUPPIE FLU: A History of Chronic Fatigue Syndromes

ROBERT A. ARONOWITZ

Every busy internist treats a variety of patients who refer themselves with chronic weakness, recurrent headaches, and other vaguely defined yet deeply felt symptoms. Medical attitudes toward such men and women may be sympathetic, but also skeptical and condescending. Such patients inhabit a diagnostic—and moral—gray area. But this is hardly a novel phenomenon; physicians have always dealt with such ills, in part by conferring an explanatory diagnosis. A century ago such patients might, for example, have been termed neurasthenic or chlorotic. Just as patients expect a prescription, so they expect a diagnosis; it is an anticipated and necessary part of the healing ritual, bestowing a certain legitimacy on the patient's malaise and at the same time undergirding the diagnostician's claim to social authority.

In contemporary America, such symptom patterns are sometimes attributed to the lingering effects of chronic virus infection. To skeptical clinicians this seems no more than a handy residual category, a wastebasket designation for patients with nagging, probably "functional"—and thus illegitimate—symptoms. The diagnosis has the virtue of a reassuring somaticism that holds the promise of being reduced to operational immunological terms. As Aronowitz argues, the original articulation of Epstein-Barr virus infection indeed grew out of just such status-imparting laboratory findings. Acceptance of a clinical entity such as Epstein-Barr disease, or its successor, chronic fatigue syndrome, imposes a coherent etiological order on an otherwise bewildering assortment of symptoms. Moreover, this presumed somaticism categorically distinguishes the pathological consequences of a long-term virus infection from a more ambiguous—and equally hypothetical—origin in psychic, emotional, or intrafamilial factors.

Aronowitz also emphasizes the field-dependency of both physicians and patients. We all see what we are prepared to see. In the case of localized epidemics at the Los Angeles General and Royal Free Hospitals, he contends, perceptions, experience, and responses were conditioned by a heightened awareness of polio. In the Epstein-Barr syndrome, a prior concern with other long-term virus infections predisposed at least some physicians to frame an admittedly elusive symptomatology in just such terms.

This case study illustrates another kind of social negotiation as well. Once a putative clinical entity has been widely and publicly articulated and disseminated, it plays an active role in a complex series of life course dramas—in families, at work, and in the doctor-patient relationship. The cultural availability of the newly legitimate entity termed chronic fatigue syndrome has turned some patients into advocates, aggressive defenders of "their" ailment's legitimacy, in their quest for the consolation of a discrete somatic diagnosis to encapsulate their elusive yet dismaying feelings.

—C. E. R.

EVERY PRACTICING CLINICIAN confronts a recurrent dilemma in the form of patients who suffer from vague and poorly defined ailments; if they can be diagnosed as suffering from disease, it is typically an entity with no agreed-upon etiology or pathophysiological basis. Almost every branch of medicine makes use of such "functional" diagnoses. Gastroenterologists, for example, routinely diagnose irritable bowel syndrome when diarrhea or constipation cannot be attributed to an infectious agent or anatomical abnormality. Because the somatic basis of such ills is questionable, physicians themselves are skeptical about the legitimacy of these diseases, and patients may feel stigmatized. Often both physician and patient are left with the uneasy feeling that organic disease has been missed or that important but difficult to manage psychosocial issues have been sidestepped. Not infrequently, physicians offer these diagnoses only with reluctance and only to patients who demand them. Critical medical appraisal of functional ailments has been bogged down in methodological and epistemological conundrums. Without distinctive objective abnormalities, these syndromes are not easily distinguished from ubiquitous background complaints. There is no well-accepted way to determine whether associated psychological symptoms are effects or causes.

Chronic fatigue syndrome is perhaps the most characteristic functional disease of the 1980s. A public and medical debate quickly arose over its legitimacy. Is its cause in the patient's mind or body? Does its epidemiology reflect biological events or social phenomena such as abnormal illness behavior and altered medical perception?

Participants in the debate have frequently made historical comparisons to support their view. Proponents of the disease's legitimacy have argued that the long history of diseases similar to chronic fatigue syndrome demonstrates the disease's somatic basis. Antagonists have cited the same history to demonstrate the tendency of patients to seek remission of personal responsibility for emotional problems and of doctors to invent accommodating diseases. For example, antagonists have compared chronic fatigue syndrome to neurasthenia, the nineteenth-century disease of "nerve exhaustion," whose hypothesized somatic basis seems fanciful in retrospect.[1] Both sides have frequently compared chronic fatigue syndrome to a cluster of epidemic diseases of the past fifty years thought to result from novel viral infections, most commonly grouped as myalgic encephalitis.

I have taken the cue from these casual citations to compare and contrast the histories of myalgic encephalitis and chronic fatigue syndrome. As their biological basis remains unclear, I have relied on more readily observable nonbiological factors to explain the definition and subsequent course of these diseases. This analysis has identified a set of such factors, which, I will argue, have shaped the epidemiology of these diseases, the controversies that have surrounded them, and the perception that they are the same or similar disease(s). While this historical comparison cannot resolve the debate about the legitimacy of these diseases, it does elucidate some characteristic problems in medical and lay conceptions of disease definition and legitimacy.[2]

Myalgic Encephalitis

I will focus the discussion of myalgic encephalitis on the oldest epidemic typically included under this heading, an outbreak of "atypical poliomyelitis" among health-care workers at the Los Angeles County General Hospital (LAC) in 1934. This outbreak figures in almost all historical comparisons between chronic fatigue syndrome and myalgic encephalitis. For example, a recent call for a conference on the postviral fatigue syndrome states that the "post-viral fatigue syndrome has disturbed and intrigued both the medical and general public since the first documented epidemic occurred in the Los Angeles General

Hospital.[3] Since 1934, when the medical and nursing staff were devastated during that epidemic, post-viral fatigue syndrome has become an increasing cause of disability in North America."[4]

The LAC epidemic was part of a larger polio outbreak in Los Angeles. According to the wisdom of the time, the declining polio incidence in California in the three years preceding 1934 foreshadowed a particularly severe epidemic to come. Fulfilling the prediction, 2,648 polio cases were reported in California before July 1934, over 1700 in the Los Angeles area,[5] a higher incidence than any previous epidemic in southern California.

Even at the outset, it was apparent that cases reported to public health authorities suffered unusually mild symptoms. Mortality was low and persistent paralysis rare or nonexistent.[6] Other unusual features were the high percentage of adult cases, the absence of spinal fluid abnormalities, the early onset in late winter, and an increased frequency of psychoneurotic complaints attributed by some to the mild encephalopathy known to occur in polio.[7]

Doctors had only their clinical judgment to distinguish polio from other diseases. No explicit case definition was available for clinical or reporting purposes. "Space does not permit a discussion of the problem of differential diagnosis," one Los Angeles physician elaborated; "Unbelievable as it seems, we have seen 55 different conditions sent in as poliomyelitis, which proved to be something else, some of them far removed from central nervous system disease."[8]

These diagnostic difficulties, along with other nonbiological factors, contributed to the high incidence rates. Public health officials, fearing an upsurge of polio and believing that effective intervention depended on early diagnosis, launched a campaign to raise public awareness about the protean nature of the disease and the need for early recognition. Patients were thus more prone to detect symptoms and seek medical attention. Clinicians, subject to the same public health campaign and sharing their patients' fears, were likely to have a lower threshold for suspecting polio in the increasing number of patients seeking care.

Public health officials tried to balance measures they felt were effective, such as early isolation of active cases and administration of convalescent serum, against the risk of encouraging public hysteria. "There is a well founded fear of this disease," said Dr. Dunshee, head of the California Department of Public Health, "and there is also an unfortunate terror that is wholly unnecessary."[9] At the outset of the epidemic, public health officials decided to prohibit school assemblies and fairs and enforce hygienic procedures at beer parlors, but decided against closing schools, theaters, and "other places where people gather together."[10] These public health practices not only shaped the way the epidemic was generally understood, but the practices themselves were influenced by widely held lay attitudes. "The public almost forces us to require a specific isolation period," one clinician noted, "and I think it is of practical importance that we prepare to give some sort of arbitrary period."[11]

Throughout the epidemic, public health authorities offered advice about changes in life-style that might slow transmission, noting for example that "dust is a germ carrier . . . where possible housewives should use vacuum cleaners for sweeping rather than the old fashioned broom."[12] The head of the California Department of Health stated what must have seemed obvious: "the most important thing [is] the avoidance of overfatigue and the adherence to a proper personal hygiene."[13]

The low fatality rate and near absence of paralysis were repeatedly cited by public health officials as evidence of the success of their education campaign, isolation of acute cases, and serum treatment. They also claimed success in calming public hysteria by "the systematic manner in which the situation had been handled."[14] Nevertheless, health authorities were criticized for exaggerating the risks and extent of the epidemic.[15]

In June, Los Angeles officials petitioned Simon Flexner of the Rockefeller Institute to send a team of specialists to study the epidemic.[16] The group was welcomed as potential saviors, even though their goal of isolating polio virus from live cases was not the kind of help the public understood or expected.[17]

LAC received more suspected polio cases than other area hospitals, 2,499 before the end of 1934. Physicians stood guard at the hospital entrance and questioned all incoming patients and visitors. A suspected case of polio was immediately dispatched to the contagion ward. Over 100 patients per day were isolated during the most intense part of the epidemic.[18] The new "acute" hospital had just opened in April, and the influx of cases required reopening recently closed wards.[19]

Staff on the contagion ward were required to have their temperature taken every day. Interns were often left unsupervised by attending physicians, who preferred to consult on the telephone. One doctor who worked on the contagion ward was made to feel unwelcome in many private homes by would-be hosts who feared polio.[20]

Health-care workers at LAC began taking sick in May, and by December a total of 198 of them were reported as polio cases to the public health authorities. This represented an overall attack rate among hospital employees of 4.4 percent, ranging from 1 percent for the osteopathic unit to 16 percent for student nurses. No deaths were reported.[21] Concern that the hospital epidemic would cause a health-care worker shortage led the hospital to administer convalescent serum to the entire hospital staff.[22]

Clinical presentations of LAC workers were as varied and atypical as in the community. One summary of clinical findings highlighted the "scarcity of the usual."[23] None of the 25 cases randomly chosen for an appendix to a report of a U.S. Public Health Service investigation of the hospital epidemic presented with definite paralysis or spinal fluid abnormalities. The unusual clinical features were also exemplified by the impossibility of calculating the ratio of paralytic to nonparalytic cases, a traditional polio statistic, because what was recorded in medical charts as paralysis was typically only minor motor impairment detected after vigorous neurological screening.[24]

Gilliam, author of the Public Health Service report, nevertheless concluded that the LAC epidemic represented the person-to-person spread of an infectious agent, "probably poliovirus."[25] His conclusion followed from the similarity of the hospital cases to those of the larger community and the high attack rates among workers in the contagion ward. The community epidemic itself was thought to be polio because a few autopsies showed typical neuropathic changes, and the Rockefeller investigators were able to isolate the polio virus from the nasal secretions of one patient.

These conclusions about the LAC epidemic were at odds with what was then known about polio epidemiology. Acutely ill cases were rarely the source of contagion, and the high attack rate among adults was unprecedented. Only one institutional polio epidemic had previously been recorded. The issue of mass hysteria, raised by some observers as a competing explanation, was dealt with only indirectly in medical publications.[26] "An epidemic like this is almost like a battlefront," Rockefeller investigator Leslie Webster wrote to his wife; "in the one case, 'shell shock' develops, in the other 'queer polio'. . . ." Probably I see and am in contact with 100—200 polio patients a day—but remember hardly any of them are sick. . . . There is hysteria of the populace due to fear of getting the disease, hysteria on the part of the profession in not daring to say a disease isn't polio and refusing the absolutely useless protective serum."[27]

Although the majority of LAC cases fully recovered, a subset suffered from prolonged and recurrent symptoms. A group of nurses claimed permanent disability and complained about their treatment by the hospital administration. Newspapers criticized the transfer of these nurses to other hospitals, with headlines such as "County Scraps 150 Heroic Nurses."[28] A grand jury investigation of the disposition of these nurses kept the hospital epidemic in public view for the next few years.

In the 1950s, the LAC epidemic featured prominently in a number of reviews, which noted that several epidemics had many features in common. In editorials entitled "A New Clinical Entity"[29] and "Not Poliomyelitis,"[30] a new syndrome was defined, most commonly called "benign myalgic encephalitis."[31] This label has been the most durable and is so named because no one died (benign), diffuse muscle pains (myalgia) were prominent, and subjective symptoms were thought to be secondary to brain infection (encephalitis). Links were established to epidemics in South Africa,[32] Australia,[33] Great Britain,[34] Florida,[35] and other places.[36] The two most frequently cited myalgic encephalitis epidemics are Icelandic disease and Royal Free disease, named after their locations. As the years passed, increasing attention was given to sporadic, nonepidemic cases of fatigue and other nonspecific symptoms, whose recognition contributed to the growing interest in myalgic encephalitis and further distinguished it from polio. Such cases presently constitute the majority of patients who receive this diagnosis.

Chronic Fatigue Syndrome

The brief history of chronic fatigue syndrome spans less than a decade. It began with the publication in the early 1980s of a few case series that described a lingering viral-like illness, manifested as fatigue and other largely subjective symptoms, that appeared to be associated with serological evidence of recurrent or prolonged infection with the Epstein-Barr virus (EBV).[37] The idea of recurrent EBV infection was not new, isolated cases having been reported over the preceding forty years, but it was not recognized as a widespread phenomenon.[38]

The later reports seemed plausible on a number of grounds. First, EBV, like other herpesviruses, persists in the body after acute infection and thus might cause recurrent or continual clinical disease. Second, the major clinical syndrome associated with EBV, acute infectious mononucleosis, shared fatigue and many other protean clinical features with the chronic syndrome being described.[39]

From the beginning, it was difficult to attribute symptoms such as chronic fatigue to EBV infection on the basis of antibody tests because most adults have been exposed to the virus and thus have antibody. One of the first case reports claimed that symptomatic patients had an elevated level of a class of antibody (IgM) to the EBV capsid antigen, usually present only in acute infection, thus adding plausibility to the EBV reactivation hypothesis.[40] However, subsequent studies did not reproduce this finding. The other case series that received wide circulation showed EBV antibodies apparently fitting a pattern of continued or reactivated infection.[41]

During the same year these studies were published (1985), the Centers for Disease Control (CDC) sent a team to Lake Tahoe to investigate an outbreak of a prolonged viral-like syndrome in over a hundred patients. Tests by local doctors, aware of recent case reports, had found many of these patients to have high antibody titers to various EBV antigens. *Science* reported the epidemic under the headline "Mystery Disease at Lake Tahoe."[42]

The subject stirred considerable local controversy. According to one popular report, other Tahoe physicians were skeptical that an epidemic was occurring. One said that "they [Peterson and Cheney, the Tahoe internists who reported the epidemic] think they notice something, then they start seeing it everywhere."[43] Another doctor noted that he saw none of these patients in his practice and concluded "there has to be something wrong with Peterson and Cheney's diagnosing procedure." In response, one patient said "Peterson and Cheney believed we were sick, that's why they got all these patients."[44]

The CDC investigators, following standard epidemiological practice, created a case definition and intensively studied fifteen Tahoe patients who met their criteria. Although they observed serological abnormalities similar to those reported, they also noted substantial overlap with their controls, as well as serological evidence of other viral infections, including cytomegalovirus. The CDC group concluded: (1) A sensitive and specific laboratory test was still needed to define a group of patients who might have a somatic basis for their symptoms. (2) Reported symptoms were too vague for a proper case definition. (3) There was too much clinical overlap with normals. (4) The EBV serological tests were not reproducible enough to be reliable.[45]

In April 1985, the National Institute of Allergic and Infectious Diseases (NIAID) organized a consensus conference on chronic Epstein-Barr virus infection. Among the fifty attendants were several patients. Conference participants agreed that only those patients with severe lymphoproliferative or hypoplastic disorders and very high levels of Epstein-Barr antibodies "most likely have chronic Epstein-Barr virus infection."[46]

Despite the skepticism of the CDC investigators and expert medical opinion, chronic Epstein-Barr virus infection had been launched—in both lay and medical worlds. Popular journals reported the results widely, private laboratories promoted EBV blood tests, and patients began arriving at doctors' doors. During the next three years interest in chronic Epstein-Barr virus infection spread among clinicians, and medical journals generally treated it as a credible entity. For example, an editorial in a leading journal of allergy and immunology declared that "the syndrome of chronic Epstein-Barr virus infection exists."[47]

Hints of skepticism appeared in some of the initial medical publications. For example, one study noted an "excessive risk for educated adult white women," a wry comment given the implausibility of a biological explanation for this susceptibility. The same study described symptoms as "woes," suggesting that these patients might have been more troubled than ill.[48] However, these initial medical publications did not explicitly consider that many patients with the diagnosis of chronic Epstein-Barr virus infection suffered from psychiatric problems.

By 1988 a skeptical trend became evident in the medical literature. Signifying this change of opinion was another consensus conference whose results were published in early 1988.[49] The most important development was a call to rename chronic Epstein-Barr virus infection "chronic fatigue syndrome." The new consensus stressed the minimal diagnostic utility of the EBV serological tests and questioned the etiologic relationship of symptoms to EBV. Conference participants proposed an alternative "Chinese menu" approach to diagnosis: the presence of symptoms meeting two major plus any eight of fourteen minor criteria. The new definition was criticized as arbitrary, in part because its sensitivity, specificity, and predictive value were not known.[50] The consensus group authors countered that such information was not knowable for this disease, which had no "gold standard" diagnostic test (one that definitively identifies a disease as a biopsy confirms cancer).[51]

The importance of pathobiological mechanisms in defining and legitimating diseases is illustrated by the fact that chronic fatigue syndrome gained notice as a new disease only as a result of attention given to the apparent correlation between abnormal EBV scrologies and a vague viral-like illness. Although the causal link between EBV and chronic fatigue was later severed, the purely clinical syndrome, relabeled chronic fatigue syndrome, began to take on a life of its own.

A series of studies soon appeared casting doubt on the somatic basis of the syndrome. A few demonstrated that patients with chronic fatigue also have a very high prevalence of psychiatric disease.[52] In one study, researchers evaluated patients presenting to a "fatigue clinic." When given a structured psychiatric interview most of these patients "tested out" as having a psychiatric disorder. The authors then went on to judge which of their patients had symptoms ascribable to the identified psychiatric disorder and concluded that the proportion was two thirds.[53]

While selection bias and the lack of control groups marred this study, a more fundamental dilemma was the direction of causality. Other studies have consistently shown that chronically ill patients have high psychiatric comorbidity, especially as measured by psychometric tests.[54] The patients in this study, who were suffering from an as yet undiagnosed chronic medical condition for a mean of thirteen years, might have been expected, by analogy to other chronic diseases, to have had a good deal of psychiatric comorbidity. Moreover, by judging for themselves when fatigue had a psychiatric basis, the authors in effect predetermined the relationship they set out to "study." Their conclusions suffered from this circular logic.

Another debunking study that received much attention was undertaken by the National Institutes of Health group that had previously published one of the first chronic Epstein-Barr virus infection case series. This time they performed a randomized, double-blind, placebo-controlled test of the effect of acyclovir (an antiviral drug active against herpesviruses) on patients who met the CDC criteria for chronic fatigue syndrome and had very high levels of antibodies associated with EBV reactivation.[55] Acyclovir had no advantage over placebo. Although the study was small, and subjects were not typical of most patients with chronic fatigue syndrome, the negative results were offered as additional evidence against the EBV hypothesis and the disorder's legitimacy in general.[56]

More recently, researchers (including one of the original Tahoe physicians) found DNA fragments similar to parts of the HTLV-II (a retrovirus in the same class as HIV, the putative etiological agent of AIDS) genome in patients suffering from chronic fatigue. While their report received front page coverage in newspapers across the country, it was greeted with skepticism by scientists at the meeting announcing the results.[57] Only further detailed laboratory and epidemiological work will be able to prove whether this correlation is a valid indication of an etiological relationship or, as is more likely, a spurious finding.

In summary, scientists and clinicians at first correlated serological evidence of chronic Epstein-Barr infection with a vague clinical syndrome, but the correlation was not confirmed. The symptom cluster was reformulated as chronic fatigue syndrome and developed its own momentum among patients and the general public. Recent medical studies have tried to demonstrate that the syndrome is better thought of as psychiatric, but these studies are methodologically weak. Doctors' apparent growing skepticism about chronic fatigue syndrome— despite continued public interest, media attention, and new scientific "breakthroughs"—is probably related not to published studies but to other considerations, such as the syndromes's problematic nonetiologic definition, its determined lay advocates, and other factors to be considered in the next two sections.

Nonbiological Determinants of Disease Definition and Identity

These historical sketches of myalgic encephalitis and chronic fatigue syndrome point to a set of nonbiological factors that these diseases share and that account for their controversial status. I will discuss six: (1) attitudes and beliefs about disease; (2) disease advocacy; (3) media coverage; (4) ecological relationships; (5) therapeutic and diagnostic practices; and (6) economic relationships.

Lay and medical attitudes and beliefs have played important roles in the definition and appearance of these diseases. The widely held conclusion that LAC workers contracted polio was considered plausible despite the presence of many atypical features because individual susceptibility to polio was believed to be a matter of "overwork and excess fatiguability."[58] "A surprising number of doctors and nurses came down," Dr. Parrish, the Los Angeles Public health chief surmised, "because they have worked themselves to the point of exhaustion and their resistance has broke down."[59]

Individual susceptibility was of special concern in polio as it was increasingly apparent that the denominator of silent infection was large, perhaps including whole populations in some epidemics.[60] The question of individual susceptibility to polio catalyzed the transformation of George Draper from prominent Rockefeller biomedical scientist to leader of "constitutionalism," a protopsychosomatic movement.[61] Factors that might explain the "Why me?" and "Why now?" of this disease were the subject of both lay speculation and scientific research.

Another attitudinal factor in the LAC outbreak was health-care workers' fear of contagion from hospitalized cases despite epidemiological evidence, especially the work of Wickman in Sweden and Frost in the United States, that polio was almost exclusively spread by contact with persons with inapparent infection.[62] "It was as if a plague had invaded the city," noted prominent polio researcher and historian John Paul, "and the place where cases were assembled and cared for [LAC] was to be shunned as a veritable pest house."[63]

In chronic fatigue syndrome, lay and medical disease theorizing has focused on a disruption of immune regulation as the reason that particular individuals succumb. One lay advocacy group is tellingly called "the Chronic Fatigue and Immune Dysfunction Syndrome Association" (CFIDS). On occasion, stress is linked to immune dysfunction, providing a pseudobiological link between an inciting virus, life events, and disease.

Vigorous advocacy has played a determining role in the development of these diseases and their controversy. In both the LAC and Royal Free epidemics, the very fact that patients were themselves health-care professionals was an important source of disease advocacy. When health-care workers, especially doctors, are patients, their traditional authority to define disease lends legitimacy to a questionable syndrome. A concrete example of the nexus of different roles was Mary Bigler, head of the LAC contagion unit, who took ill in June 1934 and later co-authored one of the epidemiological reviews of the outbreak.[64] The similar part played by lay advocacy groups in chronic fatigue syndrome is detailed in the next section.

Media coverage has shaped the way these diseases have been understood and propagated. In Los Angeles, newspapers were skillfully manipulated by the public health authorities on whom the papers depended for statistics and commentary. Newspaper reports probably had a major influence on the lay and medical interpretation of ubiquitous background complaints and self-limited illness as due to polio. In May and June 1934, when the public health authorities were committed to the existence of a polio epidemic, articles almost daily compared the high current incidence of the epidemic to prior years. Later in the summer, when the authorities were criticized for exaggerating the risks and extent of the epidemic, more reassuring articles about the disease's mildness and decreasing incidence took over.[65] The media similarly helped to spread an awareness and interest in chronic fatigue syndrome, especially with regard to the Lake Tahoe epidemic.

Ecological relationships, by which I mean the interdependence among prominent diseases at any particular time and place, have strongly shaped the characterization of these diseases. Myalgic encephalitis would not have captured attention on clinical grounds alone. It needed polio as a vehicle for recognition, even though the link between the two was later severed.

A good deal of evidence indicates that the perception of the chronic fatigue syndrome was affected by the contemporary AIDS epidemic. As a newly described viral infection of the 1980s, localized in the immune system and preferentially attacking specific social groups, chronic fatigue syndrome invited such a comparison. The stress is not so much on biological similarities as on related controversies. What patients and medical critics keep pointing out is the marginal status of chronic fatigue syndrome patients: deserving of sympathy, if not disability benefits, but injured cruelly less by the disease itself than by medical and lay attitudes that rob them of their dignity by impugning the legitimacy of their symptoms. The stigma of the "psychosomatic label" suffered by chronic fatigue syndrome patients is akin to the stigma of being in an AIDS risk group. AIDS is also cited to argue that new protean infectious diseases are possible and can effect hundreds of thousands before their etiology is understood.

Specific therapeutic and diagnostic practices also have shaped these epidemics.[66] Speaking of the LAC epidemic, Paul charged that the elaborate orthopedic treatment of polio cases "gave the poliomyelitis ward the appearance of a ward occupied by patients who suffered extensive trauma inflicted in a disaster area, whereas in actuality very few patients turned out to have any paralysis at all."[67]

A major factor contributing to the high incidence and unusual features of the LAC epidemic was the absence of a case definition for clinical or reporting purposes. Analogously, elevated EBV antibodies, thought to indicate the chronic fatigue syndrome but later found to be prevalent in the general adult population, helped to launch the syndrome's notability.

Finally, economic factors such as business interests, disability concerns, and labor relations all played a role in shaping these diseases. Early in the Los Angeles epidemic, public health authorities used the stark incidence figures to justify increased public health spending. Later, these same authorities became increasingly sensitive to the economic implications of public hysteria, especially for southern

California's tourist industry. Combining a growing sobriety about the mildness of the epidemic with advocation of the protective values of fresh air and uncrowded mountains, they distributed a series of bulletins about the safety of southern California's tourist attractions. Nevertheless, publicity about the epidemic appeared to threaten the tourist industry, prompting U.S. Surgeon General Cummings to declare that the epidemic was "not serious."[68] At the very onset of the hospital epidemic, disability was a contentious labor issue for hospital employees, to be followed by a prolonged battle over workmens' compensation for student nurses whose pre-disability income was so low that ordinary compensation was woefully inadequate.

Disability has also been a central concern in chronic fatigue syndrome. The CFIDS association issued a pamphlet entitled "you can win your rightful benefits." Another important economic factor has been the promotion of serological tests by private laboratories, which resulted in more testing and thus more diagnoses.

Controversy over Disease Legitimacy

These common nonbiological determinants demonstrate why both detractors and supporters of the legitimacy of myalgic encephalitis and chronic fatigue syndrome have perceived them, in the absence of common etiological or objective clinical abnormalities, as being the same disease. Even more striking, however, are similarities in the controversies these diseases have inspired. It is useful to discern two distinct levels of controversy. At the most explicit level, for example the typical presentation in medical publications, chronic fatigue syndrome and myalgic encephalitis are measured against conventional criteria of disease specificity. At a more ideological level, the debate over disease legitimacy reflects tensions and contradictions between lay and medical attitudes toward medical authority, phenomenological versus objective definitions of disease, and the social construction of disease.

In medical publications, the explicit debate concerns whether the diseases are caused by a virus or hysteria. Antagonists have typically focused their critique on such standard epidemiological problems as the vagueness of case definitions, the lack of measurable abnormality, atypical epidemiological features, and the substantial overlap with psychiatric disease.

Supporters of these syndromes have taken a largely defensive posture in the published medical debate. They have argued that patients have exhibited little premorbid psychiatric disease[69] and that long-term follow-up has shown only minor psychiatric morbidity but continued physical impairment.[70] Moreover, proponents have pointed out the discrepancy between patients' mental symptoms and any model of mass hysteria.[71] While allowing that some fraction of those with the diagnosis had exaggerated or hysterical illness, "it would be manifestly erroneous to consider as hysteria the emotional instability associated with this illness in all of the cases in which it was present."[72] Lastly, proponents have used the aggregate mass of myalgic encephalitis epidemics and chronic fatigue syndrome cases to suggest that only a somatic etiology would adequately explain their recurrence and prevalence.

The nonbiological determinants discussed earlier offer some clues to the ideological basis of antagonism toward these syndromes. Physicians have objected to vigorous lay advocacy, especially when patients demand diagnoses doctors are reluctant to confer. Beliefs about disease susceptibility, however important, have not agreed well with reigning reductionist models of disease etiology. Legitimacy is more likely to be questioned if a disease is defined largely by its treatment and if its incidence is tied to arbitrary public health activities and economic factors. In short, these syndromes are too transparently socially constructed. Although many among researchers, clinicians, and the public at large are aware of the importance of social factors in the definition of "legitimate" diseases such as tuberculosis and AIDS, they nevertheless tend to view such factors as delegitimating in the absence of a specific biological explanatory mechanism.

Another ideological objection to the legitimacy of these diseases is the concern that their diagnoses are controlled more by the patient than the physician. The patient-centered CDC criteria for chronic fatigue syndrome, for example, are perceived as permitting the patient, rather than the doctor, to define the disease. Intensifying this concern is the widespread belief that the stressed and mentally ill are an immense market for somatic diagnoses, making these syndromes vulnerable to abuse. Journalistic accounts have used the terms "yuppie flu" and "Hollywood blahs" for chronic fatigue syndrome, suggesting it may result from the stress of being an ambitious young professional. Cleveland Amory made this explicit. After listing the CDC clinical criteria for chronic fatigue syndrome, he wrote that it "left out number twelve—a six figure income by age thirty. Now that's tiring."[73] One medical observer characterized myalgic encephalitis patients as having "four-star abilities with five-star ambitions."[74]

More than merely blaming the patients, these mean-spirited remarks take aim at a perceived misuse of the sick role. Antagonists see the typical myalgic encephalitis or chronic fatigue syndrome patient as claiming the benefits of the sick role without being particularly sick. "The seasoned clinician," one physician wrote, "will recognize the current epidemic in diagnosis of the EB syndrome to be a manifestation of the current narcissistic and hedonistic society in that there is an ever-increasing tendency for people to blame something or someone else rather than look at themselves."[75] The "epidemic in diagnosis" requires not only patient abuse of the sick role but doctors as willing accomplices. Critics have pointed out the prejudicial influence of the "doctor as patient" in these syndromes and have argued that those with the disease should not be allowed to investigate it.

Less caustic critics have maintained that attributing an etiological role to psychosocial factors is not necessarily stigmatizing and should not deny the patient the benefits of being sick. For example, the authors of the "mass hysteria" hypothesis for the Royal Free epidemic argued that

> Many people will feel that the diagnosis of hysteria is distasteful. This ought not to prevent its discussion, but perhaps makes it worthwhile to point out that the diagnosis of hysteria in its epidemic form is not a slur on either the individuals or the institution involved. Whereas it is true that sporadic cases of hysterical disability often have disordered personalities, the hysterical reaction is part of everyone's potential and can be elicited in any individual by the right set of circumstance.[76]

One lay observer of the chronic fatigue syndrome controversy made a similar point:

> Illness is illness. . . . Psychological research has shown us that individuals with psychological disorders are no more responsible for their illness than people with physical illness and that both can have an organic cause.[77]

Despite the appealing rhetoric, both sides of the debate have acted as if the onus of responsibility falls much more severely on the patient whose disease does not have a specific somatic etiology.

While the ideological basis for antagonism to the legitimacy of these syndromes has typically been sub rosa in medical publications, the protagonists' ideology is often more straightforwardly presented in the popular press and publications of lay groups (especially in chronic fatigue syndrome). The basic argument repeatedly made by lay advocates equates the legitimacy of these syndromes with the reality of patient suffering. Not accepting their legitimacy means not believing the patient. A common form of the argument is to contrast physician skepticism with a patient's certain knowledge of suffering something real." While some doctors dismiss it as a faddish disease," one lay observer noted, "the EB syndrome sufferer must face the future knowing that any debate will not change the fact that something is very wrong."[78]

Lay accounts emphasize the harm caused by medical skepticism. One report related how a patient diagnosed with chronic EBV infection later turned out to have cancer. After undergoing surgery and chemotherapy, she remained more afraid of a recurrence of the EBV infection and the pain of not being believed than of a recurrence of her cancer.[79]

Proponents repeatedly emphasize the legitimacy of the patients' illness experience. Medicine is portrayed as overly reliant on objective tests while denigrating the patients' experience. One chronic fatigue syndrome sufferer questioned the hubris of scientists who equate "not known" with "not real": "How does one explain a disease that doesn't show, can't be measured, and as a consequence, is erroneously attributed to the patient's willful gloom?"[80] "The difference between crazed neurotic and a seriously ill person is simply a test," one myalgic encephalitis patient observed, "that would allow me to be ill."[81] The implicit argument is that the patient's phenomenological experience of sickness and suffering should be as or more important than medicine's "objective" criteria in defining and diagnosing disease.

Narrative accounts of myalgic encephalitis and chronic fatigue syndrome characteristically reflect this conflict between medical and lay authority to define disease, and the harm inflicted on the individual whose claim to have a disease is medically rejected. A successful young woman suddenly develops a mysterious debilitating illness. Because her physicians are unable to make a precise diagnosis, they become inpatient and suggest that the problem is psychological. Friends and family become more frustrated with the patient's situation and begin to lose interest and sympathy.[82] When all hope appears to be gone, the patient is diagnosed as suffering from myalgic encephalitis or chronic fatigue syndrome either because she discovers the diagnosis herself or meets a knowledgeable and compassionate doctor. With the disease named and some time lapsed, the patient begins to recover, often delivering a moral lesson in the process.[83]

The rhetoric of this advocacy and the emphasis on the patient's subjective experience of illness often evoke a bitter struggle between patients and the biomedical establishment. "The rigid mindset of those who have tried to submerge the illness as a clinical entity, discredit the physicians who have stood by us, and demean those of us who have CFIDS," one lay advocate warned, "must never be forgotten or covered up."[84]

A characteristic feature of such rhetoric has been to depict medical opposition to the syndromes as a conspiracy. One report claimed that the CDC investigation for new viruses has been impeded by suspicious National Institutes of Health (NIH) researchers who kept their research to themselves. One advocate complained that "there's a national epidemic of immune system dysfunction and viral disorders in progress which until recently the CDC has been more interested in covering up than doing something about."[85] One of the two Tahoe physicians who requested that the CDC investigate the epidemic has been depicted as being run out of town by economic interests worried that publicity about the disease would damage the Tahoe tourist industry.[86]

The underlying power struggle is also seen in the tactics of lay advocacy groups. One chronic fatigue syndrome group publishes summaries of relevant scientific and popular reports and pragmatically assesses their usefulness for the cause. "This was not the most sympathetic of articles, in regard to the way it treated chronic fatigue syndrome," a typical summary goes, "but it helped to keep the syndrome in the public spotlight."[87] This same group maintains an honor roll of physicians loyal to the cause, and organized a drive to have the NIH dismiss prominent researcher Steven Straus after he published two studies viewed as antithetical to the disease's legitimacy.[88]

Although lay advocates stridently criticize medical hegemony, they typically stop short of categorically advocating the primacy of the patient's phenomenological experience of illness or of articulating an explicitly relativist critique of conventional modes of categorizing disease. Biomedical science is still looked on as the ultimate arbiter of disease legitimacy. Even the staunchest critics of medical authority desire to have these syndromes accepted by doctors as "ordinary" diseases.

The reconciliation of this apparent paradox—a vigorous attack on medical authority and the desire for its approval—takes the form of compromise. Some proponents have argued that medical knowledge is provisional and the lack of an objective test for myalgic encephalitis or chronic fatigue syndrome is only a temporary limitation of medical technology and competence.[89] Others claim that objective evidence already exists, in the form of measurable physiological abnormalities, distinctive clinical signs, and etiological agents, and fault doctors and researchers for their inability to form a consensus around this evidence.[90] Studies that failed to find distinctive abnormalities are criticized on all levels: study design, inferred conclusions, and the ideological motivations of the authors.[91]

The tension between the critique of medical authority and the desire for its acceptance of the patient's condition as a normal disease is exemplified by one patient's complaint to a medical journal over the change in name from chronic EBV infection to chronic fatigue syndrome, which argued that "instead of affirming the infectious nature of the illness, [the change] reinforces its psychiatric nature . . . these implications feed right into the alternative healing misinformation mill."[92] This patient went on to suggest names like "chronic viral syndrome" and "chronic mononucleosis syndrome" as being "appropriately vague" while still connoting an infectious origin. The crux of this argument is that doctors should provisionally accept the somatic basis of the chronic fatigue syndrome in order to keep patient care within the legitimate medical establishment.[93]

In summary, the controversy over the legitimacy of myalgic encephalitis and chronic fatigue syndrome pits those who find the social construction of these diseases debunking, and who see in them both a threat to medical control over diagnosis and a potential source of abuse of the sick role, against those who believe that greater weight should be given to the patient's experience of sickness in disease definition, and who more generally challenge medical hegemony. At the same time, both sides share assumptions about the proper priority of specific biological mechanisms in disease definition and the greater burden of responsibility on the individual who suffers from a disease that so far lacks such mechanisms.

Conclusion

Rosenberg makes the point that with few exceptions "in our culture a disease does not exist as a social phenomenon until we agree that it does."[94] The controversy surrounding myalgic encephalitis and chronic fatigue syndrome exposes the conflict often inherent in such agreements and the process through which they are reached.

The initial agreement as to the nature of the LAC epidemic turned on the shared belief that an unusual polio outbreak had occurred, while chronic fatigue syndrome was first proposed as a novel but plausible consequence of a specific etiological agent (EBV). Even from the beginning, however, disease specificity was a tenuous rationale for these diseases. Clinicians and researchers at LAC were always aware that the clinical presentations of health-care workers were atypical for polio, just as chronic fatigue syndrome's brief history included considerable early biomedical skepticism about the etiological role of EBV.

Given these weak beginnings, the rise of interest in these syndromes might reasonably be seen as resulting from a commonality of interest among medical scientists, doctors, and patients. In chronic fatigue syndrome, for example, medical scientists benefited from being able to correlate abnormalities detected by laboratory technology with a clinical problem, especially one resulting in a new disease.[95] Clinicians thus found a solution to many problematic encounters with patients in which a disease could not be diagnosed. Patients received absolution from responsibility, relief from uncertainty, and the promise of effective therapy. The conception of chronic fatigue as a disease also placed limits on doctors' paternalistic judgments about the role of psychological factors in causing symptoms.

The ensuing controversy had as much to do with the different types of equity each group held in these diseases as with a failure to form a biomedical consensus around specific etiological agents or physiological derangements. The issue of specific disease mechanisms has nevertheless divided skeptical scientists and physicians from patients, lay advocacy groups, and fellow-traveling physicians who believed in, and had something to gain from, the recognition and acceptance of these diseases—whether or not explanatory mechanisms actually existed.

At the same time, the controversy reveals how much both sides have been bound by shared notions concerning the centrality of disease as the rationale for individual doctor-patient relations, and health policy more generally. Even as lay groups have strongly advocated patient autonomy and rights, they have accepted that the tools, conceptual framework, and authority to legitimate disease rest ultimately and appropriately with scientists and clinicians. Radical epistemological critiques of biomedical models of disease have been rare. Biomedical failure to confirm these syndromes as diseases is seen as temporary, provisional, and correctable by further research. Skeptical doctors, even when dismissing the status of these syndromes, attributed them to yet another category of disease—the psychological.[96]

The nonbiological factors and ideological considerations that form the basis of the social construction of these diseases and their ensuing controversies are almost always omitted or remain between the lines in biomedical publications. Explicit statements of prominent researchers' "politics" concerning the status of these diseases are to be found only in popular accounts or in private correspondence.[97] Published reviews of the controversy over myalgic encephalitis typically end in hopes of a biomedical solution. "So far, then, there is no definite answer as to what causes this perplexing syndrome," one reviewer concluded, "but further controlled trials and the application of gene probes and monoclonal antibodies may provide one."[98] And thus, characteristically, we again avoid addressing any of the epistemological issues raised by these diseases.

Paralleling such constraints in biomedical discourse are limits on what constitutes acceptable biomedical investigation. The Public Health Service report on the LAC epidemic and the CDC study of the Lake Tahoe case cluster did little to resolve the controversy that surrounded these diseases. The social factors underlying chronic fatigue syndrome's acceptance, its spread among select populations, its declining interest for doctors, and the influence of its determined lay advocates have all contributed to the syndrome's incidence and distribution, that is, its epidemiology, yet such factors are typically outside the gaze of clinical or epidemiological research.

In concluding, I would like to stress that the issue of disease legitimacy in medical practice, while brought into dramatic relief by chronic fatigue syndrome and its borderland antecedents, has general significance. Lyme disease, whose somatic ontogeny was methodically uncovered over the last fifteen years, is increasingly diagnosed as a chronic condition. The relationship of chronic constitutional symptoms to prior infection and abnormal laboratory findings presents a configuration of problems to researcher, doctor, and patient very similar to that of myalgic encephalitis and chronic fatigue syndrome.[99]

In an era in which the financial burden of health care coupled with a perception of declining or static health gains is generally recognized as the major problem facing modern medicine, more attention might be paid to the costs and health effects of the creation of new diseases that emerge from the detection of novel abnormalities and clinical-laboratory correlations. It may not be farfetched to expand the issues raised in the debate over chronic fatigue syndrome and myalgic encephalitis to include the increasing number of new diseases rendered discoverable by our advanced technological capacity. For example, we can now observe individual differences in serum cholesterol levels and continuous electrocardiographic monitoring that can lead to diagnoses such as hypercholesterolemia and silent cardiac ischemia, "diseases" that have no corresponding phenomenological basis until a patient is found or constructed by screening tests. Similarly, the expanding knowledge of the human genome will undoubtedly lead to the construction of new diseases based on correlations between individual genetic variation and clinical states. What configuration of nonbiological factors explains the appearance of such diseases at a particular moment in time? In whose interest is it to view diseases as legitimate? What is the effect of a new label on the patient and the doctor-patient encounter? These questions remain at the center of medical practice.

ACKNOWLEDGMENTS

I would especially like to thank Sankey Williams, Charles Rosenberg, and Rosemary Stevens for comments on earlier drafts of this manuscript.

NOTES

1. See S. B. Straus, "The Chronic Mononucleosis Syndrome," *Journal of Infectious Disease* 157, no. 3 (1988):405–412, for such a comparison. Skeptics about chronic fatigue syndrome's legitimacy would probably readily agree with William Osler's depiction of the problematic doctor-patient relationship in neurasthenia, *Principles and Practice of Medicine*, 3d ed. (New York: Appleton, 1898), p. 1130: "in all forms there is a striking lack of accordance between the symptoms of which the patient complains and the objective changes discover-

able by the physician. . . . As has been said, it is education more than medicine that these patients need, but the patients themselves do not wish to be educated." Space limitations do not allow a detailed consideration of neurasthenia. C. Rosenberg examines the social construction of neurasthenia in "George Beard and American Nervousness," in *No Other Gods* (Baltimore: Johns Hopkins University Press, 1976). B. Sicherman discusses the negotiations between doctors and patients over the use and meaning of the neurasthenia diagnosis in "The Uses of a Diagnosis: Doctors, Patients and Neurasthenia," *Journal of the History of Medicine* (Jan. 1977):33–54. From the nineteenth century on, feminist social critics portrayed neurasthenia and its treatments as tools of male domination over women, e.g., Charlotte Perkins Gilman, *The Yellow Wallpaper* (New York: Feminist Press, 1973).

2. An obvious problem with this approach is that the "social construction" of chronic fatigue syndrome might be understood as arguing against its somatic basis. This is neither my intention nor my conclusion. Social factors have played an important role regardless of whether or not a culpable virus is ultimately discovered.

3. This term, used mainly by British clinicians, encompasses chronic fatigue syndrome, myalgic encephalitis, and other diseases.

4. Advertisement for The First World Symposium on Post-Viral Fatigue Syndrome, *New England Journal of Medicine* 321:19 (1989).

5. Severe polio epidemics were thought to occur after three or four years of declining incidence. The numbers of reported cases in California during the four years preceding 1934 were: 903, 293, 191, and 170. J. D. Dunshee and I. M. Stevens, "Previous History of Poliomyelitis in California," *American Journal of Public Health* 24, no. 12 (1934):1197–1200.

6. According to one contemporary epidemiologist, G. M. Stevens, "The 1934 Epidemic of Poliomyelitis in Southern California," *American Journal of Public Health* 24, no. 12 (1934):1213, there was "never an epidemic with so low a death rate, with so many of the spinal type, so many with only paresis, or neuritis, so many of the straggling or recurrent variety."

7. Referring only to the Los Angeles area, another epidemiologist, A. Dower et al., "Clinical Features of Poliomyelitis in Los Angeles," *American Journal of Public Health* 24, no. 12 (1934):1210, noted that "The high adult morbidity was itself a variation. The symptoms were, for the most part, milder than usual, and the sequelae less crippling. . . . The degree and duration of muscle pain, tenderness and severe cramping were out of proportion to the motor phenomena. . . . Even more striking was the rapid, and apparently complete, recovery of some cases which appeared early to be doomed to extensive paralyses. An unusual number of cases showed no increase in spinal fluid pressure."

8. Ibid., p. 1211.

9. "Poliomyelitis Wane Foreseen," *Los Angeles Times,* June 17, 1934, p. 18.

10. "Drive Started on Paralysis," *Los Angeles Times,* June 1, 1934, p. 3.

11. From *Poliomyelitis: Papers and Discussions Presented at the First International Poliomyelitis Conference* (Philadelphia: Lippincott, 1949), p. 123.

12. "Pomeroy To Air Plans on Play Camps," *Los Angeles Times,* June 6, 1934, p. 5.

13. "Paralysis Scare Held Unjustified," *Los Angeles Times,* July 8, 1934, p. 20.

14. "Paralysis Situation Improving," *Los Angeles Times,* June 24, 1934, p. 23.

15. Responding to accusations of inflated incidence statistics, Health Commission president Baxter replied that "the records at the General hospital are open for anyone to ascertain the facts" ("Experts Study Child Disease," *Los Angeles Times,* June 22, 1934, p. 10).

16. J. Paul, *A History of Poliomyelitis* (New Haven and London: Yale University Press, 1971), pp. 212–224.

17. Media attention—and controversy—also focused on the Rockefeller investigators because they used monkeys in their virological work. One antivivisectionist characteristically complained that "experimenting is not for the cure of the disease but to advance knowledge and satisfy curiosity" (letter to ed., "Healing or Experimenting?" *Los Angeles Times,* July 7, 1934, p. 4).

18. G. M. Stevens, "The 1934 Epidemic of Poliomyelitis in Southern California," p. 1214. The hectic atmosphere at LAC is captured in this account: "At the height of the epidemic there were 21 wards with 364 nurses caring for 724 patients, 360 of whom had poliomyelitis. . . . Doctors, nurses, orderlies, maids, ambulance drivers, and all others worked overtime, often for 24 to 48 hours without letup. Fatigue, loss of sleep, and constant exposure to poliomyelitis in its most infectious stage was common to all."

19. H. E. Martin, *The History of the Los Angeles County Hospital (1878–1968) and the Los Angeles County–University of Southern California Medical Center (1968–1978)* (Los Angeles: University of Southern California Press, 1979), p. 121.

20. J. Paul, *A History of Poliomyelitis,* p. 223.

21. A. G. Gilliam, *Epidemiological Study of an Epidemic Diagnosed as Poliomyelitis, Occurring among the Personnel of the Los Angeles County General Hospital during the Summer of 1934* (U.S. Treasury Department, Public Health Service), Public Health Bulletin no. 240, April 1938.

22. A. Bower et al., "Clinical Features of Poliomyelitis in Los Angeles." Prophylactic serum therapy was probably ineffective and was not without risks. A two-year-old child died after injection of serum by her physician/father ("Stricken Child Dies Following Serum Injection," *Los Angeles Times,* June 25, 1934, p. 3).

23. M. F. Bigler and J. M. Nielsen, "Poliomyelitis in Los Angeles in 1934," *Bulletin Los Angeles Neurological Society* 2 (1937):48.

24. A. G. Gilliam, *Epidemiological Study.*

25. Ibid.

26. For example, Paul (*A History of Poliomyelitis,* p. 224) felt that the Los Angeles polio epidemic illustrated "the part that the character and attitude of the community can play in distorting the accepted textbook picture of a common disease—almost beyond recognition."

27. Ibid., p. 219.

28. "County Scraps 150 Heroic Nurses," *Los Angeles Times,* Oct 10, 1935, cited in H. E. Martin, *The History of the Los Angeles County Hospital,* p. 123.

29. "A New Clinical Entity," *Lancet* 1 (1956):789.

30. "Not Poliomyelitis," *Lancet* 2 (1954):1060–1061.

31. Syndromal definitions like that of myalgic encephalitis challenged the notion that a particular set of symptoms is uniquely caused by a specific infectious agent. D. N. White and R. B. Burtch, "Iceland Disease: A New Infection Simulating Acute Anterior Poliomyelitis," *Neurology* 4 (1953):515, made this explicit when discussing a myalgic encephalitis epidemic: "It would seem possible that a wide spectrum of viruses may be responsible for a wide variety of clinical conditions including acute infectious polyneuritis, acute lymphocytic choriomeningitis, as well as the encephalitides and poliomyelitis."

32. "The Durban 'Mystery Disease'," *South African Medical Journal* 29 (1955):997–998.

33. A. A. Pellew, "A Clinical Description of a Disease Resembling Poliomyelitis Seen in Adelaide, 1949–1951," *Medical Journal of Australia* 1 (1951): 944–946.

34. A. D. Macrae and J. F. Galpine, "An Illness Resembling Poliomyelitis Observed in Nurses," *Lancet* 2 (1954):350–352.

35. D. C. Poskanzer et al., "Epidemic Neuromyasthenia: An Outbreak in Florida," *New England Journal of Medicine* 257, no. 8 (1957):356–364. Although CDC investigators and subsequent authors linked the outbreak to myalgic encephalitis, the CDC group concluded cautiously—and revealingly—that "the illness epidemic in Punta Gorda illustrates problems inherent in an attempt to define a nonfatal illness with a broad array of symptoms that blend into the host of psychoneurotic and minor medical complaints endemic in a community" (p. 362).

36. For a detailed review of the clinical and epidemiological features of these epidemics see E. D. Acheson, "The Clinical Syndrome Variously Called Benign Myalgic Encephalitis, Iceland Disease and Epidemic Neuromyasthenia," *American Journal of Medicine* 26 (1959):569–595.

37. S. E. Straus et al., "Persisting Illness and Fatigue with Evidence of Epstein-Barr Virus Infection," *Annals of Internal Medicine* 102, no. 1 (1985):7–16; J. F. Jones et al., "Evidence for Active Epstein-Barr Virus in Patients with Persistent, Unexplained Illnesses: Elevated Anti-early Antigen Antibodies," *Annals of Internal Medicine* 102, no. 1 (1985):1–7; and M. Tobi et al., "Prolonged Atypical Illness Associated with Serological Evidence of Persistent Epstein-Barr Virus Infection," *Lancet* 1 (1982):61–64.

38. R. Issacs, "Chronic Infectious Mononucleosis," *Blood* 3 (1948):858–861; R. E. Kaufman, "Recurrences in Infectious Mononucleosis," *American Practice* 1, no. 7 (1950):673–676; and C. E. Bender, "Recurrent Mononucleosis," *Journal of the American Medical Association* 182, no. 9 (1962):954–956.

39. Another point of similarity was that acute infectious mononucleosis had attracted much psychosomatic speculation, e.g., in S. V. Kasl, A. S. Evans and J. C. Niederman, "Psychosocial Risk Factors in the Development of Infectious Mononucleosis," *Psychosomatic Medicine* 41, no. 6 (1949):445–466.

40. M. Tobi et al., "Prolonged Atypical Illness Associated with Serological Evidence of Persistent Epstein-Barr Virus Infection."

41. R. E. Dubois et al., "Chronic Mononucleosis Syndrome," *Southern Medical Journal* 77, no. 1 (1984):1376–1382; J. F. Jones et al., "Evidence for Active Epstein-Barr Virus"; and S. E. Straus et al., "Persisting Illness and Fatigue with Evidence of Epstein-Barr Virus Infection."

42. D. M. Barnes, "Research News: Mystery Disease at Lake Tahoe Challenges Virologists and Clinicians," *Science* 234 (1986):541–542.

43. W. Boly, "Raggedy Ann Town," *Hippocrates* (July/Aug 1987):35.

44. Ibid., p. 36.

45. "Chronic Fatigue Possibly Related to Epstein-Barr Virus," *Morbidity and Mortality World Report* 35, no. 21 (1986):350–352.

46. "Chronic Epstein-Barr Virus Disease: A Workshop Held by the National Institute of Allergy and Infectious Diseases," *Annals of Internal Medicine* 103, no. 6 (1985):951–953.

47. R. C. Welliver, "Allergy and the Syndrome of Chronic Epstein-Barr Virus Infection," editorial, *Journal of Allergy and Clinical Immunology* 78, no. 2 (1986):278–281.

48. S. E. Straus et al., "Allergy and the Chronic Fatigue Syndrome," *Journal of Allergy and Clinical Immunology* 81, no. 5 (1988):791.

49. G. P. Holmes et al., "Chronic Fatigue Syndrome: A Working Case Definition," *Annals of Internal Medicine* 108, no. 3 (1988):387–389.

50. D. A. Matthews, T. J. Lane, and P. Manu, "Definition of the Chronic Fatigue Syndrome," *Annals of Internal Medicine* 109, no. 6 (1988):512.

51. G. P. Holmes et al., "In Response: Definition of the Chronic Fatigue Syndrome," *Annals of Internal Medicine* 109, no. 6 (1988):512.

52. M. J. Kruesi, J. Dale, and S. E. Straus, "Psychiatric Diagnoses in Patients Who Have the Chronic Fatigue Syndrome," *Journal of Clinical Psychiatry* 50, no. 2 (1989):53–56, and P. Manu, T. J. Lane, and D. A. Matthews, "The Frequency of the Chronic Fatigue Syndrome in Patients with Symptoms of Persistent Fatigue," *Annals of Internal Medicine* 109 (1988):554–556.

53. P. Manu, D. A. Matthews, and T. J. Lane, "The Mental Health of Patients with a Chief Complaint of Chronic Fatigue: A Prospective Evaluation and Follow-up," *Archives of Internal Medicine* 148 (1988):2213–2217.

54. J. E. Helzer et al., "A Controlled Study of the Association between Ulcerative Colitis and Psychiatric Diagnosis," *Digestive Disease Science* 27, no. 6 (1982):513, and T. Pincus et al., "Elevated MMPI Scores for Hypochondriasis, Depression, and Hysteria in Patients with Rheumatoid Arthritis Reflect Disease Rather than Psychological Status," *Arthritis and Rheumatism* 29, no. 12 (1986):1456–1466.

55. S. E. Straus et al., "Acyclovir Treatment of the Chronic Fatigue Syndrome: Lack of Efficacy in a Placebo-controlled Trial," *New England Journal of Medicine* 319, no. 26 (1988):1692–1696.

56. M. N. Swartz, "The Chronic Fatigue Syndrome—One Entity or Many?" *New England Journal of Medicine* 319, no. 26 (1988):1726–1728.

57. J. Palca, "Does a Retrovirus Explain Fatigue Syndrome Puzzle?" *Science* 249 (1990):1240–1241.

58. A. Bower et al., "Clinical Features of Poliomyelitis in Los Angeles," p. 1211.

59. "Light Shed on Disease," *Los Angeles Times*, July 15, 1934, p. 18.

60. Individual susceptibility to poliovirus was also thought to result from specific attributes such as having had a tonsillectomy (leading to removal of one defense against infection).

61. George Draper's views of susceptibility are especially striking and important. Draper was one of Flexner's chief Rockefeller polio investigators and wrote extensively on polio, e.g., *Infantile Paralysis* (Philadelphia: P. Blakiston's, 1917). Draper felt that polio exemplified the problem of disease susceptibility, with its few sick patients among the many infected. He advocated a theory that correlated susceptibility to different diseases with characteristic morphological types (faces and other bodily features), which appears unfounded and simplistic by today's standards. But the importance of facial morphology in Draper's theories has been exaggerated, and his work can be viewed as a reasonable attempt to reinject the individual into reductionistic, postgerm-theory models of disease causation; e.g., see *Disease and the Man* (New York: Macmillan, 1930). The American psychosomatic movement of the 1930s and 1940s grew quite directly out of Draper's "Constitution Clinic" where Murray formed the psychosomatic hypothesis for ulcerative colitis. See R. Aronowitz and H. Spiro, "The Rise and Fall of the Psychosomatic Hypothesis in Ulcerative Colitis," *Journal of Clinical Gastroenterology* 10, no. 3 (1988):298–305.

62. I. Wickman, *Beitrage zur Kenntnis der Heine-Medinschen Krankheit (Poliomyelitis acuta und verwandter Erkrankungen)* (Berlin: Karger, 1907) and W. H. Frost, "Acute Anterior Poliomyelitis (Infantile Paralysis); A Precis," "Public Health Bulletin no. 44, 1911.

63. J. Paul, *A History of Poliomyelitis*, p. 212.

64. M. F. Bigler and J. M. Nielson, "Poliomyelitis in Los Angeles in 1934."

65. "Drive Started on Paralysis," *Los Angeles Times*, June 1, 1934, p. 3 and "Sharp Drop in Malady Estimated," *Los Angeles Times*, July 16, 1934, p. 12. Throughout the summer, statistics on new cases were reported like today's Dow Jones averages; "sudden drops" one day, "slight increases" the next.

66. Specific therapies are themselves shaped by widely held beliefs about disease. For example, serum therapy was conveniently rationalized by both modern concepts of specific humoral immunity as well as vaguer notions of acquired resistance. Acquired resistance received explanations such as that "mankind is building up within itself the kind of immunity that jungle men, for example, build up against jungle fevers. And the fact that no one over 61 has succumbed indicates that older people are already immune. Immunity begins that way—usually with old folks" ("Light Shed on Disease," *Los Angeles Times*, July 15, 1934, p. 18).

67. J. Paul, *A History of Poliomyelitis*, p. 222.

68. "Surgeon General Cummings Minimizes Outbreak," *Los Angeles Times*, July 7, 1934, p. 3.

69. K. G. Fegan, P. O. Behan, and E. J. Bell, "Myalgic Encephalomyelitis—Report of an Epidemic," *Journal of Royal College of General Practioners* 33 (1983):336, noted that "the fact that they [cases] were all known to have good pre-morbid personalities, made us consider an organic cause for their illness."

70. E. D. Acheson. "The Clinical Syndrome Variously Called Benign Myalgic Encephalitis, Iceland Disease and Epidemic Neuromyasthenia."

71. E.g., "Dr. A. M. Ramsey told a meeting of the Myalgic Encephalitis Study Group in London [May 1988] that nurses affected in the 1955 Royal Free Hospital epidemic had been so ridiculed by some physicians that they have refused to talk about the long-term consequences of the disease. Such denial is the opposite of what one would anticipate from a patient with hysteria," quoted from B. Hyde and S. Bergmann, "Akureyri Disease (Myalgic Encephalitis), Forty Years Later," *Lancet* (1988):1191–1192.

72. E. D. Acheson, "The Clinical Syndrome Variously Called Benign Myalgic Encephalomyelitis, Iceland Disease and Epidemic Neuromyasthenia," p. 575.

73. C. Amory, "The Best and Worst of Everything," *Parade*, Jan. 1, 1989, p. 6.

74. Attributed to Dr. Stuart Rosen, a research fellow at Charing Cross Hospital, by D. Jackson, "In the Library," *CFIDS Chronicle* (Spring 1989):70.

75. R. Holland, "Is It Nobler in the Yuppie Mind to Suffer the Slings of EBV," *Medical Post* (Canada), Nov. 1, 1988, p. 12, cited in D. Jackson, "In the Library," p. 70.

76. C. P. McEvedy and A. W. Beard, "Royal Free Epidemic of 1955: A Reconsideration," *British Medical Journal* 1 (1970):10.

77. D. Edelson, "In Defense of Stephen E. Straus, MD," letter to the editor," *CFIDS Chronicle* (Spring 1989):45.

78. "Stress and a New Plague," *Maclean's* 49 (1986):75.

79. H. Johnson, "Journey into Fear," *Rolling Stone* (July 16–30, 1987):58.

80. Ibid., p. 57.

81. A. J. Church, "Myalgic Encephalitis 'An Obscene Cosmic Joke?'" *Medical Journal of Australia* 1 (1980):307–309.

82. E.g., "My friends became leery of my situation. At first they were sympathetic, but after a while they didn't understand why I couldn't get better," L. Heumerdinger, "A Baffling Syndrome, Perhaps Born out of Stress, Leaves a Would Be Screenwriter Sick and Tired," *People Weekly* 29 (1988):131.

83. A good example of this genre is L. Marsa, "Newest Mysterious Illness: Chronic Fatigue Syndrome," *Redbook* (April 1988):120. The lesson for the chronic fatigue syndrome sufferer in this vignette is "I no longer feel I need to do it all."

84. "Fresh A.I.R.E. (Advocacy, Information, Research, and Encouragement)," *CFIDS Chronicle* (Jan/Feb 1989):9.

85. W. Boly, "Raggedy Ann Town," p. 39.

86. According to this physician, "Maybe we violated some law of nature that says one does not do research projects on viruses in resort communities," Ibid., p. 40.

87. D. Jackson, "Comment from Recent CFIDS Items in Popular Magazines/Newspapers," *CFIDS Chronicle* (Jan/Feb 1989):67.

88. Some representative comments: "Sound your desire to have Dr. Steven Straus removed from his current position of 'C.F.S. expert at the N.I.H.,'" "Fresh A.I.R.E. (Advocacy, Information, Research and Encouragement)," *CFIDS Chronicle* (Jan/Feb 1989):11. "We will not rest (figuratively speaking) until CFIDS achieves the legitimacy and attention it merits and ultimately, is conquered. And we will speak and print the truth, even when it is a king (e.g., a noted researcher) who is caught wearing no clothes," Ibid., p. 9.

89. E.g., "Too often because our illness does not 'fit the rules' we have been victims of an arrogant medical doctrine that holds 'if I can't understand or diagnose your illness (and my technology can't detect it), it must not exist—or you must be psychoneurotic,'" "Fresh A.I.R.E. (Advocacy, Information, Research and Encouragement)," *CFIDS Chronicle* (Jan/Feb 1989):9.

90. A striking example of how poorly rationalized laboratory abnormalities help to legitimize these diseases was given for a school-based epidemic by G. W. Small and J. F. Borus, "Outbreak of Illness in a School Chorus," *New England Journal of Medicine* 308, no. 11 (1983):632–635. A urinary abnormality found in all cases significantly contributed to the definition and acceptance of the outbreak. The abnormality later proved to exist in the urine container, which made the competing explanation of mass hysteria appear more compelling.

91. The acyclovir trial in chronic fatigue syndrome was subject to all these criticisms in a series of articles in *CFIDS Chronicle* (Spring 1989).

92. C. Radford, "The Chronic Fatigue Syndrome," letter to *Annals of Internal*

Medicine (July 15, 1988):166. The apparent role reversal in which otherwise holistic critics of medicine admonish doctors for their mind-body speculation and lobby for a somatic conception of chronic fatigue syndrome is not lost on some advocates. One patient reflected, "In his tiny office, he [doctor] offered me hot coffee and chocolate donuts, which I turned down regretfully. I mused on a world where the healers thrive on sugar and coffee while the sick limp through their days with infusions of vitamins, mineral and Evian water. Nothing seemed fair or reasonable anymore," H. Johnson, "Journey into Fear," p. 46.

93. Another critic likened the new label "chronic fatigue syndrome" to renaming diabetes "chronic thirst syndrome," N. Walker, "Welcoming Speech to The San Francisco Chronic Fatigue Syndrome Conference, April 15, 1989," *CFIDS Chronicle* (Spring 1989):2. The unintended irony is that diabetes is a Greek syndromal label for "chronic urination syndrome."

94. C. Rosenberg, "Disease in History: Frames and Framers," *Milbank Quarterly* 67, Supp. 1 (1989):1–15.

95. What significance many of these correlations might have is increasingly recognized as a general problem for modern medicine. See, for example, T. E. Quill, M. Lipkin Jr., and P. Greenland, "The Medicalization of Normal Variants: The Case of Mitral Valve Prolapse," *Journal of General Internal Medicine* 3, No. 3 (1988):267–276.

96. It is remarkable and illuminating to note the many continuities in the way psychological diseases, and various formulations of psychological factors to explain somatic disease, have figured in standard medical practice over the last century. Mid- to late-nineteenth-century neurologists argued that all forms of madness would eventually be understood as resulting from observable pathology. While their classic example was the dementia of syphilis, they most frequently treated patients labeled as having neurasthenia, a disease with many similarities to myalgic encephalitis and chronic fatigue syndrome. While psychosomatic theories of disease gained some marginal acceptance with the ascendancy of Freudian psychiatry, their acceptance was shortlived, especially among psychiatrists whose research paradigm increasingly called for a somatic cause for most psychiatric illness. Patients with complaints not explainable by an existing somatic disease need to be provided with one, creating a continuous stimulus for functional diagnoses in different medical generations.

97. E.g., H. Johnson, "Journey into Fear," part two, Aug. 13, 1987, 44, reveals prominent chronic fatigue syndrome researcher Anthony Komaroff's "politics."

98. J. Dawson, "Royal Free Disease: Perplexity Continues," *British Medical Journal* 294 (1987):328.

99. R. Aronowitz, "Lyme Disease: The Social Construction of a New Disease and Its Social Consequences," *Milbank Quarterly* 69, no. 1 (1991):79–112.

NEGOTIATING DISEASE:
The Public Arena

9   THE ILLUSION
OF MEDICAL CERTAINTY:
Silicosis and the Politics
of Industrial Disability, 1930–1960

GERALD MARKOWITZ and DAVID ROSNER

The give and take surrounding the definition of occupational disease is a very real and unavoidably political process, yet at the same time a metaphor for the way society deals with disease more generally. In the case of an entity like silicosis, we can see how a variety of interested groups struggled to fashion particular outcomes—in terms of both policy and accepted pathology. In the arena of occupational health these seemingly disparate spheres are inevitably linked; the outcome of particular negotiations often turns on a consensual willingness to accept the existence and occupation-related etiology of a specific—and thus legitimate—clinical entity.

Silicosis in twentieth-century America provides an illuminating case in point. As Markowitz and Rosner emphasize, the syndrome's very *visibility*—as well, of course, as its clinical characteristics and accepted course—was negotiated and renegotiated, decade by decade. The negotiators were a varied lot: unions, industry, public health officials, industrial physicians, politicians, insurance executives, state, and—by the 1930s—federal administrators. The debate over the etiology and pathological trajectory of silicosis was not simply an academic discussion among radiologists and pulmonary specialists; the imposition of clinical judgments had consequent policy implications. And the establishment of an administrative procedure for making such judgments was a necessarily political action. In such a context, the power to diagnose was the power to legitimate economic claims. The practical stakes were significant and the questions to be negotiated complex. Would workman's compensation be due men unable to work as a consequence of chronic lung disease? And how was silicosis related to particular environmental circumstances and other, supervening, ills, most importantly tuberculosis? Was ability to work a viable and appropriate indicator of marginal—and potential—pathology? Who was to be the gatekeeper, the designator of incapacitating and occupationally derived sickness: a pulmonary specialist? an expert in occupational illness? a lay jury in a civil court? or any competent general practitioner?

The history of silicosis in twentieth-century America also illustrates another sort of negotiation, between human bodies and their surroundings. Foundries, mines, and quarries presented very different (and constantly changing) environments. Market relationships, technical change, and social custom all interacted to produce particular work careers and particular levels and kinds of dust in the air breathed by miners or foundrymen. The complexity and interrelatedness of these interactions underline another axiom as well: the artificiality of our distinction between the social and the biological. A complex and ever-changing configuration of cultural, economic, and technical factors builds and shapes mines and foundries, alters modes of production—and introduces ventilators, fans, and OSHA inspectors along with high speed drills and sanders. Markowitz and Rosner show how the constantly renegotiated entity called silicosis became visible, then central to the depression-era debate over welfare and occupational health, and then gradually faded to the margin of social perceptions after the Second World War.

It should also be emphasized that the social negotiations determining the comparative attention paid particular ailments is an ongoing phenomenon—hardly banished by increasingly sophisticated levels of medical understanding. One might refer to the comparative invisibility of lead poisoning in late-twentieth-century America as contrasted with a perhaps overwrought public awareness of the dangers associated with even the smallest atmospheric concentration of asbestos fibers. We see and fear what we have been prepared to see.

—C. E. R.

Silicosis is a chronic lung disease caused by the inhalation of silica dust. At various times throughout the twentieth century it was perceived as a major health problem affecting millions of American workers in mining, tunneling, foundries, quarrying, sandblasting, and other extractive industries. How is it that this disease became the focus of intense attention at a particular time and then faded from view? The introduction to this book states that "disease does not exist until we have agreed that it does." Here, we hope to shed light on that process of agreement. Professional groups were not alone in framing the disease. Rather, government officials, insurance executives, and labor representatives all played critical roles in defining and accepting silicosis as a legitimate disease category.

During the Great Depression a deluge of lawsuits by unemployed workers claiming disability from silicosis forced major industries, insurance companies, and government and labor officials to address the relationship between occupational disease and disability. The relationship was marked by questions rather than agreed upon solutions. What is an industrial disease? How can health problems related to occupation be distinguished from other, nonindustrial ailments? How should liability for risk be assigned? Should a worker be remunerated for physical impairment or loss of wages due to occupational disability through the workmen's compensation system? Should industry be held accountable for chronic illnesses whose symptoms appear years and sometimes decades after exposure? At what point in the progress of a disease should compensation be paid? Is diagnosis sufficient for compensation claims or is inability to work the criterion? Who defines inability to work—the employee, the government, the physician, or the company? In the case of silicosis, labor, management, industry, and insurance representatives argued—in nontechnical terms—over the power to define what the medical community would later call "latency," "time of onset," and "disease process." The heated debates of the 1930s became the basis for revision of state and federal workmen's compensation systems for occupational disease.

As the lawsuits produced a "silicosis crisis," representatives of large insurance companies and the foundry and metal-mining industries pressed local and federal government officials to integrate silicosis into the existing workmen's compensation system. Spokesmen such as F. Robertson Jones and Henry D. Sayer, both of the Association of Casualty and Insurance Executives, worried about the impact of lawsuits on the stability of the insurance industry. Anthony Lanza, the well-known industrial hygienist of the Metropolitan Life Insurance Company, and R. R. Sayers, of the U.S. Public Health Service, were prominent public health experts who defined the medical criteria for diagnosis. Robert J. Watt, of the Massachusetts Federation of Labor, was one of the most active labor representatives in the silicosis controversy.

In the post–World War II era few unions other than the radical International Union of Mine, Mill, and Smelter Workers and even fewer industry and government representatives continued to consider silicosis a major health problem. In part, this was due to the financial and political resolution of the silicosis crisis of the 1930s. Incorporation of silicosis into many state workmen's compensation systems stifled public debate, and the decline in disability lawsuits removed the silicosis controversy from the forum of the courts. By the 1960s and 1970s national attention shifted to other industrial lung conditions as workers organized to demand special protection: Coal miners agitated for special protective legislation against black lung, and new political, medical, and labor constituencies brought public prominence to other lung conditions such as brown and white lung.

The Great Depression
and the Silicosis Crisis

The introduction in the early twentieth century of sandblasting, pneumatic tools, and other mechanical devices dramatically increased workers' exposure to silica dust in a wide variety of industries. Between the First World War and the Crash of 1929, silicosis had devastated many mining and quarrying communities. By the beginning of the Great Depression, however, a number of factors coincided to create a "silicosis crisis." As many industrial workers were thrown out of work or denied employment, some workers focused on the role of disease and disability in creating their plight. Unable to find work, the unemployed sued former employers and insurance companies for damages due to occupational disease, creating what industry and insurance spokesmen termed a "liability crisis."[1] During good times, workers would have been able to find employment despite varying degrees of discomfort and disability. During the Depression, an enormous pool of dependent and disabled workers sought to gain redress from current and former employers through the courts.

By 1936, government officials and business leaders believed that "silica dust is probably the most serious occupational disease hazard in existence today" and that it "typifie[d] the whole occupational disease" problem.[2] As a result of the silicosis liability crisis, insurance and business representatives argued that decisions reached by juries and courts were dictated more by sympathy for or antipathy against the claimants than by "objective" science and law. They, along with labor unions and reform politicians, suggested that occupational disease be incorporated into the workmen's compensation system. The system would provide "reasonable" economic protection for workers while limiting the financial liability of insurance companies and industry. By taking decisions regarding culpability out of the hands of lay juries, insurance and business representatives maintained that predictability and expertise would substitute for randomness and subjectivity.[3]

But serious issues had to be addressed before workmen's compensation could be used to manage the silicosis crisis. Insurance companies, industry, labor, and government needed to decide on how to integrate chronic disease into a system that had evolved around injuries and accidents on the job. Traditionally, workmen's compensation systems paid a dollar amount for the type and severity of any particular injury. Loss of a limb, an eye, or a life were all given a price, and workers or their families could count on receiving the applicable established payments. These "scheduled" payments might differ between states, but the principle of concrete awards for predictable events was the basis of these systems.

Now many industrial diseases, and silicosis in particular, challenged the compensation system because the effects of disease were rarely obvious and clear-cut. Unlike accidents and acute poisonings, silicosis was a chronic condition that might or might not show symptoms at any particular time. Medical opinion was divided as to how it progressed, whether disability and impairment were inevitable, and the length of time between exposure and first symptoms. In light of the enormous medical uncertainty regarding the nature of silicosis, legislators, insurance company representatives, industry spokesmen, public health officials, and labor faced the political problem of defining criteria for its compensation, whose financial implications were largely unknown.

The debate of the 1930s centered on two differing conceptions of disability. The first held that compensation should be treated simply as a response to accidental injury. If a worker lost a limb or was disfigured in any way, even with no consequent specific loss of income, the injury was compensable. An insult to a worker's body was sufficient reason for payment. In the 1920s, for example, compensation boards commonly held that an employee "may become permanently partially disabled by the loss of some member of his body without suffering a loss in earning capacity" but was still entitled to compensation for suffering. Despite the fact that the worker could continue to work, "the majority of the states have provided in their laws for a schedule of such injuries."[4] But the extraordinary number of lawsuits for silicosis made extending such scheduled uncontested payments to cover the disease extremely expensive: "[W]e apparently think of every case of pulmonary fibrosis as requiring compensation, whether disabled or able to work," complained Roy Jones of the U.S. Public Health Service. He worried that the implementation of such a policy could bankrupt tottering industries.[5]

By the late 1930s, insurance, business, and public health officials pressed for a second, more restrictive, definition of compensable disease, using decreased earning capacity as an objective "criterion" to measure disability.[6] Insurance, medical, and industry spokesmen advocated defining disability due to a compensable disease narrowly in terms of lost income, not impaired function.[7]

The narrowing of compensation board criteria coincided with a more restrictive definition of silicosis within the public health community. In 1917, for example, a pathbreaking study documented the importance of silicosis in an American mining community by noting that "the first stage [of silicosis] is characterized with slight or moderate dyspnea on exertion." The study maintained that workers at all stages of silicosis, whether early or late, suffered inhibited breathing. By 1935, however, the Public Health Service no longer mentioned shortness of breath as characteristic of the disease in its early stages. Now, the focus of concern was on the relationship between silicosis and decreased earning capacity: "The term disability . . . may be defined as a decreased capacity to do the work required of the individual in the course of his usual occupation and/or an increased susceptibility to respiratory infection causing a loss of time from work which may reasonably be considered as primarily the result of the pulmonary fibrosis."[8]

Relating their physical condition to their ability to work further limited workers' ability to gain compensation. Insurance and public health experts believed that early-stage silicosis, uncomplicated by tuberculosis, did not decrease a person's ability to work. For example, one public health expert argued that "simple silicosis . . . causes relatively little severe disability,"[9] and Anthony Lanza claimed even further in 1936 that "disability in silicosis is seldom due to the silicosis itself."[10] Such medical opinion reinforced the restrictive views of compensation boards. A Wisconsin court ruled that a compensation board's refusal to provide money to a diseased worker was appropriate because "medical disability, does not, in the absence of an actual wage loss, entitle one to compensation."[11]

What Is an Occupational Disease?
The Problem of Tuberculosis

Negotiations over the occupational disease of silicosis was complicated by its association with tuberculosis, assumed to have a nonindustrial origin. By the mid-1930s public health officials and statisticians recognized that silicotic workers tended to contract tuberculosis at a much higher rate than did the general population. In light of this greater associated risk, labor, government, and insurance officials agreed that compensation was due victims of silicotuberculosis.[12] Despite this agreement, however, insurance company representatives and industry spokesmen maintained that the compensation system should not bear the liability for tuberculosis, a disease not generally regarded as specific to a particular occupation. Business rejected the idea that it should pay for "secondary" illnesses, claiming that tuberculosis was an infectious disease associated with poverty and living conditions rather than the workplace. Since tuberculosis was not an occupational disease, and silicosis itself was rarely severely disabling, the worker must bear the primary burden of silicotuberculosis, however disabling. The insurance industry in particular held that when tuberculosis was a complicating factor in creating disability "there should be some provision for *reduction* [our emphasis] of compensation benefits."[13]

Who Should Determine Disability?

The business community maintained that medical experts could end the contentious political debate. But even these experts were unable to arrive at a definitive diagnosis of "pure" silicosis, despite the availability of the x ray, stress tests, lung function tests, and other technologies.[14] Because doctors were divided in the years before the integration of silicosis into workmen's compensation, juries were left to judge for themselves the adequacy, honesty, and reliability of individual medical experts. A spokesman for the foundry industry complained that "boards of laymen, after hearing the partisan opinions of physicians selected by the disputants, regardless of the weight of medical opinion" make decisions on highly technical medical matters. If "difficult diseases such as silicosis . . . [were] . . . brought under the compensation law," he protested, "the results would be chaotic and probably ruinous." Rather than allow overt political interests to define criteria for compensation, he called for the "establishment of competent and impartial medical tribunals, free from political influences, to decide all medical questions involved in controverted cases."[15] Some labor and New Deal officials saw this move to give medical professionals control over the definition of disease as nothing more than a political use of "objective science." Despite the rhetoric of objectivity, they charged, the call for change was designed to disenfranchise labor by stripping away its right to participate in decisions affecting workers' lives.

As early as 1925, industry, insurance, and medical professionals began to advocate the establishment of an official medical advisory panel. They rationalized that such a panel was necessary to determine "objectively" the outcome of accident and disease cases brought before the compensation system. Such a process would "reduce the possibilities of political influence to a minimum and would give assurance that only properly qualified men would be considered."[16] In the 1930s such suggestions gained greater urgency, given "the desirability of removing this type of case from the sphere of ex parte medical testimony" in the courts.[17] The prestigious Committee on Pneumonoconioses of the American Public Health Association declared its support for such a scheme in 1933: "Without some form of medical control, the management of compensation for a disease such as silicosis would be difficult." It held this position despite the lack of a medical consensus regarding the mechanism by which silica dust affected lung tissue, the course of silicosis once diagnosed, or even the degree of disability associated with various stages of the disease. The Committee acknowledged that silicosis was "a disease in which the definition of disability is obscure," but went on to urge that "medical advice is necessary to determine whether the worker's health is impaired to a degree which constitutes disability."[18]

It was at this point, in 1936, that the Department of Labor held a national conference to bring together the various constituencies in the silicosis debate to address the highly political medical issue. But despite the Department's hope that labor, management, government, and public health professionals would reach a consensus, real power at the meeting devolved onto the insurance and public health professionals. As a result, the conference concluded that efforts should be redoubled "to search for impartial, objective, quantitative measures by means of which the subjective complaints may be evaluated." In addition to bacteriological cultures of sputum and medical evaluations of x rays, the Committee advocated the use of "various physiological tests to determine respiratory efficiency." Such tests would include measurements of pulmonary capacity and ventilation during muscular work. It would be the medical board, allegedly removed from the politics of workplace struggle and the real-life pressures of unemployment during the Depression, that would become the arbiters of workers' complaints. There should be established "a medical advisory board of experts on dust disease . . . as an official, impartial body of the State." This body would have absolute authority in all disputed cases involving medical uncertainty. "The findings of this board in cases of medical fact should be final."[19] Thus, the appearance of medical and scientific objectivity served to cloak the reality of uncertainty, controversy, and differing political and social perspectives.

Labor representatives at the Conference rejected the attempt to take the discussion of silicosis out of the public arena and place it into the hands of professionals. Labor viewed this use of experts itself as driven by politics rather than science. Martin Durkin, an official from the Illinois Department of Labor who would later serve briefly as Secretary of Labor under Eisenhower, objected to the substitution of the findings of a medical board for the judgment of courts and juries of citizens, pointing out that "medicine is not an exact science, and doctors, even experts, have been mistaken." He disagreed "that medical boards, composed of experts, should be the final arbiters of disputed medical questions. . . . Such a procedure violates every known idea of American justice and denies a person the right to a trial of the important issue."[20]

Underlying this dissent were ongoing negotiations between labor officials and medical and public health experts, two groups with fundamentally different approaches to the issue of industrial hygiene. Broadly speaking, state and federal labor administrators believed that medical and public health opinion regarding industrial disease was biased in favor of industry rather than the workforce, despite its claim of scientific objectivity. New Deal administrators in the U.S. Department of Labor and some state departments saw themselves as allies of organized labor during the turbulent years of the 1930s and generally adopted labor's distrust of medicine.[21] Throughout the 1920s companies had used physical examinations to discriminate against workers. Taking the silicosis issue out of a political arena in which workers' influence mattered would enable a coalition of other interests to control the definition of "disease."

Different Perspectives on Disability

By 1939 the most thoughtful industrial physicians recognized that these differing perspectives undercut an unambigous conception of occupational disease. When Anthony Lanza was asked how he might try to define occupational disease for the purpose of writing workmen's compensation legislation, he admitted that "at one time, I felt that I knew what was an occupational disease, but I no longer feel that way."[22] Ludwig Teleky, an internationally renowned authority on industrial lung diseases, noted in 1941 that differing views of the relationship of industry to disability and disease were embodied in two competing compensation schemes. The first, "blanket" coverage, was all-inclusive, encompassing all diseases associated with employment. Even diseases like pneumonia, if contracted as a result of work conditions, could be classified as an occupational disease. The second was much narrower, including only diseases "peculiar to a certain occupation," which would have to be clearly and definitively enumerated in a schedule in the workmen's compensation laws. Pneumonia, a disease not specifically associated with a particular industry, would not be compensated, even if conditions in a particular plant predisposed workers to this illness.[23]

Insurance carriers and industry representatives, not surprisingly, pressed hard for ways to limit liability claims and decrease costs. They proposed that each state develop "a schedule of the diseases to be deemed 'occupational diseases'" peculiar to that state in the opinion of medical authorities, and that only those specified "diseases" be compensated. Disease causes would have to "be traced to origins in 'trade risks'—i.e., risks, not of ordinary life, but created by special practices or processes in industrial occupations."[24] Henry D. Sayer, another representative of the carriers, remarked on the divergence of opinion regarding compensation legislation. "We all start out with the proposition that all diseases fairly chargeable to an industry should be compensated by the industry," he began. "We differ, however, in the method of coverage, and the difference in method means a vast difference in the rule of liability and the burden imposed upon industry." Sayer opposed "blanket" coverage as applying a vague definition of occupational disease that would allow virtually any disease to be the basis of compensation claims. He wrote that "great danger lies in the fact that such general and vague language will lend itself to the inclusion of any and every sort of illness and disease to which human flesh is heir." He concluded with the derisive comment, ironic in today's hindsight, that such absurd notions could lead to the inclusion of seemingly "natural" diseases such as tuberculosis, pneumonia, and cancer as occupational diseases: "We surely do not think of colds, pneumonia, tuberculosis . . . and cancer as occupational diseases."[25]

Sayer believed that blanket coverage would do nothing to resolve the liability crisis and would provoke a deluge of lawsuits and compensation claims along with massive discontent unless the definitions were closely controlled. The all-inclusive method "will give rise and lead certainly to a great volume of litigation, all looking to court interpretations as to what is and what is not an occupational disease."[26] Others added new dimensions to the arguments in favor of the schedule method.

Sayer saw beyond the immediate crisis over silicosis and sought to protect the insurance industry from assuming the liability for broader social and health problems. He thought industry should pay when it was directly responsible for a workers' ailment, but "no such obligation should be placed on industry" for society's general ill-health.[27] Many in the insurance industry feared that blanket coverage would tend to obscure the distinctions between workers' compensation and a more general system of relief during the Depression. F. Robertson Jones, general manager of the Association of Casualty and Insurance Executives, summarized the fears and political goals of the insurance industry, which worried that workmen's compensation would become a tool of reformers seeking to shift the costs of social welfare benefits for unemployment from the public to the private sector: "The chief trouble today is that we have confused compensation with relief. If we can keep these two ideas separate and can restrict the tendency to turn the compensation system into a universal pension system having no particular relation to employment, we shall have accomplished something."[28] Anthony Lanza, who only a few years later would wonder whether there were any "objective" criteria for measuring occupational disease, reinforced this position. In 1936 he said a major reason for opposing blanket coverage and favoring the schedule method was that a schedule could be framed to include "only true occupational disease," making it possible "to estimate with fair accuracy what will be the liability that the employers and the compensation carrier have to face."[29]

Labor Perspectives

Organized labor, like the insurance industry, the public health community, and management, had a variety of positions regarding disability caused by silicosis. Spokesmen for the American Federation of Labor developed a rhetoric of dissent and strongly objected to the position of insurance carriers and industry. They argued that "a disease may grow out of the employment and be caused by it, and yet not be 'characteristic' of it, or 'peculiar' to it." The schedule method of payment limited employees' access under the workmen's compensation law and prevented them from gaining restitution for legitimate injuries to their health and well-being. Schedules would "not include any new disease until long after its discovery and after considerable harm has been done to the worker."[30] They disagreed that schedules were the only reasonable means of organizing coverage and rejected the idea that compensation should supersede in importance factory inspection and regulation of the workplace. Finally, they proposed a system of federal grants-in-aid that would establish national standards for compensation and federal regulation of occupational disease hazards.

The American Federation of Labor and some of its affiliates opposed most of the fundamental assumptions of the insurance and industry representatives. Labor wanted an expanded federal role in workplace regulation to complement the compensation mechanism and advocated establishing federal standards for the compensation system itself. Rather than breaking off the issue of compensation from that of workplace regulation, inspection, and control, they called for an integrated approach coordinated by the federal Department of Labor rather than state or federal health authorities.[31]

The Department of Labor and
Development of the Murray Bill

Both labor and Department of Labor officials believed that the broad problem of chronic industrial disease and disability would emerge as a critical issue in later decades. In early 1939 Verne E. Zimmer of the U.S. Department of Labor drafted a bill, for Senator James E. Murray of Montana, to address the issues raised by the American Federation of Labor during the silicosis crisis.[32] Although strongly supported by Secretary of Labor Frances Perkins, the bill gave little indication that it was an administration measure. Based on the belief that industrial disease posed long-term challenges requiring federal interventions and that neither federal nor state public health authorities were willing to undertake serious activities, the bill sought to move authority for occupational disease programs into the Department of Labor. In a memo to Secretary Perkins, Zimmer identified two major purposes in drafting the legislation: first, to provide financial assistance to the states, through the Secretary of Labor, for control of silicosis in industry, and second, to provide funds to state compensation systems specifically to enable "full benefits to claimants for silicosis."[33] Senate Bill 2256, introduced April 27, 1939,[34] was the first attempt at establishing federal regulation of safety and health conditions other than the more limited application of the Walsh-Healey Act, which authorized federal labor officials to impose regulations at the worksites of industries doing business with the federal government.[35]

At the Senate hearing on the bill, Dr. Walter N. Polakov, a physician working with the United Mine Workers of America, then part of the new and militant Congress of Industrial Organizations, argued that such federal intervention in workplace regulation was necessary because of a direct relationship between increasing disability and poverty. American workers who earned $1,000 or less a year were more than twice as likely to be disabled than those who earned $5,000 or more. Unlike representatives of management and industry, who used the term "disability" to describe physical impairments of individual workers, Polakov saw disablement differently. He pointed out that if disability represented inability to find work, then all workers excluded from employment because of a suspicious chest x ray or other medical finding were, technically, disabled. He also sought to broaden the definition of disability to include "any organic or functional disorder the source of which may be traced to harmful working conditions or environment." He argued that simply listing occupational diseases was inadequate since it would not "include the occupational hazards resulting from the tempo of the work and from the nervous strain of maintaining continuous sustained attention, correct perception, and prompt reaction in the environs of general nervous tension in the work and great responsibility in modern mass production, where a slight mistake in touching the wrong button may kill a number of people to say nothing about damage, of course." Control of these hazards should not be the responsibility of medical and insurance personnel; insurance personnel especially were untrustworthy because "industrial hygiene and safety are dangerous to insurance companies' profits since as the risk is lessened, so is the volume of business."[36]

Despite support from organized labor, two different versions of the Murray Bill suffered defeat. In the closing months of the New Deal, the attention of the White House turned away from domestic legislation and toward preparation for war. With the Public Health Service opposed to a bill that threatened to strip it of authority with respect to industrial disease, the White House, never fully behind the bill, gave it little political support. Both bills died in committee.

Disability and Silicosis:
The Case of New York

Many instances can testify to the political and economic nature of the extremely heated debates over silicosis during the 1930s. Martin Cherniak, for example, relates the story of the Gaulley Bridge incident and the congressional hearings that followed the discovery of hundreds of bodies of workers who had died from silicosis.[37] Elsewhere we have recounted the story of the struggles over silicosis among lead and zinc miners.[38] Here, we will trace the conflict over the compensation system. Government, industry, insurance, and labor leaders in every state all took part in the mid-1930s in shaping the compensation laws to include silicosis, but the case of New York State perhaps best exemplifies the ambiguities of their various roles in defining the disease. In this highly industrial state, "coverage" of the disease under a new compensation law first passed in 1934 was so limited that few workers obtained any redress through the system.

The liability crisis that led to the inclusion of silicosis in the compensation legislation was described by the state's Industrial Commissioner, Elmer Andrews, in early 1936. Andrews described how "certain industries in this State, particularly in the up-State areas, suddenly developed an intense interest in silicosis" when New York workers filed many suits in the early 1930s under the common law for damages due to silicosis exposure. Despite the fact that industry had fought coverage of silicosis or occupational disease since the inception of workmen's compensation, Andrews pointed out, "immediately the attitude of twenty years was reversed." He noted that "employers who had opposed inclusion of silicosis under the Workmen's Compensation Law came running to the State pleading for the inclusion of silicosis under the Workmen's Compensation Act so that they would be protected against the unlimited and terrifying common law damage suits which were being filed against them."[39]

In response to these suits, and to the growing financial crisis, a bill sponsored primarily by the foundry industry and their insurance representatives was introduced to the New York State legislature in March 1934 to add silicosis to the list of occupational diseases covered under workmen's compensation. Although the bill passed, Governor Herbert Lehman vetoed it, saying that he favored a blanket bill covering occupational diseases.[40] Within six months, at Lehman's initiative, the workmen's compensation law was amended to include all occupational diseases, including silicosis. Fearing a rash of compensation claims, the insurance industry demanded that companies institute compulsory physical examinations at the worksite to insure that all workers in the

foundry industry in particular were free from silicosis. It feared specifically that workers laid off during the Depression might claim disability
for diseases that were either not disabling or were incurred at earlier
worksites. In light of the idiosyncratic course of silicosis, the industry
sought to protect itself from previously encountered risks. It went so
far as to propose that employers be prohibited from participating in
the workmen's compensation system after September 1, 1935, the date
the new law took effect, unless all their employees had been screened.[41]
At the same time, the insurance industry in New York announced a rise
in rates for workmen's compensation insurance in anticipation of increased claims, sometimes by as much as 400 percent.[42]

Within six months all the major parties in New York State—labor,
industry, insurance, and state government—agreed that the new law
was not working. Employees were no better off since insurance industry demands to fire or not hire silicotic workers forced many plants to
face "the threat of shutdowns which would put hundreds of skilled
workers on the street and add many to the relief rolls." The demand
for physical examination of workers was a special hardship, which "resulted in the elimination of many old and experienced workers not
solely due to silicosis but for other possible physical defects that could
be found."[43] In an undiplomatic moment, Commissioner Andrews
succinctly characterized the reaction of the insurance industry: "They
insisted that the working force be 'dry cleaned.'"[44] In summary,
Andrews said, "faced with these rate increases, closed plants and
unemployed workers, matters were in a critical condition."[45] In the
short space of a half year, blanket coverage for occupational diseases in
the new compensation law had effectively alienated insurance carriers,
industry, and labor alike. Given the troubled state of the economy during the Depression, no one wanted to further disrupt the State's
economy.

In response to this crisis, a new bill was introduced on behalf of the
foundry and insurance industries that was even more limited in scope
than before the amendment and virtually made it impossible for
workers to quality for compensation. The bill provided no compensation for partial disability and limited compensation for total disability
to $3,000. Moreover, compensation for disablement or death that oc-
curred during the first calendar month in which the Act became
effective would not exceed $500 and would increase only $50 each
month thereafter. Compensation for silicosis was further restricted by
a provision that precluded any payment if the last exposure preceded
September 1, 1935.[46]

The issue of compensation forced labor to choose between health
and jobs, and the New York State Federation of Labor sought to protect
jobs.[47] At the legislative hearing, Industrial Commissioner Elmer
Andrews acknowledged that the bill "is not an ideal one," but defended
it as "a compromise solution of an emergency situation and a difficult

problem." The bill aimed to keep men at work by keeping compensation rates low and plants open, to encourage dust prevention, and to
discourage pre-employment physicals. Both labor and industry appeared in support of the bill, with George Meany, president of the New
York State Federation of Labor, defending the compromise as necessary, if not ideal. At the least, it provided some assurance that workers
could continue to work at their trades.[48]

Others were not nearly as accepting of the need to compromise
workers' health. Most notably, two of the leaders of labor reform efforts
in the twentieth century, Secretary of Labor Frances Perkins and John
B. Andrews, Secretary of the American Association for Labor Legislation, both saw the bill as a dangerous and destructive precedent in labor
legislation. In late February 1936, Andrews wrote to Verne Zimmer of
the U.S. Department of Labor, enclosing a summary of his objections to
the bill and asking Zimmer for his opinion. Shortly after, on March 4,
1936, during hearings on the bill in Albany, Labor Secretary Perkins
herself responded to Andrews's letter with a detailed critique of the bill,
depicting it as a dramatic step backward in occupational disease legislation. First, she pointed out, the bill placed a "definite limitation on
workmen's compensation benefits payable for total disability and
death." Second, it imposed a "drastic limitation on medical benefits."
Third, it excluded "any liability whatsoever for partial disability regardless of extent or duration."

It was obviously particularly painful for Perkins, who had been involved in the movement for workmen's compensation in New York
State and had administered the state's labor department under Governor Roosevelt, to witness the "complete reversal" of the progressive
features of New York's law, previously "among the most beneficial measures of its kind in the country." Not only were its provisions inadequate
for claimants, but the bill provided little or no protection for workers
threatened with dismissal. It gave no assurance "as to the retention of
silicotic workers in industry through the abolition of medical examinations," she complained. In fact, the bill's provision fixing liability on the
last employer "seems to invite the continuance of pre-employment examinations as a protection against accrued liability." She was particularly disturbed by the section of the bill that prohibited the use in
compensation claims of information on industrial conditions gathered
through the offices of the Industrial Commissioner. This, she protested, completely undermined the chances for a claimant to achieve "a
fair and equitable disposition of a pending compensation claim." Her
objections were so strong that she ended by asking for its complete rejection. "So restricted and meager are the benefits under this proposed
amendment that it offers little to workers as a substitute for the common law remedy available previous to enactment of the all-inclusive
occupational disease act effective last September." She concluded, "I
would prefer to this weak palliative the frank elimination of silicosis

from coverage until such time that a suitable and acceptable compensation plan can be devised."[49]

Despite Perkins's appeals to the governor and an intensive lobbying campaign against the bill by the American Association for Labor Legislation, it passed and was signed by the governor in June 1936.[50] The Industrial Commissioner of New York State summed up the position of state officials by acknowledging the weaknesses in the bill and asserting that the state had no other options without federal legislation. Appealing to the National Conference on Silicosis in mid-April 1936, Commissioner Andrews asked for assistance "in bringing about, in the very near future, some measure of compensation coverage for silicosis in the principal industrial, mining and quarrying states. The almost total lack of such coverage is what has made necessary the appallingly low maximum compensation now proposed in the New York State bill."[51] In 1940, compensation payments were raised modestly.[52]

Disability Policy in the Postwar Years

During the fifteen years following World War II, attention to silicosis declined among all the constituencies that had heatedly debated the issue during the Depression. All agreed to a more restrictive definition of the problem. Labor turned its efforts toward wages and fringe benefits, while insurance companies and industry in general succeeded in limiting their liability through the development of scheduled workmen's compensation payments. Postwar hostility to organized labor among many in Congress, state government, and industry combined with comparatively low unemployment to produce a decrease in disability claims for disease and a more restrictive definition of disability itself. The medical and public health community's interest in silicosis also dramatically waned. A review of articles on silicosis in *Index Medicus* for the quarter century following the mid-1930s political crisis over the disease reveals a sharp drop in their number.

George W. Wright, along with Leroy U. Gardner, his colleague at Saranac Laboratories, wrote a number of important articles on silicosis that help explain the medical community's changing perception of the silicosis problem. Historically, he wrote, physicians sought to determine if patients suffered decreased capacity in their everyday life. Present disease and disability were measured against earlier abilities, and a doctor's responsibility was to restore full health as far as possible. But the recent experience with silicosis had changed physicians' traditional responsibilities. The compensation system now defined disability "in terms of lack of ability to earn wages and not in terms of a diminution of capacity to breathe or exercise." Physicians should no longer act "as one usually does in terms of loss of ability," but rather to evaluate a workers' ability to earn a living. "The [physicians'] usual approach to problems of health [should] be changed by deemphasizing the question whether or not the man has suffered an injury, and directing attention to determining whether or not the claimant still possesses sufficient physical capacity to earn wages as stipulated under the compensation act."

The medical community had to face two problems, Wright maintained. First, was there an impairment that prevented a person from earning a living? Second, was that impairment caused by work? Wright argued that evaluation of an impairment was an extremely difficult and highly technical issue. Since the patient's own subjective evaluation of impairment was unreliable and might be motivated by "frank malingering," physicians had to rely on their own experience and interpretation of the data. Furthermore, the data also were highly suspect. Wright complained that an x-ray showing of "an anatomic alteration of the lungs or heart is still commonly used as evidence that these organs must of necessity be functioning abnormally . . . and the extent of the anatomic change is frequently considered an index of the degree of functional impairment." But, Wright argued, repeated experiments had shown little correlation between the medical evidence and measurements of a patient's capacity to do work. Thus, only functional impairment could be used to diagnose silicosis.[53]

With the experience of the Depression still fresh, Wright feared, there was "a grave danger" that "physicians may be inclined to ascribe all the pulmonary ailments of men who have been exposed to dust or fumes to the inhalation of those foreign substances." In the past, local physicians had too much power in making critical decisions. It was far better, Wright argued, to depend upon the testimony of "the experienced plant physician."[54] He called for more control by industrial physicians, who were more likely to read and follow articles about industrial disease in technical journals and would hence be able to distinguish between diseases of industrial and nonindustrial origin. "Several diseases of a nonindustrial origin are especially prone to mimic the symptomatology of industrial pulmonary disease and also to lead to physiologic alterations that cause an incompetency to earn wages." Implicit in Wright's argument was that neither workers, their advocates, nor their community practitioners could adequately understand, much less diagnose, silicosis. Industrial disease had to be defined by the medical specialist.

Conclusion

Because silicosis is a chronic condition, its history raises broader questions regarding chronic diseases and disability in general. As we have shown, political, economic, and scientific arguments were intertwined throughout the negotiations over the nature, course, etiology, and treatment of diseases rooted in industrial society. The long period between exposure to toxic materials and resulting disability, the uncertainty of clinical and roentgenological diagnosis, the problematic measurement of disability, and the ambiguity inherent in assigning liability all made decisions about the degree or even the existence of disability part of a continuing negotiation among labor, government, and public health and medical communities. In case of silicosis, it was the circumstances of economic depression, unemployment, changing methods of industrial production, and legal debates over compensation that defined the crisis of the 1930s. By the 1960s, new constituencies entered into the process of socially constructing chronic and industrial disease. Coal miners pressed for compensation in cases of "coal workers' pheumoconiosis" according to symptoms reported by patients and tests by internists; textile workers and their advocates helped define the public response to byssinosis; the legal profession, some physicians, and the asbestos workers union forced the problem of asbestos-related lung cancer, mesothelioma, and asbestosis onto the national agenda.

As the study of silicosis indicates, it is impossible to understand the emergence of industrial and chronic health issues without studying the specific historical circumstances in which these issues are framed. Professionals, political interests, and economic constituencies all play important roles in interpreting the events in peoples lives that are called industrial disease and disability. As circumstances change, changing interpretations become the new bases for claims about latency, onset, incidence, prevalence, morbidity, and mortality. Policymakers, leaders of interest groups, and, most importantly, workers themselves take considerable risk when they assume that science-based claims are, by definition, objective.

NOTES

This article appeared in somewhat different form in *Milbank Quarterly* 67, suppl. 2, part 1 (1989):228–253. The authors would like to thank the Milbank Memorial Fund, the National Endowment for the Humanities, and PSC-CUNY for their financial support of this project. David Rosner would like to acknowledge the support of the Guggenheim Foundation. The authors would also like to thank the participants in the Milbank Roundtable and Dan Fox and David Willis, in particular for their comments.

1. Anthony Bale, "Compensation Crisis: The Value and Meaning of Work-related Injuries and Illnesses in the United States, 1842–1932," Ph.D. dissertation, Brandeis University, 1986.

2. U.S. Department of Labor, "'Silicosis,' some Pertinent Facts about this Occupational Disease for Use by the Secretary at Joplin, Missouri," April 26, 1940, NA RG 174, Office of the Secretary, Joplin, Missouri," April 26, 1940, NA RG 174, Office of the Secretary, Joplin, MO.

3. Bale, op. cit.

4. M. Frinke, "'Loss of Use' or the Impairment of Function," *Monthly Labor Review* 10 (August 1920):121–130.

5. R. R. Jones to E. O. Jones, December 10, 1935, National Archives, Record Group 100, 7-2-1-5-1.

6. O. A. Sander, "A Practical Discussion of the Silicosis Problem," U.S. Department of Labor, Division of Labor Standards, *Bulletin No. 10* (Washington, DC: Government Printing Office, 1936), p. 261.

7. M. Kossoris and O. A. Freed, "Experience with Silicosis under Wisconsin Workmen's Compensation Act, 1920–1936," *Monthly Labor Review* 44 (May 1937):1097.

8. R. R. Sayers, "Relationship of Asbestosis and Silicosis to Disability," U.S. Department of Labor, Division of Labor Standards, *Bulletin No. 4* (Washington, DC: Government Printing Office, 1935), p. 71.

9. T. C. Waters, "Legislative Control and Compensation," *Third Symposium on Silicosis*, ed. B. E. Kueckle (Wausau, WI: Employers' Mutual Liability Insurance Co., June 21–25, 1937), p. 245.

10. A. J. Lanza, "Health Problems in Mining," *Mining Congress Journal* 22 (Nov. 1936):26–27.

11. "Occupational Disease—Disease, Silicosis-Disability," *The Weekly Underwriter* 134 (Feb. 22, 1936):425.

12. T. C. Waters, "Formulation of Legislation," *Fourth Saranac Laboratory Symposium on Silicosis*, ed. B. E. Kuechle (Wausau, WI: Employers' Mutual Liability Insurance Co., 1939), pp. 320–333.

13. R. N. Caverly, "Dust Diseases as a Legislative Problem," *Insurance Counsel Journal* 4 (January 1937):30.

14. Tri-state Conference, "Proceedings," National Archives, Record Group 100, 7-0-4(3), April 23, 1940, pp. 20–21.

15. F. R. Jones, "Problems of Compensation for Occupational Diseases." Address delivered at the National Convention of the National Founders Association, Nov. 15, 1934. American Association for Labor Legislation Manuscripts, Occupational Disease Pamphlets, 1933–1934, Industrial and Labor Relations Archives, Cornell University.

16. National Industrial Conference Board, "Uniform Medical Provisions for Workmen's Compensation Acts in the United States," *Special Report No. 31*, New York City, 1925, p. 17.

17. A. J. Lanza, "Health Problems in Mining," *Mining Congress Journal* 22 (Nov. 1936):26–27.

18. American Public Health Association, *Workmen's Compensation for Silicosis*, Report of the Committee on Pneumonoconioses (R. R. Sayers, A. J. Lanza,

A. R. Smith, E. R. Hayhurst). New York: Association of Casualty and Surety Executives. American Association for Labor Legislation Manuscripts, Occupational Disease Pamphlets, 1933–1934, Industrial and Labor Relations Archives, Cornell University.

19. U.S. Department of Labor, Division of Labor Standards, "National Silicosis Conference, Report on Medical Control," *Bulletin No. 21*, part 1 (Washington, DC: Government Printing Office, 1938); U.S. Department of Labor, Division of Labor Standards, "National Silicosis Conference, Report on Regulatory and Administrative Phases," *Bulletin No. 21*, part 4 (Washington, D.C.: 1938).

20. U.S. Department of Labor, Division of Labor Standards, "National Silicosis Conference, Report on Regulatory and Administrative Phases," *Bulletin No. 21*, part 4 (Washington, DC: Government Printing Office 1938), pp. 6–7.

21. David Rosner and Gerald Markowitz, eds., *Dying for Work, Workers' Safety and Health in Twentieth Century America* (Bloomington: Indiana University Press, 1987).

22. A. J. Lanza, "The Insurance Carrier," in *Fourth Saranac Laboratory Symposium on Silicosis*, ed. B. E. Kuechle (Wausau, WI: Employers Mutual Liability Insurance Company, 1939), p. 316.

23. L. Teleky, "The Compensation of Occupational Diseases," *Journal of Industrial Hygiene and Toxicology* 23 (Oct. 1941):357–358.

24. Association of Casualty and Surety Executives, *Perils in Measures for Compensation for Occupational Diseases* (NY: Feb. 28, 1935). American Association for Labor Legislation Manuscripts, Occupational Disease Pamphlets, 1935–1936, Industrial and Labor Relations Archives, Cornell University.

25. H. D. Sayer, *Occupational Diseases—Real and Supposed*, n.d. American Association for Labor Legislation Manuscripts, Occupational Disease Pamphlets, Industrial and Labor Relations Archives, Cornell University.

26. Sayer, n.d.

27. "Tendencies in Workmen's Compensation," *The Weekly Underwriter* 131 (August 1934):289–290.

28. F. R. Jones, "Problems of Compensation for Occupational Diseases." Address delivered at the National Convention of the National Founders Association, Nov. 15, 1934. American Association for Labor Legislation Manuscripts, Occupational Disease Pamphlets, 1933–1934, Industrial and Labor Relations Archives, Cornell University.

29. A. J. Lanza, "Health Problems in Mining," *Mining Congress Journal* 22 (Nov. 1936):26–27.

30. J. A. Padway, "What We Expect under Workmen's Compensation and What We Are Getting," U.S. Department of Labor, Division of Labor Standards, *Bulletin No. 36* (Washington, DC: Government Printing Office, 1939), p. 31.

31. R. J. Watt, "Draft of Report, Supplementing Report of the Economic, Legal and Insurance Committee—National Silicosis Conference," 1938, National Archives, Record Group 100, 7-0-4(1).

32. V. A. Zimmer to The Secretary, June 7, 1939, National Archives, Record Group 174, Office of the Secretary, Folder: Bills, Miscellaneous.

33. Zimmer to Perkins, 1939.

34. U.S. 76th Congress, First Session, Senate Bill 2256, April 27, 1939.

35. T. C. Waters, "A Critical Review of Occupational Disease Legislation," *Mining Congress Journal* 25 (Dec. 1939):34–36.

36. U.S. 76th Congress, 3d Session, May 13,14,16, *Senate Hearings before a Subcommittee of the Committee on Education and Labor on S.3461* (Washington, DC: Government Printing Office), pp. 65–70.

37. Martin Cherniack, *The Hawk's Nest Incident: America's Worst Industrial Disaster* (New Haven: Yale University Press, 1986).

38. Gerald Markowitz and David Rosner, "Street of Walking Death," *Journal of American History* (September 1990).

39. E. Andrews, New York State Department of Labor news release, March 31, 1936, "Testimony of Elmer F. Andrews before New York State Senate Labor and Industries Committee," National Archives, Record Group 100, 7-2-1-5-1.

40. "Three Liquor Bills Signed by Lehman," *New York Times*, March 30, 1934, p. 7; "Guaranteed Liens Suspended to 1936," *New York Times*, May 16, 1935, p. 8.

41. "Pitts Insurance Test for Silica Workers," *New York Times*, July 21, 1935, p. 7.

42. E. Andrews, New York State Department of Labor news release, March 31, 1936, "Testimony of Elmer F. Andrews before New York State Senate Labor and Industries Committee," National Archives, Record Group 100, 7-2-1-5-1.

43. New York State Federation of Labor, "Hearings Held March 10th and March 11th in Senate and Assembly," *Bulletin* (March 23, 1936), pp. 3–4.

44. E. Andrews, New York State Department of Labor news release, March 31, 1936, "Testimony of Elmer F. Andrews before New York State Senate Labor and Industries Committee," National Archives, Record Group 100, 7-2-1-5-1.

45. New York State Federation of Labor, 1936, pp. 3–4.

46. E. Andrews, "Memorandum for Dr. Greenburg," April 13, 1936, National Archives, Record Group 174, Office of the Secretary, Folder: State Labor Department, New York.

47. "Silicosis Problem in State at 'Crisis,'" *New York Times*, April 15, 1936.

48. New York State Federation of Labor, 1936, pp. 3–4.

49. F. Perkins to John B. Andrews, March 4, 1936, National Archives, Record Group 174, Folder: Secretary Frances Perkins, Labor Standards, January–April 1936, Box 59.

50. American Association for Labor Legislation press release, "Workmens' Compensation Law Threatened," March 4, 1936, National Archives, Record Group 174, Folder: Office of the Secretary, Labor Standards, January–April 1936; F. Perkins to Herbert H. Lehman, March 4, 1936, National Archives, Record Group 174, Folder: Secretary Frances Perkins, Labor Standards, January–April, 1936, Box 59.

51. E. Andrews, "Memorandum for Dr. Greenburg," April 13, 1936, National Archives, Record Group 174, Office of the Secretary, Folder: State Labor Department, New York.

52. "Silicosis Pay Rise Signed by Lehman," *New York Times*, April 18, 1940, p. 18.

53. G. W. Wright, "Disability Evaluation in Industrial Pulmonary Disease," *Journal of the American Medical Association* 141 (Dec. 24, 1949):1218–1222.

54. Wright, 1949, pp. 1218–1222.

10   THE LEGAL ART OF PSYCHIATRIC DIAGNOSIS: Searching for Reliability

JANET A. TIGHE

Society constructs its response to disease in the particular. But the word construct implies a monolithic consistency, a unified imposition of definitions and policies, the existence of a blueprint, and the presence of a builder. As we have argued, however, the construction process is more accurately seen as one of negotiation—with emphasis on the term *process*. Negotiation, moreover, implies diversity of interest and perception as well as process.

The microcosm of law and, in particular, the need to determine the responsibility of criminal defendants (or the makers of wills) illustrate this in stark, almost caricatured form. Every trial represents a kind of sociodrama, an acting out of the process of social negotiation. Unlike the debates of academics in symposia and scholarly journals, it is a drama that must have a resolution. Lawyers and physician/witnesses play their roles as individuals but inevitably also represent corporate interests—acting out the values of the professions that socialized and credentialed them. Judges, jury members, and defendants represent still other interests. As Janet Tighe emphasizes, law and medicine have rather different needs and socialized perceptions; even within medicine itself specialists in mental illness may have a very different orientation from that of the majority of their fellow practitioners.

Individual disease concepts must also be seen as protagonists. Disease categories can—and are made to—serve as mediating factors in the resolution of such judicial encounters. And thus Tighe's focus on that much-contested entity, "moral insanity." As we can see in retrospect, moral insanity was a nonviable social category; it certainly failed to mediate between the several contesting disciplines in the arena of the courtroom or in the more generally conflicted discourse that has both divided and united law and medicine.

Moral insanity intensified rather than reduced conflict. In some sense this was entirely—and ironically—predictable. Although the substance of the concept (that legitimate mental disease could exist without marked delusion, hallucination, or extreme alterations in mood) reflected an important clinical observation and proved useful to practitioners, the concept itself did not provide the discrete, reassuringly constrained category demanded by lawyers and nonprofessionals. It seemed, in fact, antithetical to their assumptions, raising the alarming spectre of an illness whose characteristic pathognomonic feature was the very antisocial behavior the court was required to judge.

As Tighe stresses, however, the demand for certainty can never be consistent with the elusive and idiosyncratic phenomena that face the clinician—or judge and jury in a criminal trial. Only in a textbook or nosological table can one expect to find the particularity of experience embodied in coherent and neatly ordered categories. No alteration in the rules of law defining responsibility or the role of expert testimony can provide a definitive solution to this fundamental dilemma.

—C. E. R.

I T IS NO LONGER novel or provocative to begin an analysis of psychiatric thinking with comments about the "social construction" of mental illness. An enormous literature delineating the various sociocultural forces that have shaped definitions of mental disease already exists. In this literature psychiatric diagnosis is treated as a complex and subtle process of negotiation involving a varied cast of characters. This cast includes not only doctors and patients but patients' families, nurses, social workers, insurance companies, the courts, and legislators.[1] Scholarly interest in the roles played by many of these negotiators has been uneven. One of the most neglected members of the cast, and the subject of this chapter, is the American legal profession; the law and lawyers have played a subtle, but often significant, role in "framing" disease.

The part played by American lawyers in the drama of psychiatric diagnosis has taken many different forms. We examine two of the most prominent: the law's roles in leading public debate over mental disease definitions supplied by the psychiatric profession, and in creating an elaborate set of expert testimony rules. The question dominating the exploration of these two legal strategies for shaping definitions of mental illness is simple: What does the law want from a psychiatric diagnosis? Although different in orientation and purpose, the two legal strategies examined here demonstrate the law's interest in a specific type of disease definition. As we shall see, this legal definition is often at odds with the diagnostic criteria employed by psychiatrists in their pursuit of clinical goals. Our main analytic focus will be on this gap between medical and legal definitions of mental illness and the struggle of the two professions to bridge it.

An analysis of the American legal profession's contributions to the debate over the proper definition of mental illness must begin with a recognition of the troubled nature of this debate in medical and lay circles—prior to any contribution by the law. No medical specialist's ability to diagnose accurately has been so persistently and publicly challenged as that of American psychiatrists. Such challenges have been aimed at both the individual physician and the discipline as a whole. They have come from both inside the medical profession and outside observers and have consisted of limited critiques of specific disease concepts as well as broad denunciations of psychiatry's legitimacy as a scientific medical specialty.[2] Much of this criticism arises from the fact that psychiatry is a specialty that has historically treated individuals whose major symptoms of disease are behavioral. Unfortunately for psychiatry, when the specialty came into being in the nineteenth century the most prestigious and believable models of disease were somatic and mechanistic. If an illness lacked a clear underlying pathophysiology and a few well-marked physical symptoms—as did so many of the conditions treated by psychiatrists—its legitimacy as a disease was questioned. To put the matter simply, medical science's most compelling "disease-framing options" were not applicable to most of the conditions over which psychiatry claimed expertise.[3]

No group has been more acutely aware of the diagnostic shortcomings of psychiatry than the physicians who practiced it. The current controversy over the "statistical reliability" of the "multiaxial diagnostic" categories found in the American Psychiatric Association's official *Diagnostic and Statistical Manual of Mental Diseases* is only the most recent manifestation of a debate dating back to the early years of the nineteenth century.[4] Although the two groups speak quite different languages, America's first medical experts on mental disease—the asylum superintendents of the 1840s—and today's statistically oriented psychiatrists faced the same problem. How could they give their disease definitions compelling authority?

American lawyers offered an answer to this question almost as soon as they began to use physicians specializing in mental illness as expert witnesses in cases involving insanity. In the United States this development occurred in the second quarter of the nineteenth century. By this

time various legal inquiries were making regular use of medical assess-
ments of mental function. The most notable of these inquiries were
commitment hearings, habeas corpus suits, criminal cases involving the
insanity defense, and testamentary capacity cases. These forms of litigation touched upon some of the most fundamental rights of
American citizens, including protection from unlawful incarceration,
the guarantee of due process, and trial by a jury of one's peers, as well as
the freedom to manage and dispose of one's property according to
one's wishes. During these proceedings doctors were asked to give their
opinion of the mental status either of a patient they had treated or, as
experts in mental diseases, of an individual whose case record the court
or one of the opposing counsels had asked them to review. In both situations they were asked to make a diagnosis in a forum that was quite
different from their normal clinical setting. The legal forum was highly
public and had its own peculiar set of rules and rituals, buried among
which were a set of specifications for the definition of mental illness.[5]

Attorneys interested in mental disease were not hesitant about com-
municating their likes and dislikes to the psychiatric profession. By
1850 lawyers had produced a sizable body of commentary on mental
illness. Legal opinions about psychiatric disease were expressed at
trials, especially during cross-examination of hostile medical witnesses,
at professional meetings, in journal articles, legal textbooks, lay newspapers, and periodicals. The disparate natures of the sources in which
this commentary can be found make it evident that practitioners of the
law wanted to reach an audience of doctors and nonprofessionals, as
well as their own colleagues. One of the sentiments legal commentators
were most anxious to communicate to this broad audience was a profound suspicion of all disease entities affecting primarily the will
and/or emotions. This suspicion was voiced most expansively in the
nineteenth century when the concept of what was termed moral insanity was introduced into legal inquiries.[6]

To appreciate the nature of American jurists' role in the rise and fall
of the disease designation of moral insanity it is necessary to have some
sense of the concept's importance in psychiatric thinking. Moral insanity was a visible symbol of a powerful, generalized movement in
American psychiatry—the expansion of the definition of insanity to include a class of mental disorders affecting an individual's emotional
and volitional capacities rather than reason or intellect. It was the
French who did the most influential early work in this field, describing
a variety of new disease conditions, the most prominent of which were
manie sans délire, monomania, *folie raisonnante*, and a group of impulse
disorders such as kleptomania and erotomania.[7] British and American
physicians quickly showed an interest in the idea that there were nonintellectual, partial insanities and began to use these diagnostic categories
themselves. One British physician, James C. Prichard, published a particularly influential treatise on these conditions in 1835 in which he

applied the term moral insanity to such cases.[8] Although Prichard used
the term in a very specific sense, in the United States its meaning was
quickly expanded. By midcentury, moral insanity in the United States
had become a "catch all for many forms of mental illness in which intellectual powers seemed to remain partially or completely intact."[9] It had
also become the single most controversial psychiatric disease definition
presented in American courts.

No other doctrine of disease stimulated as much legal interest and
bitter criticism as the notion of mental disorder confined primarily to
the mind's nonintellectual faculties, stemming in part from its conflict
with traditional legal ideas and images of insanity. In most definitions
of insanity used by the law during the nineteenth century, disturbances
of the reason or intellect were the most recognizable features. The so-
called M'Naghten Rule, enunciated by British judges in 1843, is a good
example of this popular legal assumption. The rule basically prescribes
a "test" to determine whether an individual accused of a crime suffers
from a mental illness believed by the courts to preclude responsibility
for the criminal action. According to the M'Naghten test:

> To establish a defense on the ground of insanity it must be clearly proved
> that at the time of committing the act, the accused was labouring under such
> a defect of reason, from disease of the mind, as not to know the nature and
> quality of the act he was doing, or if he did know it, that he did not know he
> was doing what was wrong.

This criterion of a knowledge of right and wrong was widely used in the
United States, even before its formalization by the judges in the
M'Naghten case.[10]

The moral insanity doctrine's clash with the law's traditional reason-
focused model of mental illness was not the only basis for legal objections to it. Many lawyers were dismayed by the threat it posed to one of
the law's most basic tenets—individual responsibility. The principle
that individuals are accountable for their actions was clearly enunciated, widely discussed, and the keystone of an elaborate system of legal
rules and procedures. Furthermore, in the minds of many jurists it was
this principle that gave the law its coercive power over the individual,
and thus its force in society. Responsibility, especially as defined at the
time of the moral insanity controversy, assumed that each individual
was a free moral agent. All human action, including crime, was thus the
product of free and deliberate choice.[11]

The mentally ill constitute one of the few classes exempted from the
responsibility imposed on a free moral being, and according to the law
their irresponsibility is based on a disease-induced incapacity to conceive criminal intent. In legal terms a mentally ill individual was
functioning under "duress" and therefore could not have the *mens
rea*—guilty mind—necessary to make an action criminal. The lawyer's

task in such cases was to trace the chain of causation from the crime to the mind of the individual accused of committing it. Tests like the M'Naghten Rule were the law's primary means of establishing this causal link. The "material fact" in each case, from the legal point of view, was not merely whether the individual was mentally ill, but whether the consequence of the mental illness was a "defect of reason" and thus a "subjugation of his moral freedom." Based as they were on the idea that all human action arose from the purposive choice of a free moral being, the "tests" assumed that the chain of causation behind all crime began with the individual.[12]

In nineteenth-century science and medicine, however, deterministic explanatory models had become increasingly fashionable. Medical speculation as to the etiology of mental illness often included such remote causes as an unfortunate home life, bad companions, poor heredity, or "youthful masturbation." Such "causes" were largely irrelevant in the legal universe of reason and knowledge tests. Furthermore, many of the new psychological theories like moral insanity were cast in materialistic terms that reduced all mental activity to functions of the brain and nervous system. Such determinism was increasingly allied with reductionist theories of hereditary taint and anthropological degeneration.[13] These deterministic conceptions presented the law with a highly troublesome set of questions. How could an individual who was a victim of brain malformation or hereditary degeneracy be held responsible for a crime? If a criminal suffered from a physically diseased moral sense, what became of free moral choice? What did knowledge of right and wrong have to do with "uncontrollable impulses" or diseased moral faculties?[14]

The threat to fundamental legal principles and practices implied in these questions inspired much of the legal campaign against moral insanity. But technical arguments about *mens rea* and the right-and-wrong tests were only a small part of the campaign. Lawyers also cited the broadly based nineteenth-century discussion of the moral insanity doctrine's generally deleterious effect on social mores. At the core of this discussion was the fear that psychiatry was being used to shield people who were evil rather than ill, which grew out of the fact that the main symptom of many supposed instances of moral insanity was antisocial or criminal behavior. One of the leading proponents of the moral insanity doctrine, the physician Isaac Ray, described typically afflicted individuals as "unable to offer the slightest resistance to the overwhelming power that impels" them to such antisocial behavior as lying, stealing, sexual perversions, incendiarism, and even homicide. Driven by what Ray called "blind automatic impulse," they were compelled to act "by a kind of instinctive irresistibility" even though they retained their power to reason and might even be conscious of the "impropriety and enormity" of their conduct. One nineteenth-century lawyer observed dryly that a much older and more succinct designation described such individuals—"bad men." Much of the general public agreed with his judgment and regarded moral insanity as a dangerous doctrine.[15]

By the 1860s a sizable chorus of concerned ministers, moral philosophers, politicians, ordinary citizens, and not a few physicians joined the legal profession in denouncing moral insanity. The outspoken asylum superintendent John P. Gray said moral insanity had a tendency "to tempt men to indulge their strongest passions, under the false impression that God had so constituted them that their passions or impulses are not generally governable by their will or their reason, and that, therefore, there is not punishable guilt in indulging them."[16] As Gray's comment makes clear, moral insanity opponents saw the doctrine as a direct threat to the social order because it undermined the basic religious and legal principles, like free will and individual responsibility, that made the "social order as well as personal safety" possible. Any suggestion that the will could be overcome by some "irresistible impulse" or uncontrollable passion endangered the law's, and by implication the public's, ability to control violent individuals. Turning medicine to the task of "tak[ing] the guilt out of sin and convert[ing] crime into innocence" was a trend they believed had to be stopped.[17]

The legal campaign to stop physicians from diagnosing moral insanity was not well organized and directed, but it was widespread and persistent. Leading legal textbooks, for example, regarded the idea of moral or emotional insanity with skepticism. Even legal scholars such as Francis Wharton who originally accepted the moral insanity doctrine had, by the 1870s, come to reject it. Stating that the doctrine "had not support . . . either in psychology or law," Wharton's influential textbook declared it "unphilosophical," "vague," "impractical," and absolutely insupportable as a basis for legal action.[18] A similar negative opinion appeared in other popular legal textbooks of the era, such as those by John Elwell, William Guy, John Ordronaux, and Alfred Taylor.[19] Lawyers also felt compelled to express their distaste for the moral insanity doctrine in the pages of such widely circulated legal and medical journals as the *Criminal Law Magazine*, the *Medico-Legal Journal*, the *Journal of Nervous and Mental Disease*, and the *American Journal of Insanity*.[20]

Jurists often saved their most biting comments on moral insanity and other disease definitions that claimed inhibition of volition for statements during trials. The trial of President Garfield's assassin, Charles Guiteau, was an especially rich opportunity for American lawyers bent on destroying moral insanity. The vexed concept was a central issue at the trial itself and in the related discussion that raged afterward. Yet no medical witness, especially among those testifying for the defense, voluntarily called Guiteau morally insane. Using many names, including "moral monstrosity," "moral imbecility," and "primaire verrucktheit," defense experts carefully avoided the term moral insanity. It was the prosecuting attorneys who actually introduced it into the trial, as part of a carefully conceived plan to undermine the credibility of defense witnesses. The prosecution's stratagem reveals how the doctrine by the mid-1880s had become more of a legal tool to bedevil psychiatrists than a useful clinical device.[21] The moral-insanity diagnosis was even derided at a dinner celebrating the founding of the New York Medico-Legal Society. At the dinner *New York Tribune* reporter and attorney Whitelaw Reid made a toast urging that the doctrine of "emotional" or "moral insanity" be abolished "forever from the face of the earth."[22]

Not all American jurists shared Reid's view. A small minority, which included attorney Walter Cocke, explicitly defended the notion of moral insanity. However, the defending statements had a skeptical quality that did little to improve moral insanity's reputation. Cocke, for example, while pleading that such concepts as moral insanity and "instant emotional homicidal insanity" are "valuable instruments of justice" for a defense attorney, also admitted that they are "an absurdity" to "all save mere alienistic doctrinaires." Their usefulness to defense lawyers like Cocke stemmed from the fact that they gave a jury a convenient "pretext" for evading "technical rules of evidence." Armed with moral insanity and similar doctrines, a jury could, Cocke argued, decide a case "on the universal sense of right" rather than on the "letter of the law." To illustrate his point Cocke cited a recent case in which such a plea, although an admitted "ruse," had allowed a jury to acquit a husband who had slain his wife's seducer. While Cocke's characterizing moral insanity and related psychiatric diagnoses as "legal fictions" was a bit extreme, it does make a telling point.[23] By the second half of the nineteenth century, testifying in court to a moral insanity diagnosis, even when elicited by an attorney such as Cocke, did little for the reputation of the testifying physician. Thus, the underlying message in both Cocke's article and the Guiteau prosecutors' stratagem was the same: The law was more than willing to use "foolish" diagnoses as "a means of bending a rigid rule to the requirements of equity and policy." Their willingness to do so, it is important to note, in no way changed their opinion of its absurdity.

This legal consensus about moral insanity's "absurdity" did little to eliminate the disease category until physicians began to discredit it in medical circles. Fought in the reports of asylum superintendents and the pages of medical journals, the moral insanity war had many distinctively medical elements. Physician opponents of the doctrine, such as John Gray, David Reese, Andrew McFarland, and W. S. Chipley, loaded their articles with case histories and clinical data. Other physicians, including Isaac Ray, John Tyler, and Charles Nichols, who believed moral insanity to be a legitimate disease entity, were just as concerned to couch their arguments in terms of biomedical theory and clinical observation. This often led to a battle between case histories, theoretical explanations of brain physiology, and autopsy findings.[24]

Yet, what is curious is how these "medical" arguments became closely intertwined with moral and legal arguments. Despite repeated assertions by contesting doctors that their presentations were "medical" and "scientific," the discussions usually contained impassioned defenses of such legal and theological notions as free will and individual responsibility. Moreover, many of moral insanity's medical opponents expressed concern that the doctrine was undermining social order and the rule of law. Some, like Gray, even went so far as to champion the legal tests of responsibility as a bulwark against the "cavil and subtleties" of moral insanity. Quoting freely from judges and attorneys who had attacked the moral insanity doctrine in court, these doctors were happy to ally themselves with the law.[25] Moral insanity's proponents were just as attentive to legal arguments, only their aim was to refute them, as evidenced by a central theme in all five editions of Isaac Ray's seminal text on the medical jurisprudence of insanity.[26] Ray's arguments and the myriad of similar ones appearing in nineteenth-century medical journals make it clear that whatever position a physician embraced in the conflict over moral insanity, by the 1880s it necessarily included at least a sidelong glance at the law.

As American physicians concerned with mental illness peeked over their shoulders at the law, they began to see some disconcerting things. One of the most troubling was a growing legal perception that the moral insanity controversy revealed a basic weakness in all psychiatric knowledge. If doctors could not agree among themselves over this diagnosis, how could they speak with authority about any disease. These doubts about psychiatry's reliability found their way into such widely circulated legal textbooks as Francis Wharton's *Treatise on Mental Unsoundness* and Alfred S. Taylor's *Manual of Medical Jurisprudence*.[27] Both authors saw in the medical controversy over moral insanity evidence of the physicians' lack of "certain knowledge" about mental illness and unreliability as expert witnesses. As these authors realized, the warring parties in the moral insanity conflict were arguing about far more than the existence of a single disease entity. Some of psychiatry's basic tenets—regarding the human capacity for purposive action, the elements of an individuals emotional make-up, the role of heredity in human development, and the relationship between physical and psychological symptoms—became problematical in the debate over moral insanity. To many legal practitioners the knowledge base upon which nineteenth-century psychiatrists were attempting to build their specialty now seemed fundamentally flawed.

The growing legal skepticism about the value of psychiatric diagnoses was deeply disturbing to late-nineteenth-century American psychiatrists. They expressed their concern in a number of ways. One faction within the Asylum Superintendents' Association (the forerunner of the American Psychiatric Association) even attempted to force their organization to adopt some generally acceptable—and prudent—

public stand on moral insanity. The primary champion of this faction, Andrew McFarland, declared that "It becomes us to settle the question of moral insanity . . . and I am satisfied that whatever is the opinion of this body, will become the accepted position of the courts, and the accepted opinion of the higher minds of the country."[28]

Despite McFarland's assurances, the effort to bring even the small and close-knit Association to a consensus on moral insanity failed, and debate over the doctrine raged on for years. Finally, a few superintendents suggested that the Association stop discussing the matter altogether, for it brought only discord and public censure.[29] Although inconclusive, the Asylum Superintendents' Association's debate over taking a public stand is revealing. What it demonstrated most clearly was the fact that the psychiatric profession itself was losing its hold on the moral insanity doctrine. Moral insanity by the end of the nineteenth century was no longer merely a disease category determined by physicians but an explosive social and legal issue that lawyers felt they had the right and responsibility to address.

No longer a viable clinical category, moral insanity became the symbol of a profound problem—the legal profession's growing mistrust of all psychiatric diagnoses. Even as the term moral insanity was fading into disuse, legal concern heightened over the ideas that had inspired it. Concerned American lawyers recognized that psychiatry's abandonment of moral insanity did not translate into an abandonment of the underlying theory that mental illness did not necessarily affect ability to reason. Emotions and the workings of the human will were, in fact, becoming an increasingly significant part of all psychiatric definitions of disease at the end of the nineteenth and the beginning of the twentieth centuries. The development in these years of new schools of psychiatric thought, such as psychoanalysis and psychobiology, demonstrates the discipline's deepening commitment to exploring the emotions and thus the human capacity for purposive action. The search for a general disease model at odds with that of the law was accompanied by another disturbing tendency. Many psychiatrists were expanding their view of what behavior constituted symptoms of disease. Since these broader definitions of disease encompassed an increasingly large group of deviant behaviors, the categorical distinction between sick and bad, so important to the law, became increasingly elusive.[30]

The American legal profession was also disturbed by inconsistencies and contradictions in psychiatry's expanding nosology. Definitional conflicts, similar to the one that had swirled around moral insanity, had by no means diminished. In the early twentieth century no standard psychiatric nomenclature yet fulfilled the clearly recognized need for some common system of definitions psychiatrists could use in discussions of patients. A nosological cacophony was making it increasingly difficult for doctors to speak with clarity to one another, much less to the outside world. Judging by the volume and frequency of nosological arguments in American psychiatric journals of that era, the problem evidently promised no easy solution.[31]

The unwillingness or inability of American psychiatrists to resolve these problems prompted several members of the legal profession to take action. Most of these lawyers confined themselves to diatribes about the need for doctors to reform their classification scheme; New Yorker Clark Bell went further. In 1885 Bell, a recognized insanity law expert with a long history as a medicolegal reformer, launched a campaign to produce the first uniform "classification of mental diseases," an audacious step for an "outsider." Here was a man of the law who believed he had the right to intrude directly into the world of psychiatry to elicit the kind of disease definitions he believed society needed. Admittedly, Bell was an unusual outsider. A guiding member of the United States' first nationally organized medicolegal society, he was also the editor of America's first interdisciplinary medico-legal journal. These activities gave him a credibility that few in the legal or medical communities could dispute. But even these credentials did not guarantee his being taken seriously, especially by physicians. Only the psychiatric community's growing recognition that their definitional problems were attracting adverse legal attention can explain Bell's success in securing medical cooperation.[32]

Bell's attracting a medical audience is even more noteworthy, given his view of the nosology's purpose. What he wanted was a system of disease definitions "simple and practical" enough to end psychiatry's accelerating slide into public disrepute. Bell's emphasis on repairing psychiatry's damaged reputation is important. It shows that, from a legal perspective, a nosology's primary function is to improve the quality of legal testimony. While Bell clearly foresaw other functions of a nosology, such as providing a basis for the compilation of statistics and the more efficient organization of treatment facilities, it was its legal application that concerned him the most.[33]

A clear precise nosology, Bell reasoned, would be an invaluable tool for evaluating expert psychiatric testimony. It would represent the best thinking in the field and leave little room for the "squabbles" between doctors that "so disturbed" the courts in recent years.[34] Lawyer Bell's belief that a systematic nosology would help end psychiatrists' diagnostic conflicts reinforces the impression left by the legal response to the moral insanity controversy: What the law wanted most from psychiatry was consistency.

Bell's pressuring of American psychiatrists to reform their nosology was ultimately unsuccessful. Although the failure of this legally engineered effort can in some measure be traced to theoretical confusion in psychiatry at that time, the inappropriateness of Bell's reform method also played a part. He began with a plea to all interested parties to send him their ideas as to a proper basis for a classification system. After collecting suggestions he planned to have a committee of "classification referees" review them and choose the "best." Unfortunately for Bell, such a democratic procedure did not produce the results he had hoped for. His grandiose scheme to develop an authoritative nomenclature progressed no farther than the collection of an odd assortment of possible classification models. They were all dutifully published in the *Medico-Legal Journal* but were so contradictory that all efforts toward a consensus on the "best" failed.[35]

This failure probably came as no surprise to the psychiatrists who participated in Bell's campaign; they had already experienced defeat in similar efforts at nosological reform.[36] Although they had responded to Bell's call with a mixture of hope and doubt, most would have agreed with prominent psychiatric administrator Pliny Earle when he told Bell that "I greatly fear that the day is far distant at which a classification universally or even generally satisfactory . . . will be attained."[37] History has proved Earle's prophetic powers to be stronger than even he could have imagined. Not only did Bell fail to win widespread support for his reform, but all subsequent attempts by others have been beset by controversy.[38] Not even the computer sophistication and statistical pyrotechnics of the American Psychiatric Association's third and most recent edition of its *Diagnostic and Statistical Manual* have ended the debate over the reliability of psychiatry's disease definitions.[39]

Surprisingly, the persistence of diagnostic disagreements has not diminished the amount of psychiatric testimony in American legal proceedings. On the contrary, it has increased.[40] This does not signify the law's abandonment of efforts to secure a legally suitable definition of mental illness. It does, however, suggest that the law has relied on other strategies for shaping psychiatric thinking. One of the most important, but often overlooked, is the American system for handling expert testimony.

The American legal system had evolved a fairly elaborate set of rules governing expert witnesses by the time medical specialists in mental illness became a regular part of legal proceedings. The rules have continued to expand and change over the last two centuries, creating complex and often confusing guidelines for psychiatrists. This confusion is compounded by the fact that individual states and the federal government have developed their own distinct rules. Most jurisdictions include provisions for courts to decide if a given expert is qualified, for subjecting expert testimony to cross-examination, and for establishing the evidentiary basis of an expert's opinion. These provisions, particularly with regard to evidentiary basis, are a means to make detailed items of evidence cited by an expert, as well as the general state of knowledge about this evidence, clear to the jury. If enforced strictly, and admittedly enforcement has varied markedly in U.S. courts, these provisions can dictate not only who is allowed to diagnose in a given case, but under what circumstances and in what terminology.[41]

Needless to say, reforming the system of expert testimony has been a popular topic among forensically active psychiatrists. Since the middle of the nineteenth century they have complained about the detrimental effect that many of these rules, particularly those making the physician a partisan, have had on testimony about disease. Chafing under legal constraints, they have lamented the damage done to their ability to produce, in the words of one twentieth-century physician, a "sound, scientific diagnosis." As these doctors realized, legal rules for guiding expert testimony were merely a means of controlling the flow of psychiatric information in American courts.

One of the clearest discussions of how the rules control psychiatric information came from nineteenth-century lawyer Francis Wharton. Wharton was a leading authority on the "medical jurisprudence of insanity" and wrote a textbook on the subject that went through four editions between 1855 and 1882. With succeeding editions, Wharton's thinking on expert psychiatric testimony underwent some revealing changes that stand as a blueprint of what many American lawyers, even in the twentieth century, would like from a psychiatric diagnosis. Wharton shifted from a devout believer "that the testimony of medical men is on medical questions to be received by the court and jury as authoritative" to a cautious skeptic who constructed an elaborate scheme for improving the quality of badly flawed medical thinking. What changed Wharton's mind was the fierce battle over moral insanity that engulfed the psychiatric profession during his lifetime. The battle, discussed earlier, convinced him that psychiatric knowledge had not yet reached the level of "certainty" the law required. It was Wharton's proposals for achieving the desired certainty that make his writing so useful in the search for the law's idea of a proper definition of mental disease.[42]

Wharton began by asserting that the expert knowledge of nonlegal professionals is vital to the common-law system of justice. The common law, he wrote, "May more properly be treated as the precipitate of the wisdom of all ages—all professions—all countries." After thus affirming his openness to knowledge systems other than that of the law, Wharton went on to qualify his remarks. What the law needs is not merely knowledge, but "certain" knowledge. "Certainty," claimed Wharton, "is the only proper foundation for court decisions about guilt and innocence, reward and punishment." "This certainty," he continued, "can be no other than that which bears the seal of technical science." Medical experts, like "experts in all other brands of science," were to be the source of this "certainty" in appropriate cases. "Whether a man is really or only apparently deranged, is a question which cannot be decided with the certainty belonging to science except by a physician; nor is it possible without a thorough knowledge of psychological medicine, to pronounce upon the influence exercised by specific forms of disease upon given action." Science and the physician's position as a scientist gave the doctor authority to speak as an expert in court.[43] But what if scientific experts did not all speak with the same voice? What happened to certainty when medical scientists disagreed?

These questions did not trouble Wharton in the first edition of his book, although he did note that "occasional conflicts of opinion among medical witnesses" did occur when "insanity was at issue." He deplored these "conflicts" but did not indicate they were a major problem in 1855. By the third edition (1873), intraprofessional differences of opinion had, it seemed to Wharton, reached crisis proportions; the "occasional conflicts" of 1855 had, by 1873, become common occurrences. What disturbed him most was their destructive effect on the medical profession's reputation. Quoting cases exemplifying such conflicts among experts, he argued that "doubt" and "distrust" had come to replace the "respect" formerly accorded the medical expert. To Wharton, the most alarming aspect of these confrontations was their demonstration that "there is no theory so absurd but that it has found some philosopher by whom it is maintained." Even though Wharton was sure that "the sober practical thought of the great body of alienists rejects these extravagances [definitions of insanity]," "this evil" went on unabated. In the face of this evil Wharton was forced to recommend a major reform in the way the courts elicited expert testimony. His reform scheme made explicit his idea of what makes a knowledge system certain. "Certainty" demanded only one answer to every question. In Wharton's mind, and those of the great majority of his legal colleagues, what would make doctors "certain" and "reliable" witnesses was their agreement on a diagnosis.[44]

What Wharton and most subsequent legal scholars of expert testimony wanted from the psychiatric profession was the same thing that Bell and the nosology reformers wanted—an authoritative consensus on diagnostic matters. Physicians called to testify should, claimed Wharton, state not simply their own personal views but the "general sense" of their profession.[45] As the previous discussion of the American psychiatric profession's nosological problems makes clear, this "general sense" was often very difficult if not impossible to ascertain.

Nonetheless, authoritative consensus was precisely what most American legal experts wanted, and they widely shared Wharton's ideas about the need. One of the most vocal of the jurists who agreed with Wharton was Clark Bell, whose work as a medicolegal reformer has already been discussed. In a speech urging the Medico-Legal Society to take up the issue of expert testimony reform, Bell declared that it was time to put this class of legal evidence on a "more solid and substantial basis and to make it a means of arriving at the truth, and to aggregate facts, rather than to cloud and embarrass a case in the mazes and uncertainties of the diverse and contradictory opinion of medical men." Like Wharton, Bell believed that the "scientific certainty" he wanted from psychiatry was equivalent to the most widely accepted medical theory.

Bell was much less explicit than Wharton on how psychiatrists should decide on the most widely accepted theory, but he was convinced it existed.[46]

Concerns like Wharton's and Bell's were not confined to the nineteenth century. Early twentieth-century mental health law experts, such as Edwin Keedy and Sheldon Glueck, were just as convinced of the destructive nature of psychiatry's diagnostic conflicts and the need to minimize them. Although these lawyers found Wharton's reform plans unacceptable, the ones they developed were similar in spirit. To Glueck and Keedy and most other twentieth-century medicolegal theorists, psychiatry was a useful but troubled science. These men of the law were not satisfied with psychiatric definitions of mental illness but were not ready to abandon these categories altogether. Instead they developed legal devices to filter and modify the definitions when presented in court.[47]

Such devices have essentially equated diagnostic reliability with psychiatrists' ability to reach agreement, an equation that is not universally accepted by modern American psychiatrists. Some psychiatrists, and physicians in other specialties as well, question whether the search for "agreement" and consensus is an appropriate model for a scientific endeavor.[48] Forensically active nineteenth-century psychiatrists, most notably Isaac Ray, had also been troubled by a legal emphasis on expert unanimity. Ray believed that advocates who wanted psychiatrists to speak with "one voice" exaggerated not only the frequency of psychiatric disagreements but the ability to end them, and displayed a poor understanding of the nature of scientific knowledge. To Ray they appeared to be saying that differences of opinion between physicians called as experts were the "fruit of a mercenary or partisan spirit." He asserted instead that "on questions of natural science, men may differ in their opinions, without being influenced by unworthy motives, and without being ignorant or superficial. Medical science presents no exception to the general rule." Legal reform schemes aimed at "unanmity" were, from Ray's vantage point, trying to reform an aspect of medical practice that could not and ultimately should not be changed.[49]

Early twentieth-century psychiatrists proved even more resistant than Ray and his nineteenth-century colleagues to the idea that unanimity equaled truth in science. Given the diversity of theoretical orientations that characterized American psychiatry in that period, such resistance is not surprising. American psychiatry was a discipline that would in the twentieth century come to embrace the disciples of various psychoanalytic schools, as well as adherents to such differing approaches as psychobiology, behaviorism, and psychosurgery. Agreement in a specialty torn by such divisions was highly unlikely, as many prominent psychiatrists, including Adolf Meyer of Johns Hopkins, realized.[50] His refusal to participate in the American Psychiatric Association's effort in the 1920s to produce a uniform nomenclature, and the generally little interest shown by the psychiatric rank and file in this and similar projects, were signs that American psychiatrists did not perceive "agreement" as the hallmark of a valid scientific diagnosis.[51]

Not until after World War II did American psychiatrists manage to focus sustained attention on the issue of diagnostic consensus and reliability. Postwar concern was stimulated by outside pressures, including new demands for epidemiological data, more powerful statistical methods, the growing importance of psychiatric critics like Thomas Szasz, and the expansion of legal interest in the civil liberties of mental patients. In this contemporary debate jurists are once again the most insistent group of outsiders demanding that psychiatry provide more reliable definitions of mental illness. They are also the least impressed by such psychiatric efforts to improve reliability as the revision of the *Diagnostic and Statistical Manual*, the establishment of an American Board of Forensic Psychiatry, and the American Psychiatric Association's endorsement of a more circumspect role in the courtroom. The lawyers' skeptical response suggests that little progress had been made in the effort to construct a legally acceptable definition of mental disease.[52]

What makes this legal skepticism toward psychiatry so important is that the lawyers seem to be viewing an increasing number of other medical specialties with the same jaundiced eye. This broadening of legal suspicions is part of a more general crisis in the relationship between American physicians and their patients. Born of a fundamental loss of faith, the crisis has helped transform not only the traditional therapeutic alliance between doctors and patients, but medicine's role as a social authority. In this context, the historical debate between law and psychiatry over definitions of mental illness takes on added significance. Past debates over medical credibility, stimulated by such disease concepts as moral insanity, begin to have a hauntingly familiar sound. The words and actions of lawyers like Wharton and Bell, who struggled to mold medical definitions into legally useful forms, have a new relevance.

One of the most valuable lessons to be gained from these past efforts to reframe definitions of disease has to do with the term "reliable." As the experiences of Bell, Ray, and the others make clear, the use of such seemingly straightforward terms as "reliability" and "certainty" contains a hidden agenda. What the creators of "diagnosis related groups," insurance forms, reimbursement legislation, and legal codes mean by these terms may have little to do with their meaning to physicians. In the sometimes frantic search for disease definitions that will satisfy the requirements of public policy, it is important to remember this difference. "Medicine," the great clinician William Osler once observed, "is a science of uncertainty and an art of probability." Disease definitions born of uncertain science need fundamental alteration to fit the more "reliable" frame demanded by contemporary lawyers, businessmen, and administrators. The cost of such reframing in terms of diagnostic accuracy and therapeutic effectiveness has yet to be calculated. While such calculation is not easy, without it the price of reliability may be more than Americans are willing to pay.

NOTES

1. R. Fox, *So Far Disordered in Mind* (Berkeley, CA: University of California Press, 1978); G. Grob, *Mental Illness and American Society, 1875–1940* (Princeton, NJ: Princeton University Press, 1983); C. Rosenberg, *The Trial of the Assassin Guiteau* (Chicago, IL: University of Chicago Press, 1968); idem, "The Crisis of Psychiatric Legitimacy," in *American Psychiatry: Past, Present, and Future*, ed. G. Kriegman, R. Gardner, et al. (Charlottesville, VA: University of Virginia Press, 1975); D. Rothman, *The Discovery of the Asylum* (Boston: Little, Brown, 1971); idem, *Conscience and Convenience* (Boston: Little, Brown, 1980); A. Scull, ed., *Madhouse, Mad-doctors, and Madmen* (Philadelphia: University of Pennsylvania Press, 1981); R. Smith, *Trial by Medicine* (New York: Columbia University Press, 1981); N. Tomes, *A Generous Confidence* (Cambridge, MA: Cambridge University Press, 1984).

2. For examples of this critical literature, see E. Goffman, *Asylums* (Garden City, NY: Doubleday, 1961); T. Scheff, *Being Mentally Ill* (Chicago: Aldine Press, 1966); P. Sedgwick, *Psycho Politics* (New York: Harper & Row, 1986); T. Szasz, *Law, Liberty, and Psychiatry* (New York: Macmillan, 1963); idem, *Ideology and Insanity* (Garden City, NY: Doubleday, 1970).

3. R. Bayer, *Homosexuality and American Psychiatry* (New York: Basic Books, 1981); G. Grob, "The Origins of American Psychiatric Epidemiology," *American Journal of Public Health* 75 (1985):229–236; C. Rosenberg, "Crisis of Psychiatric Legitimacy."

4. S. Morse, "Failed Explanations and Criminal Responsibility: Experts and the Unconscious," *Virginia Law Review* 68 (1982):1046–1052; R. Spitzer and D. Klein, eds., *Critical Issues in Psychiatric Diagnosis* (New York: Raven Press, 1978); J. Ziskin, *Coping With Psychiatric and Psychological Testimony* (Beverly Hills, CA: Law and Psychology Press, 1981), pp. 138–144.

5. J. Tighe, "A Question of Responsibility," Ph.D. dissertation, University of Pennsylvania, 1983, pp. 104–125. Much of the background material for this article was drawn from "A Question of Responsibility" and several of my earlier publications. See: "'Be It Ever So Little': Reforming the Insanity Defense in the Progressive Era," *Bulletin of the History of Medicine* 57 (1983):397–411; "Francis Wharton and the Nineteenth Century Insanity Defense," *American Journal of Legal History* 27 (1983):223–253; "The New York Medico-Legal Society: Legit-

imating an Unstable Union," *International Journal of Law and Psychiatry* 9 (1986):231–243; "American Forensic Psychiatry: The Search for Persuasive Knowledge," in *The Handbook of the History of Psychiatry*, ed. T. Wallace and J. Gach (New Haven: Yale University Press, forthcoming).

6. The extensive body of nineteenth-century American legal commentary on mental illness is discussed in more detail in Tighe, "A Question of Responsibility."

7. J.E.D. Esquirol, *Mental Maladies*, trans. E. K. Hunt (Philadelphia: Lea and Blanchard, 1845); P. Pinel, *A Treatise on Insanity*, trans. D. D. Davis (Sheffield, England: W. Todd, 1806).

8. J. C. Prichard, *A Treatise on Insanity and Other Diseases Affecting the Mind* (London: Sherwood, Gilbert, and Piper, 1835).

9. The literature on moral insanity is extensive. The sources I found most helpful include: N. Dain, *Concepts of Insanity* (New Brunswick, NJ: Rutgers University Press, 1964), pp. 73–83; N. Dain and E. Carlson, "The Meaning of Moral Insanity," *Bulletin of the History of Medicine* 36 (1962):130–140; A. Fink, *Causes of Crime* (New York: A. S. Barnes, 1962), pp. 48–75; S. P. Fullinwinder, "Insanity as the Loss of Self: The Moral Insanity Controversy Revisited," *Bulletin of the History of Medicine* 49 (1975):87–101; J. Goldstein, *Console and Classify: The French Psychiatric Profession in the Nineteenth Century* (Boston: Cambridge University Press, 1987); S. Maughs, "A Concept of Psychopathology and Psychopathic Personality: Its Evolution and Historical Development," *Journal of Criminal Psychopathology* 2 (1941):329–356; A. Nye, *Crime, Madness, and Politics in Modern France* (Princeton, NJ: Princeton University Press, 1984); Rosenberg, *Guiteau*, pp. 68–70; R. deSaussure, "The Influence of the Concept of Monomania on French Medico-Legal Psychiatry, 1825–1840," *Journal of the History of Medicine* 3 (1946):365–397; R. Waldinger, "Sleep of Reason: John P. Gray and the Challenge of Moral Insanity," *Journal of the History of Medicine* 34 (1979): 163–179.

10. *English Reports* (Edinburgh: Green, Longmans, 1930), vol. 8, pp. 718, 722, 723.

11. J. Bishop, *New Commentaries on Criminal Law* (Chicago: T. H. Flood, 1892), vol. 1, sect. 287.

12. J. Ordronaux, "The Plea of Insanity as an Answer to an Indictment," *Criminal Law Magazine* 2 (1880):432–433; idem, "Judicial Problems Relating to the Disposal of Insane Criminals," *Criminal Law Magazine* 2 (1881):599–609; R. Ringgold, "Theory of Culpability," *Criminal Law Magazine* 10 (1888):641–662; F. Wharton, "Presumption in Criminal Cases," *Criminal Law Magazine* 1 (1880):32–35; idem, "Comparative Criminal Jurisprudence," *Criminal Law Magazine* 4 (1883):1–15.

13. This portrait of American psychiatric thinking is drawn from the following sources: Fink, *Causes of Crime*; Grob, *Mental Illness*; Rosenberg, *Guiteau*; idem, "The Bitter Fruit—Heredity, Disease, and Social Thought," in *No Other Gods: On Science and American Social Thought* (Baltimore, MD: Johns Hopkins University Press, 1976), pp. 25–53; B. Sicherman, *The Quest for Mental Health* (New York: Arno Press, 1980).

14. W. Hershey, "Criminal Anthropology," *Criminal Law Magazine* 15 (1893):503–504, 653, 661; Ordronaux, "Judicial Problems," pp. 591–593.

15. I. Ray, *A Treatise on the Medical Jurisprudence of Insanity* (Boston: Little, Brown, 1838), sects. 120–121, 133–134, 185–186.

16. J. P. Gray, "Moral Insanity," *American Journal of Insanity* 14 (1858): 321.

17. W. S. Chipley, "The Legal Responsibility of Inebriates," *American Journal of Insanity* 23 (1867):1–45; J. P. Gray, "Thoughts on Causation of Insanity," *American Journal of Insanity* 29 (1873):264–283; A. McFarland, "Minor Mental Maladies," *American Journal of Insanity* 20 (1864):10–26; D. Reese, "Report on Moral Insanity in Its Relations to Medical Jurisprudence," *American Medical Association Transactions* 11 (1858):727–746.

18. F. Wharton and M. Stille, *Medical Jurisprudence* (Philadelphia: Kay and Brothers, 1873), sects. 146–147, 163–189, 531–567, 576. Also see Janet Tighe, "Francis Wharton and the Nineteenth Century Insanity Defense," *American Journal of Legal History* 27 (1983):223–253.

19. J. Elwell, *Medical Jurisprudence* (New York: Voorhis, 1860); W. Guy, *Principles of Forensic Medicine* (New York: Harper, 1845); J. Ordronaux, *Jurisprudence of Medicine in its Relation to the Law of Contracts, Torts, and Evidence* (Philadelphia: Johnson, 1864); A. Taylor, *A Manual of Medical Jurisprudence* (Philadelphia: Lea, 1861, 1866).

20. Chicago Medical Society, "Minutes," *Journal of Nervous and Mental Diseases* 9 (1882):391–395; Hershey, "Criminal Anthropology," pp. 774–780; Wharton, "Presumption," pp. 37–38.

21. Fink, *Causes of Crime*, pp. 59–73; Maughs, "Concept of Psychopathology," pp. 465–499; Rosenberg, *Guiteau*; F. Wharton, "Case: U.S. v. Guiteau," *Criminal Law Magazine* 3 (1882):347–388.

22. C. Bell, *First Annual Dinner of the Medico-Legal Society of the City of New York* (New York: Thomas, 1873), p. 32.

23. W. Cocke, "Some Cases of Transitory Homicidal Mania," *Criminal Law Magazine* 7 (1886):238–239, 244–255.

24. Chipley, "Legal Responsibility," pp. 1–45; Gray, "Thoughts on Causation," pp. 264–283; McFarland, "Minor Mental Maladies," pp. 10–26; Reese, "Report on Moral Insanity," pp. 727–746.

25. Gray, "Moral Insanity," pp. 311–322; idem, "Thoughts on Causation," pp. 264–283; Reese, "Report on Moral Insanity," pp. 727–746.

26. Ray, *Treatise on Medical Jurisprudence*, 1844, 1853, 1860, and 1871 editions.

27. Taylor, *Manual of Medical Jurisprudence*, 1845 and 1861 editions; Wharton, *Medical Jurisprudence*, 1873 and 1882 editions.

28. Association of Medical Superintendents of American Institutions for the Insane (hereafter referred to as AMSAII), "Proceedings," *American Journal of Insanity* 20 (1864):63–107.

29. AMSAII, "Proceedings," *American Journal of Insanity* 23 (1867):114–122, 126–132.

30. Tighe, "A Question of Responsibility," pp. 234–500.

31. Grob, "Origins of Psychiatric Epidemiology," pp. 229–236.

32. C. Bell, "Classification of Mental Diseases," *Medico-Legal Journal* 4 (1887):49–66.

33. Bell, "Classification," pp. 49–66; idem, "Minutes of a Conference on Classification of Mental Diseases," *Medico-Legal Journal* 4 (1887):200–210; idem, "Monomania," *Medico-Legal Journal* 7 (1890):153–169; T. Meynert, "Classification of Mental Diseases as a Basis of International Statistics," *Medico-Legal Journal* 3 (1886):407–416.

34. Bell, "Classification," pp. 49–66; idem, "Monomonia," pp. 153–169.

35. Bell, "Classification," pp. 49–66; idem, "Minutes," pp. 200–210; Editorial, *Medico-Legal Journal* 4 (1887):301–310; Meynert, "Classification," pp. 407–416; "Transactions," *Medico-Legal Journal* 7 (1890):434–436.

36. Grob, "Origins of Epidemiology," pp. 229–236.

37. Pliny Earle to Clarke Bell, April 16, 1886, Pliny Earle Papers, American Antiquarian Society, Worcester, MA.

38. R. Blashfield, *The Classification of Psychopathology* (New York: Plenum Press, 1984); B. Ennis and T. Litwack, "Psychiatry and the Presumption of Expertise: Flipping Coins in the Courtroom," *California Law Review* 62 (1974): 693–752; Grob, "Origins of Epidemiology"; K. Menninger, *The Vital Balance* (New York: Viking Press, 1963); T. Millon and G. Klerman, eds., *Contemporary Directions in Psychopathology* (New York: Guilford Press, 1986); Spitzer, *Critical Issues*; L. Schwartz, "Mental Disease: The Groundwork for Legal Analysis and Legislative Action," *University of Pennsylvania Law Review* 111 (1963):389–420; Ziskin, *Coping*.

39. T. Almy, "Psychiatric Testimony: Controlling the 'Ultimate Wizardry' in Personal Injury Actions," *Forum—American Bar Association* 19 (1984):233–267; American Psychiatric Association, *Diagnostic and Statistical Manual of Mental Disorders* (Washington, DC: American Psychiatric Association Press, 1980); Blashfield, *Classification of Psychopathology*; R. Bonnie, "The Moral Basis of the Insanity Defense," *American Bar Association Journal* 69 (1983):194–197; R. Bonnie and C. Slobogin, "The Role of Mental Health Professionals in the Criminal Process: The Case for Informed Speculation," *Virginia Law Review* 66 (1980): 427–522; M. Moore, *Law and Psychiatry* (New York: Cambridge University Press, 1984); A. Stone, *Law, Psychiatry, and Morality* (Washington, DC: American Psychiatric Association Press, 1984); Ziskin, *Coping*.

40. Bonnie and Slobogin, "Role of Mental Health Professionals"; R. Bonnie, "Morality, Equality, and Expertise: Renegotiating the Relationship between Psychiatry and the Criminal Law," *Bulletin of the American Academy of Psychiatry and Law* 12 (1984):5–20; Ennis and Litwack, "Psychiatry and Presumption"; S. Halleck, "American Psychiatry and the Criminal: A Historical View," *American Journal of Psychiatry* 121, suppl., (1965):i–xx; S. Morse, "Crazy Behavior, Morals, and Science: An Analysis of Mental Health Law," *Southern California Law Review* 51 (1978);527–654; idem, "A Preference for Liberty: The Case against Involuntary Commitment of the Mentally Disordered," *California Law Review* 70 (1982):54–106; idem, "Failed Explanations"; J. Robitscher, *The Powers of Psychiatry* (Boston: Houghton Mifflin, 1980); Schwartz, "Mental Disease"; Ziskin, *Coping*.

41. For more detailed information about the development of expert testimony in the United States, see: L. Hand, "Historical and Practical Considerations Regarding Expert Testimony," *Harvard Law Review* 15 (1902):40–58; H. Pollock and E. Wiley, "A Contribution to the History of Psychiatric Expert Testimony," *American Journal of Psychiatry* 100, suppl., (1944):119–133; L. Rosenthal, "The Development of the Use of Expert Testimony," *Law and Contemporary Problems* 2 (1935):403–418; H. Weihofen, *Mental Disorder as a Criminal Defense* (Buffalo, NY: Dennis, 1954); J. Wignore, *A Treatise on the Anglo-American System of Evidence in the Trials at Common Law* (Boston: Little, Brown, 1923), vols. 1 and 4, sects. 998–999, 1078–1081, 1916–1921.

42. Wharton, *Medical Jurisprudence*, (1873), sect. 298.

43. Wharton, *Medical Jurisprudence* (1860), sects. 86–89, 91–93; (1873), sects. 338–339.

44. Wharton, *Medical Jurisprudence* (1873), sects. 171–172, 195, 279–280, 294–296, 298.

45. Ibid., sects. 275–278, 298–299; (1882), sect. 770.

46. C. Bell, "Third Inaugural [1874]," *Papers Read before the New York Medico-Legal Society* 2 (1882):498.

47. Tighe, "'Be It Ever So Little,'" pp. 397–411.

48. For an insightful discussion of the twentieth-century debate over this question, see Morse, "Failed Explanations," pp. 1021–1026.

49. AMSAII, "Proceedings," *American Journal of Insanity* 28 (1872):245–253; Ray, Medical Experts," *Contributions to Mental Pathology* (Boston: Little and Brown, 1973), pp. 411–414.

50. Adolf Meyer to Samuel Orton, April 25, 1919, Adolf Meyer Papers, Chesney Medical Archives, Johns Hopkins Medical Institutes, Baltimore, MD.

51. AMSAII, "Proceedings," (1872):245–253; Grob, "Origins of Epidemiology," pp. 232–234.

52. Morse, "Crazy Behavior," pp. 564–572; idem, "Failed Explanations," pp. 1043–1059.

MANAGING
DISEASE:
Institutions as Mediators

11 QUID PRO QUO IN CHRONIC ILLNESS: Tuberculosis in Pennsylvania, 1876–1926

BARBARA BATES

The framing of disease, in one of its dimensions, is rooted in the necessities of care and the specific biological character of particular ailments. Tuberculosis provides a useful example. The most important single cause of death (and an important factor impoverishing families) in the nineteenth century, consumption was a challenge to policy makers and welfare authorities as well as to physicians. Ubiquitous and discouraging, it constituted a problem of a very different kind than did, say, cholera, yellow fever, or typhoid. Its course was unpredictable, and the great majority of victims deteriorated gradually, hoping all the while to live and work a normal life as long as they could; when that became impossible they needed food, warmth, and care. Many hospitals would not admit them, as incurable; most families could ill afford to care for wives, husbands, and children when their symptoms became well defined.

With Koch's demonstration of the tubercle bacillus (1882) and the growing faith in sanatorium treatment, perceived options changed. Yet, the situation was still difficult. Proving that consumption was contagious posed a question; it did not provide answers. These had to come from public authorities and private charity. Only gradually did these sources provide institutional care for both the poor and the prosperous—and only gradually did the medical and nursing professions find an appropriate framework within which to organize their own response. Most physicians were reluctant to treat patients who showed little tendency to recover and if more attractive options were open, avoided the isolated, in fact dangerous, medical work of the tuberculosis sanatorium. Patients, moreover, were ordinarily loath to enter sanatorium treatment and did so only when driven by personal needs (and to spare their often financially and physically exhausted families) and by the institution's promise of remission. Chronic care has always presented grave challenges to families, to the healing professions, and to society's welfare commitment.

—C. E. R.

In 1904 DR. WARD BRINTON, working in the dispensary of the Henry Phipps Institute in Philadelphia, entered the following note in a tuberculosis patient's record:

> A cold damp day, penetrating—bad for lungs. Patient feels very badly and looks much worse—cold all the time, shivering. Sits with head bowed— Hands between his knees with swollen joints and heaving chest. Despair in every line.
>
> Is quite hopeless about himself and says he will not live until spring. . . . He gasped for breath and in his agony tore at his shirt front . . . and with his staring eyes in a leaden skin he seemed more like a subject from Dante's Hell than aught else I know.
>
> And then he babbled incoherently of wife and children and of their helplessness and almost raved against his Maker for his giving him such suffering of weakness and of dull dread unhopefulness.
>
> And nothing can be done, no, absolutely nothing that I know. And when I feel my own weakness to help him I am ashamed to face the dying man's eyes, to offer words of hope and encouragement for he knows I lie.[1]

Why did men and women take care of such chronically ill and often dying patients? What rewards did they find in the work? Why did the patients seek care? What kinds of social transactions and institutions brought patients and caretakers together so that each gained something, thus creating a viable system? What did the institutions offer the public in order to gain support? And how did these arrangements reflect the changing concepts of tuberculosis and social attitudes toward those who suffered from the disease?

The care of tuberculous patients in Pennsylvania between 1876 and 1926 sheds light on the answers to these questions. The period began with the founding of the state's first organization to care for poor consumptives. It spanned the discovery of the tubercle bacillus in 1882, the gradual recognition during the 1880s and 1890s that consumption was communicable, preventable, and curable, and the founding in 1904 of the national association to fight the disease. By 1926 the number of institutional beds dedicated to tuberculous patients in the state had risen to 4,266. Although this increase still did not meet the demand, the people of Pennsylvania had made an enormous investment.

Initial goals were religious. Clergymen hoped to give comfort to chronically ill or dying men and women, lead them to salvation, and improve society at the same time. Once prevention and cure seemed possible, however, goals gradually became medical, and the tasks of care shifted to physicians and nurses. Both the communicability of the disease and the perceived need for supervised treatment made institutionalization desirable, and the locus of care moved gradually from homes to sanatoriums and hospitals. The threat of infection helped to legitimate both private and public expenditures, and the responsibility for providing care increasingly shifted to government.

The tuberculous sick, grasping at the hope of cure, sought institutional treatment; some also wanted to protect their families from infection by leaving home. However, the personal costs of institutionalization were high: a regimented life, loss of autonomy, loss of income, separation from family and friends, and at times performing labor or permitting an autopsy after death. The relative advantages of staying at home and being treated in an institution hung in an unstable balance, changing with institutional amenities, the quality of life at home, and the apparent results of treatment. Many patients delayed entering an institution until late in the course of illness when they were destitute but when cure, an important medical goal, was no longer likely.

Four Pennsylvania organizations—a church-sponsored city mission, a research institution, a medically dominated voluntary organization, and a state health department—illustrate the changing goals in the care of consumptives and the variety of intrinsic bargains between patients and caretakers that these institutions mediated. These four examples do not exhaust the range of transactions, nor is any one of them necessarily typical of sanatoriums and hospitals for the tuberculous. In other regions, with different protagonists, geography, socioeconomic structures, or balance between voluntary and public sectors, the details undoubtedly varied.[2] Nevertheless, these examples illuminate some of the strategies society used when faced with a complicated problem that involves sickness, the risk of infection, and prolonged dependency. The patterns that evolved then are still discernible in the nation's system of care.

Care of the Chronically Ill and Dying: Traditional Models

In 1875 consumption of the lungs ranked first in the causes of death in Philadelphia, as in much of the Western world. The municipal Board of Health that year attributed 13 percent of the city's deaths to it. Yet there was no public outcry, no clamor for reform. Persons whose families seemed tainted by the disease were likely to dread the prospects of cough and wasting flesh, and some of those already sick sought treatment, but once the disease had established itself physicians had little to offer. Doctors could try to relieve symptoms and correct some of the debilitating conditions or habits thought to affect health, but, faced with a disease considered constitutional, they felt relatively helpless.

The care of chronically ill and dying consumptives was a family's responsibility. The afflicted who lacked families or had exhausted the help of relatives and friends might turn to local government. Qualified by their poverty and dependency, not their disease, they might then go to the Philadelphia Hospital, an old municipal institution that had emerged out of the city's almshouse in the mid-1830s and still functioned as such. On its medical wards, where the tuberculous mingled with other patients during the nineteenth century, consumption of the lungs was the most prevalent disease and the most common cause of death. Yet only 6 percent of the city's deaths from this disease in 1875 occurred in the Philadelphia Hospital.[3] Most consumptives died at home.

The first organization to specialize in care of the tuberculous was religious, not medical. In 1870 the Reverend William B. Stevens, Bishop of Pennsylvania, established the Philadelphia Protestant Episcopal City Mission. By preaching, visiting the sick poor, and offering both temporal and spiritual help, he hoped to Christianize the city, restore moral order, and save souls. As Mission workers made their rounds through the city, the protracted distress of destitute consumptives particularly impressed them, and in 1876 the Mission inaugurated a new department, the Home for Poor Consumptives. The Home was at first simply an administrative mechanism for giving home relief, but when the Mission acquired permanent quarters a year later, it equipped a few of its rooms upstairs as a temporary hospital. The bishop named the building the House of Mercy.[4]

Persons deemed morally worthy and medically certified as having consumption qualified for relief. If they stayed home, as most of them did, they received food, medicines, coal, other necessities, and even a little whiskey (a commonly prescribed remedy). When their homes proved too unwholesome or too poor to give them adequate care, the Mission moved them to suitable private quarters and paid for their board and nursing. The House of Mercy admitted a few. In all of these settings, physicians attended the sick as needed, and missionaries visited regularly.

Helping poor consumptives often disturbed and exhausted the missionaries but also brought emotional and spiritual satisfactions. During the winter of 1880–1881, for example, the Reverend Dr. Thomas L. Franklin, who had taken charge of the Mission's consumptives, had sixty-five of them under his care. He observed that

The work of visitation during the past winter has been severe, in bitter cold, in snow drifts, in slush . . . by day and by night, at all hours, as late as three o'clock in the morning, wending my way home from ministering to the dying. . . .

This labor, I have reason to believe, has not been in vain. Many lives were rescued that would have been lost but for the sympathy and aid of the Mission, and . . . I have been privileged to lead to the Cross souls that died in hope, and are doubtless rejoicing in the joy of their salvation.[5]

Mrs. Agar, a lay visitor, shared the rewards of the Reverend Dr. Franklin. Although appalled by the suffering she witnessed, she too felt gratified by her work. "A sick woman, in the last stages of consumption, asked me to read the Bible and pray with her," Agar reported, "and when I had done so, drew me to her and kissed me, saying, with her eyes filled with tears, 'O God, bless you and those who sent you to comfort me!'"[6]

During the 1880s and 1890s three new concepts of consumption challenged the Mission's methods. In 1882 Robert Koch reported his discovery of the tubercle bacillus. As physicians gradually accepted Koch's findings, they drew two conclusions: The disease was both contagious and preventable. The third idea came from the European sanatoriums: Consumption was curable. Fresh air, good food, and the proper amounts of rest and exercise, all supervised by a physician versed in the healing regimen, could restore a patient's health, at least if the diagnosis was made soon enough. In the early 1880s Dr. William Angney, who attended the House of Mercy, began to express concerns over the contagiousness of the Mission's patients. For the safety of families, he suggested, consumptives would be better cared for in an institution than in crowded homes. He also thought that he could claim a cure of a young consumptive woman. The next few years gave further cause for optimism; more patients improved, and a few even returned to work.[7]

The new medical goals—separating infectious patients from their families and trying to cure them—conflicted with the clergymen's beliefs and values, and two successive superintendents of the Mission resisted the doctors' urgings. One doubted the curative claims of medicine. Wherever care was provided, he believed, it was nursing, not medicine, that helped the patients. Further, institutionalization disrupted family relationships both morally and emotionally. It also cost more than care at home. Curative goals demanded new and expensive equipment and some changes in admission criteria. Physicians preferred early-stage, presumably curable cases to advanced and dying ones. They wanted to prolong and even save lives, not settle for bodily comfort and the religious guidance and solace favored by the clergymen.[8]

Despite the clergymen's concerns, medical goals gradually gained support. In 1886 the City Mission opened a new institution in Chestnut Hill, on the edge of the city. Home care dwindled, and increasing proportions of patients were hospitalized. The Mission's published records seemed to prove the wisdom of the change; death rates went down, and rates of improvement rose substantially. Analysis of the Mission's statistical reports, however, shows that these results were chiefly due to the discharge of patients from the Mission's care before they died. Almost certainly, two additional factors contributed to the apparently favorable outcome: the admission of people less seriously ill and physicians' overly optimistic assessment of patient improvement.[9] Consciously or unconsciously, leaders of many later institutions made their published statistics look better in one or more of the same ways.

Although the Mission's goals and methods grew more medical, physicians still had to accommodate themselves to the care of advanced cases. While they complained that such patients could not benefit from treatment, that they used too much of the nurses' time, that they depressed the spirits of other patients, and that they made the Mission's death rates compare poorly with those of other institutions, they found reasons to care for them. Removing the sick from deplorable home conditions was humane, they acknowledged. More important, as fears of contagion increased during the 1890s, it protected the patients' families from the disease. By the mid-1890s they had also established postgraduate education and research—two typical mechanisms by which physicians treated the poor in a way that seemed advantageous to both parties involved in the bargain. Resident physicians began to staff the Home for Consumptives at Chestnut Hill, thus gaining experience with the disease, and some of the worst cases were gathered together in a special department where Dr. Jacob Solis-Cohen, a distinguished laryngologist, could try experimental treatments.[10]

To men and women who were seriously ill and destitute, an institution offered food, shelter, personal attention, and sometimes hope. When Mrs. Josie C. Collom entered the Home at Chestnut Hill in 1900, her first impressions were highly favorable. "I like it very well," she wrote to Dr. Lawrence F. Flick, who had referred her to the Home. "Dr's Coh[e]n and Bacon say my case is a very hopefull one so I pray and hope to God that it is my throat is beaing treaty by Dr Watson he say, my voice will come back again . . . Dr Bacon and her asstant are so good and kind they try to make every thing very plasent for the pation."[11]

Institutionalization for long periods, however, disrupted patients' lives in countless ways: They lost their independence, they had to endure treatments and comply with institutional rules, and they missed their friends and families. A few objected to religious instruction. Whatever her reasons, Mrs. Collom decided to leave the Home after a three-month stay, against advice, even though she knew that her decision precluded readmission.[12] Physicians often thought they knew

what was best for a patient, but consumptives made their own judgments and choices insofar as their circumstances allowed.

Care of the Chronically Ill and Dying:
A Medical Model

The quid pro quo that evolved at Chestnut Hill in the 1890s—
treatment and care of the chronically ill and poor in exchange for medical education and research—can be seen most clearly in the Henry
Phipps Institute for the Study, Treatment, and Prevention of Tuberculosis, established in 1903. Dr. Lawrence F. Flick conceived of the
institute and directed it from 1903 to 1909. Flick was among the first
physicians in Philadelphia to argue that tuberculosis was contagious
and during the 1890s led the local campaign to prevent its spread. In
1901 he had also started the White Haven Sanatorium for poor consumptives in the Pennsylvania mountains. Flick's work attracted the
interest of Henry Phipps, a retired multimillionaire who had been the
business partner of Andrew Carnegie. Phipps, who shared Carnegie's
convictions that rich men should use their wealth to benefit society,
agreed to support Flick's ideas with a million dollars.

The primary goal of the institute, as Flick announced it in 1903, was
the extermination of tuberculosis, chiefly through preventive measures. The institute would include a hospital where patients with
advanced disease could be isolated and treated humanely. Those whose
disease was less advanced, or who would not or could not enter the hospital, would be treated and taught the most advanced preventive
methods in a dispensary. Both the hospital and dispensary would provide materials for the scientific laboratory, where physicians would
study the cause, treatment, and prevention of tuberculosis and disseminate new ideas as widely as possible. Each patient's record and each
laboratory investigation would contribute to a growing body of information, Flick believed, and the new understanding would soon eliminate the disease.[13]

Recruiting capable physicians willing to do such work in 1903 was not
easy. In addition to the discouraging aspects of treating chronically ill
and dying patients, physicians recognized the risks of infection from
repetitive contacts with the sick. For over a decade Flick and other
members of the local tuberculosis society had been emphasizing the
contagiousness of advanced consumptives, partly in an effort to arouse
public concern and stimulate corrective action. However, fear, once
aroused, was hard to control, and led some physicians and lay attendants to avoid consumptives, not to take care of them. At the turn of the
century, moreover, few physicians were experienced in diagnosing or

treating tuberculosis, and few had had training in either bacteriology or clinical pathology.

Lack of knowledge and skill, however, made work at the institute more attractive to young, ambitious physicians. They knew that postgraduate education, specialization, and institutional appointments all contributed to professional advancement, but means to acquire them were limited. The Phipps Institute offered staff physicians the opportunity to supplement their deficient medical education, become expert in physical diagnosis of chest ailments, gain experience in new laboratory procedures, and possibly publish a paper or two. They could broaden their contacts with other physicians, enhance their reputations, and thus advance their careers in a competitive medical marketplace. They also found it exhilarating to participate in the growing campaign to combat tuberculosis. In 1904 Flick was helping to organize the National Association for the Study and Prevention of Tuberculosis, and the institute's own staff was working on a new serum that seemed likely to cure the infection. With organization, research, and expanding knowledge, the doctors hoped to vanquish the ancient disease.[14]

Of all the rewards that the institute's physicians later remembered or wrote about, one predominated—the Monday evening staff meeting. And of all the subjects discussed at these meetings the correlation of physical findings with pathological changes found at autopsy evoked the most vivid memories. By a curious alchemy of scientific enthusiasm, they transmuted the ultimate failure of medicine—the death of a patient—into the gold of knowledge and fellowship. The case of every patient who died and came to autopsy at the institute was presented at these meetings. The responsible physician had to draw on a blackboard a meticulous diagram of what he had found on physical examination and then predict exactly the abnormalities that the pathologist would report. Any discrepancies between the clinical picture and autopsy results had then to be explained by the hapless physician. "No patients were more thoroughly, carefully and repeatedly examined than those in the wards of the Phipps Institute," recalled Frank A. Craig, "especially when they were seriously ill." The system, physicians believed, benefited the patients as well as the doctors. "No slipshod diagnoses, and the treatment of patients had better be good," asserted George B. Wood. "It sharpened one's wits to know that you had to give the whys and wherefores to a bunch of lions 'waiting to tear you to bits.' But with all of this give and take, the frequent getting together . . . developed a spirit of comradeship and understanding that made us united and loyal in our enthusiasm for the good work . . . and some of the personal friendships started there lasted throughout our lives."[15]

Institute policies helped to guarantee these rewards. Patients admitted to the hospital for treatment had to meet three criteria: They had to have reached an advanced stage of tuberculosis, they had to be poor, and the nearest responsible relative had to give written permission for an autopsy in case the patient should die in the institution. "Oddly enough," Wood recalled, "this [autopsy] rule did not seem to have any depressing effect on the patients themselves. . . . I remember once, as I passed through a ward, hearing one poor, scarcely alive, disease-ridden human relic say to the patient in the next bed, and with a grin on his face, 'I'll bet you ten dollars I'll beat you to the cutting-up room.'"[16] Wood did not mention the families who kept a member out of the institute because of the autopsy rule or who took a patient home when death seemed near.

The Nursing Bargain:
Training Consumptive Women

Finding suitable nurses to care for the institute's patients was much more difficult than recruiting physicians. In comparison with the doctors, almost all of whom worked part-time, nurses had even greater cause to fear infection. They spent long days in intimate contact with advanced consumptives, some completely helpless. Infectious sputum repeatedly soiled patients' bedclothes and nearby walls, floors, and furniture. The nurses had to bathe and feed the patients and keep the environment spotlessly clean. Flick believed that every particle of sputum must be removed before it dried. "The nurse may at any moment have to scrub the floor or turn washerwoman in an emergency," he observed, and "she must do it with alacrity and good humor."[17]

Turn-of-the-century training for nurses ill equipped them for the work. Over the previous two decades or so, nursing schools had proliferated to meet the demands of the burgeoning general hospitals, but many of the new schools had not sustained the standards espoused by nursing's early leaders. Because general hospitals typically excluded consumptives, few graduates were experienced with the disease. After graduation no official agency regulated licensure or practice.

The nurses whom Flick and his colleagues first employed at the institute proved disastrous. Within a few months the night watchmen began to complain: Nurses entertained gentlemen callers on night duty, made off with supplies from the kitchen and laboratory, and sometimes failed to respond to patients' calls. Two nurses who had been out drinking with male companions returned one Sunday night, laughed uproariously in the diet kitchen, and threw cantaloupe rinds at each other. The institute had been forced to hire any nurse it could and, as Flick noted sourly, it "got the scum of the profession."[18]

Flick gradually solved the nursing problem by recruiting consumptive women to do the work. The first of these, who served as ward maids, had been recovering at the White Haven Sanatorium. Because they already had the disease themselves, they were not afraid of it. Their work was satisfactory, and in the fall of 1903 Flick established a training school at the institute. The pupil nurses, all convalescent patients from White Haven, progressed through a two-year course of work and training, for which they received board, lodging, and ten dollars a month.

Although attrition at the school reached as high as 67 percent, those who managed to graduate found advantages in their new careers. Before they had become ill, almost all had worked, usually in the low-paying jobs typically available to women—in factories, domestic service, teaching, clerical work, or other white-collar positions. Some of the women, reluctant to return to conditions in which they had become sick, were looking for an alternative. Nursing paid better than their previous jobs and for some was more respectable. In addition, the growing stigma of having had tuberculosis made it more difficult for recovering patients to find other kinds of employment. Tuberculosis nursing, moreover, though often difficult and demanding, gave personal satisfactions—a sense of accomplishment and useful service. None of the nurses, as far as the records show, returned to their previous work.[19]

Training tuberculous women as nurses also benefited the institute. The women provided inexpensive labor and improved the quality of care. Flick and his colleagues, ignoring the high attrition rate in the school, promoted the model as a means of rehabilitating patients, and the system was later used widely in sanatoriums. Many of these were located in remote regions, where recruitment of staff was especially difficult, and most had trouble finding healthy workers unafraid of the disease.

Care of the Curable: The Sanatorium

Relatively few organizations focused their efforts on poor consumptives with advanced or terminal disease. In contrast, sanatoriums, which offered a hope of cure, had greater appeal to physicians, politicians, and charitable contributors. Unlike the early City Mission and the Phipps Institute, they were designed primarily for men and women who had earlier tuberculosis and a greater chance of recovery.

Another of Flick's projects also demonstrates the shift from care of advanced and dying patients toward care of the curable. In 1895, with Father John Scully, pastor of St. Joseph's Church in Philadelphia, Flick had founded the Free Hospital for Poor Consumptives, an organization that paid for the care of dying consumptives in local hospitals and promoted its cause on the grounds of humanitarianism and the prevention of tuberculosis. In 1901 it broadened its scope by opening a sanatorium for tuberculous indigents at White Haven, a village in the mountains of eastern Pennsylvania. Two goals guided the new institution: cure, when possible, and prevention, by segregating infectious patients. While Flick considered prevention more important, he believed that the hope of cure would help to attract applicants with early disease out of their homes and into the sanatorium, where they would be harmless to others.[20]

Men and women who entered White Haven Sanatorium were expected to pay for their treatment and upkeep in various ways. Like those who went to the City Mission or the Phipps Institute, they had to leave their friends and family and adapt themselves to institutional life and rules. They were also expected to work at tasks such as cleaning, doing the laundry, washing dishes, and cutting firewood. In lieu of fees, the work contributed to the sanatorium's economy and, as Flick and his colleagues explained, helped to rehabilitate the patients. Although some of the patients worried about their ability to do their jobs, and some complained of exhaustion, others considered the plan a fair bargain. They were earning the treatment and care that they otherwise could not afford.[21]

The labor system reflected medical confidence that supervised rest and exercise had therapeutic value. It also helped to balance the sanatorium's budget. Encouraged by the promise of this approach, the Free Hospital for Poor Consumptives decided to divert all its resources to its sanatorium. By early 1904 it had stopped paying for the care of dying consumptives in Philadelphia hospitals.[22]

The physicians at White Haven, however, soon recognized a serious flaw in the plan: the sanatorium was not attracting early cases. One of the reasons was clinical. Diagnosis depended almost entirely on the patient's history and a physical examination, and most general practitioners were poorly trained in physical diagnosis. They often failed to detect tuberculosis in a stage that was then considered early despite manifestations we now associate with more advanced disease: cough, weight loss, a dull percussion note over an upper lung, rales, and poor expansion of the chest.[23] Laboratory diagnosis was very limited. Bacteriological studies, when used, consisted only of sputum smears (not the more sensitive cultures), and x-ray diagnosis was controversial. Even the sanatorium did not have x-ray equipment until 1919.

Other reasons for the lack of early cases were social. Even when such a diagnosis was established, tuberculous men and women were often reluctant to leave home. Although they hoped for the cure that medical treatment seemed to promise, and some wanted to protect their families from infection, other desires and responsibilities delayed their decisions. As long as they felt reasonably well, breadwinners thought they should work to support their families, wives hesitated to leave their husbands, and mothers refused to part with their children. Prospective patients tended to delay entering a sanatorium until they were too ill to function well at work or at home. Loss of a job or lack of supportive friends and relatives often precipitated requests for admission. When the sick needed social and economic support as well as medical care, their disease was often advanced. "I am at the end of my rope," wrote V. D. Charber, who in 1904 had had tuberculosis for six years. "But while there is life there is hope and I don't want to give up the fight. I think if I could go to the proper kind of a sanatorium I might still be saved."[24]

White Haven Sanatorium tried to control the proportion of advanced cases but was not very successful. It developed a network of examining physicians to screen prospective patients, but screening was not consistent. In some instances, the physicians sympathized with the invalids, in others they hesitated to label an applicant hopeless, and in still others they feared offending a referring physician. Patients and their families sometimes persuaded their friends, clergymen, and local politicians to support their applications, and resisting these pressures took courage. Perhaps resistance was even unwise; it made political enemies. As a result of all these factors, most of the patients admitted to the sanatorium had moderately or far advanced disease. When they failed to improve after six months or so of treatment, Flick and his colleagues, trying to reduce the waiting list, discharged them back to their communities even though they were still infectious. Other patients like them took their places.[25]

The high proportion of advanced cases created unanticipated problems for the sanatorium. Patients with severe disease could not work for their upkeep and needed nursing care, more frequent medical attention, and longer periods of treatment. Increasingly elaborate facilities had to be built. The sanatorium, unable to correct the situation, had to adapt. Recurrently short of funds and unable to attract a large enough subsidy from the state government, it gradually shifted to a cash economy. One form of patient labor persisted. In 1907 the sanatorium opened a new training school for nurses and began to recruit recovering women. The two-year program legitimated the work in a bargain similar to the one devised at the Phipps Institute.

State Sanatoriums: The Limits of Government

In comparison to White Haven Sanatorium the state health department had a major advantage: The legislature funded its tuberculosis work very liberally. Many leaders of the antituberculosis movement, believing that only the government could ever meet the needs for prevention and treatment, campaigned vigorously for Pennsylvania's involvement. They justified its incursion into the practice of medicine with two arguments: The contagiousness of tuberculosis threatened the public health, and its protracted course sapped the state's economy. Intervention would protect the populace and help to return the sick to productivity. In 1907 Samuel G. Dixon, commissioner of the state health department, embarked on a massive spending program to build a system of three large tuberculosis sanatoriums and a statewide network of supporting dispensaries. Enabling legislation explicitly identified both early and advanced cases as beneficiaries: the early cases for treatment, the more advanced for their own comfort and the protection of others. Dixon was too shrewd a politician to have worded the legislation in a narrower way.[26]

In practice, however, the state's priorities tended to favor patients with less advanced stages of tuberculosis. The first sanatorium at Mont Alto, in a forest preserve in south central Pennsylvania, was far from centers of population and poorly connected to transportation routes. Initial facilities were primitive; patients lived in tents and cottages and had to walk to both the bathrooms and the dining room even during the winter. Physicians described this exercise as healthful for early cases, but Mont Alto, like White Haven Sanatorium, attracted many sicker consumptives who needed infirmary care.

Like its voluntary counterpart, the state accommodated: It built infirmaries and designed the last of its three sanatoriums for advanced cases. By 1914, however, physicians were beginning to doubt the wisdom of the plan. Over half the state's patients had far advanced disease; many seemed unsuited for care in a sanatorium. It was unfair to send dying consumptives so far from home, asserted Albert P. Francine, chief of the largest state tuberculosis dispensary in Philadelphia; local communities should care for them instead.[27]

In 1919 the state health department initiated plans to change its mix of patients. Commissioner Dixon had died, the state government was facing a fiscal crisis, and the tuberculosis program, which had become the largest state system in the country, was consuming two thirds of the health department's budget. Despite these expenditures, waiting lists for the sanatoriums remained unacceptably long, and other demands competed for limited funds. In a marked change of official philosophy, the state decided to shift the care of advanced consumptives to the counties. Care of the chronically ill and poor had long been a local responsibility, officials argued, and the counties should now accept it. The state could then focus its efforts on early and curable cases.[28]

Children figured prominently among the curables whom the state wished to attract. These were the wan, thin, and sickly boys and girls whose apparently poor resistance made them likely victims of tuberculosis in adolescence and early adulthood. Some were exposed to the disease at home. Getting them out of their homes, improving their nutrition, and correcting various physical defects, physicians believed, would boost their immunity and help carry them safely through to adulthood. Since the turn of the century, data had accumulated to show that tuberculosis developed in childhood. Autopsies revealed a rising frequency of the disease during the first two decades of life, and since 1907 a new tuberculin skin test seemed to show the same phenomenon in living children.[29]

This medical knowledge meshed nicely with contemporary interests in children's health and welfare, and out of this combination came a new institution—the tuberculosis preventorium. Its methods were modeled on those of the sanatorium, but its purpose was somewhat different: to prevent, not treat, tuberculosis. Pennsylvania's first preventorium had opened under voluntary auspices in 1913, and in 1922 the state started a similar program in one of its sanatoriums. "The State Health Department has gathered up a few of the broken bits of life from the small communities," wrote an observer for the patients' magazine at Mont Alto, "and brought them up here into the hills to be mended and healed. Two hundred youngsters, frail and pale and under weight; not sick, but just at the place where a few more days or weeks may mean the difference between illness and h[e]alth. . . . They dwell in a city of tents, living in the open; eating of the best and most nourishing food; being examined and watched by skilled physicians and so guarded against the very beginnings of serious things."[30]

By 1929 over a third of the patients admitted to the state sanatoriums were children; physicians judged that only 5 to 6 percent of them had active tuberculosis of the lungs.[31] The children took the places of chronically ill, tuberculous adults, but not because the adults were receiving care in other institutions. Most of the counties had resisted state pressures to build their own facilities, and hundreds of adult applicants awaited admission.

The children rewarded their caretakers in a way that most adults could not; they usually regained their health quickly and convincingly. The results gratified the physicians and served to justify the policies and expenditures of the state health department. Adults with advanced disease, in contrast, stayed longer, cost more to care for, discouraged their caretakers, and depressed other patients. "For many of them, nothing can be done," complained William G. Turnbull, deputy secretary of health in 1926. "The admission of these hopeless cases is an injustice to the institution, to the early cases waiting admission, and often to the patients themselves. For others much can be done," he continued, making his values clear, "and these the institution welcomes."[32]

Care versus Cure

The goal of curing patients seemed to have many advantages. For the small contributors, philanthropists, and politicians, cure was a good return for their gifts and effort. Cure pleased the patients and their relatives and won favorable comment in newspapers and magazines. It gratified the caretakers who participated in the work and enhanced their reputations and their authority. Treating curable patients in institutions, moreover, cost less per person than the protracted care of incurables, and the very presence of healthy-looking patients seemed desirable. Many men and women preferred to be treated in a place where others seemed to be getting better and where they did not have to hear the constant coughing of advanced cases or confront the likelihood of death.

The care of seriously ill and dying patients provided few such rewards. Physicians collected fees from private patients, but recoveries were too few to give them satisfaction or enhance their reputations. When patients were poor, their care could at times be linked to medical education and research, but most sanatoriums were in rural locations, far from medical schools. In most situations, care of the sick poor who had little or no hope of recovery seemed to be merely a form of welfare, and Pennsylvanians, like other Americans, had long been reluctant to encourage dependency.

The primary reason to institutionalize the tuberculous poor was preventing the spread of disease. This rationale, probably bolstered by political pressures from interested parties to accept and keep such patients in sanatoriums, proved strong enough to expand the state's care system again after some cutbacks in the early 1920s. It was not, however, forceful enough to build a system that could retain patients throughout the period of their infectiousness. After a trial of sanatorium treatment, most of the men and women, including those with far advanced and even terminal disease, went home again, even though they had not been cured and were still potentially dangerous to others.

Decades have passed since tuberculosis was a leading problem in the United States, but we have many other chronic diseases and still rely on many of the same exchanges in structuring patient care. Research institutes offer free cancer treatments to those willing to participate in experiments, and hospices run by religiously motivated people take care of patients dying of acquired immunodeficiency syndrome (AIDS) or other diseases. Hospitals no longer train and use their convalescent women as nurses, but, as the nurses at least would argue, they still often pay less than the work is worth. Medical centers still link their services to medical education and medical careers. The chronically ill still yearn to be cared for but are often reluctant to leave their homes. For this and other reasons, they may delay treatment until incapacitated and perhaps beyond medical help. Many caretakers still prefer patients who recover with treatment and have persuaded the public to pay for expensive technologies that seem to promise definitive cures.

We are now accumulating an ever larger number of persistently ill and often incurable patients. The most rapidly increasing segment of our population consists of people over eighty-five—a group especially susceptible to chronic disease and disability. The mentally ill, the retarded, and the addicted all need attention, while AIDS, like tuberculosis, has added the threat of contagion to the burden of disability. Many new treatments, when they exist at all, are halfway technologies, as Lewis Thomas labeled them; they may lengthen lives but fail to cure.[33] What new bargains between the chronically ill and their caretakers will be made? Who will the caretakers be?

Whether physicians will meet the needs of such patients—in and out of institutions—and whether they can discover suitable rewards in the care of incurables remain to be seen. Many physicians, of course, do take care of such patients; the task is an old and respected tradition in medicine. It is expressed in the words inscribed on the statute of Dr. Edward Livingston Trudeau, founder of the first successful American sanatorium and hero of the antituberculosis movement: "To cure sometimes, to relieve often, to comfort always." Ideals, however, seldom reflect the world precisely. Trudeau's own sanatorium tried to avoid the intractably ill. "I realized at once," Trudeau reported in his autobiography, "that if I was to try to obtain curative results I must confine the admission of patients to incipient and favorable cases as much as possible, and refuse to take the acute and far-advanced ones."[34]

Nurses have long provided much of the care of the chronically ill and dying. Some observers, calling attention to a cure/care dichotomy in the health-care system, have placed physicians on the cure side, nurses on the care side. Since the time of Florence Nightingale, nursing leaders have frequently expressed the special contributions that nurses can make and the somewhat different values that guide their work.[35]

Any significant change in the goals, methods, or leadership of the health-care system, however, would require some major changes in funding and a persuasive campaign to alter public opinion and public values. The antituberculosis movement could well provide a prototype for such a campaign. Starting in the 1890s, physicians and social leaders joined forces to arouse and educate the public and to persuade it to build a system to care for consumptive patients. Campaigners enlisted all major opinion-making institutions—the schools, the churches, and the press—to disseminate their propaganda. Largely as a result of their efforts, a new and expensive system of care was established, and growing numbers of the tuberculous sick left their homes and entered the institutions. The costs of sickness shifted in part from individuals to government. Although few campaigners probably intended it, the welfare system expanded substantially under the guise of medical care. While the sick hoped that institutional treatment would help them recover, they also used the system to get food, shelter, and personal attention. The institutions met the needs of these dependents, albeit partially and often reluctantly.

In retrospect, the optimistic concepts of tuberculosis as a communicable, preventable, and curable disease helped to build and shape the system of care but did not accurately characterize it. There is no convincing evidence that the available treatments had curative effects. The goal of prevention was frequently compromised. Physicians often discharged still infectious patients, and men and women with communicable disease left the institutions against advice. Instead of a system that cured and prevented disease, society had built one that met some needs of sick and dependent people, spared families some of the burdens of care at home, and reduced the public's fear of infection, if not the actual threat. These unanticipated results grew out of political, social, and economic transactions in which medical understanding of tuberculosis played only a subordinate part.

NOTES

This chapter is derived from a work in progress, tentatively titled *Bargaining for Life: A Social History of Tuberculosis, 1876–1938* (Philadelphia: University of Pennsylvania Press, in press).

1. Quoted in Frank A. Craig, *Early Days at Phipps* (Philadelphia: Henry Phipps Institute for the Study and Prevention of Tuberculosis, University of Pennsylvania, 1952), p. 41.

2. During this period Pennsylvania was a bit atypical in its generous support of voluntary hospitals as well as in its large state system of tuberculosis sanatoriums. See Rosemary Stevens, "'A Poor Sort of Memory': Voluntary Hospitals and Government before the Depression," *Milbank Memorial Fund Quarterly Health and Society* 60 (Fall 1982):552–565.

3. *Annual Message of the Mayor of Philadelphia* (1875):664–689, 714.

4. "The Philadelphia Protestant Episcopal City Mission. Plans and Appeal in Behalf of City Missions, by the Bishop of the Diocese," (1870):4–6, and *Annual Report of the Philadelphia Protestant Episcopal City Mission (PPECM)* 1 (1871) 6 (1876), 7 (1877). Archives, Episcopal Community Services, Philadelphia.

5. *Annual Report, PPECM* 11 (1881):14.

6. *Annual Report, PPECM* 9 (1879):28.

7. *Annual Report, PPECM* 11 (1881), 12 (1882), 16 (1886), 17 (1887).

8. *Annual Report, PPECM* (1881–1891) and, for the clerical resistance, 1 (1883):30–32 and 20 (1890):IV.

9. *Annual Report, PPECM* (1877–1890).

10. *Annual Report, PPECM* (1881–1900).

11. Josie C. Collom to Lawrence F. Flick, February 25, 1900, Lawrence F. Flick Papers, Historical Collections of the Library, College of Physicians of Philadelphia.

12. J. Solis Cohen to Lawrence F. Flick, May 19, 1900, Flick Papers.

13. "Henry Phipps Institute," *New York Times*, January 10, 1903, and Lawrence F. Flick, "The Treatment and Control of the Tuberculous Patient in His Home," *American Medicine* 8 (July 30, 1904):187.

14. Charles E. Rosenberg, *The Care of Strangers: The Rise of America's Hospital System*. (New York: Basic Books, 1987), pp. 166–179; Craig, *Early Days;* and Flick Papers.

15. Craig, *Early Days*, 12, 57.

16. Ibid., pp. 57–58.

17. Lawrence F. Flick, "The Hospital and the Dispensary in the Warfare against Tuberculosis," *American Medicine* 9 (May 20, 1905):825.

18. Frank Heitler to Lawrence F. Flick, July 29, 1903, Flick Papers, and Ella M. E. Flick, *Beloved Crusader: Lawrence F. Flick, Physician* (Philadelphia: Dorrance, 1944), p. 196.

19. Joseph Walsh, "The Advantage of Tuberculosis Nursing for Tuberculous Patients," *Journal of the Outdoor Life* 20 (1923):407–408. The attrition rate is calculated from lists in the *Annual Report of the Free Hospital for Poor Consumptives [FHPC]* 10 (1908):75, 11 (1909):64, and 18 (1916):32–33.

20. *Annual Reports, FHPC* (1899–1902), and Lawrence F. Flick, "The Control of Tuberculosis," *Annual Report of the Pennsylvania Board of Health and Vital Statistics* 12 (1896):642.

21. *Annual Reports, FHPC* (1904–1910), and Flick, Letters. For the English model of patients' work, see Linda Bryder, *Below the Magic Mountain: A Social History of Tuberculosis in Twentieth-Century Britain* (Oxford: Clarendon Press, 1988), pp. 54–69.

22. Lawrence F. Flick, "Work for Patients as an Economic Factor," *Transactions of the National Association for the Study and Prevention of Tuberculosis* 5 (1909):181–184, and *Annual Report, FHPC* (1904):1.

23. George W. Norris, "The Diagnosis of Incipient Pulmonary Tuberculosis," *Medical News* 85 (September 17, 1904):542–544.

24. V. D. Charber to Lawrence F. Flick, August 11, 1904, Flick Papers.

25. These generalizations summarize many scores of letters in the Flick papers. See also *Annual Reports, FHPC* (1903–1910).

26. Benjamin Lee, "State Provision for the Treatment of the Consumptive Poor," *Journal of the American Medical Association* 35 (October 20, 1900):989–990; *Proceedings of the Philadelphia County Medical Society* 21 (January 1900):1–32; and *Annual Report of the Commissioner of Health of the Commonwealth of Pennsylvania* [CHCP] 3 (1908):18–19.

27. *Annual Report, CHCP* (1908–1916), and Albert P. Francine, "The Development of the Tuberculosis Campaign in Pennsylvania, with a Discussion of Its Principles," *Pennsylvania Medical Journal* 18 (November 1914):141, 144.

28. J. D. McClean, "The Department's Future Tuberculosis Campaign," *Pennsylvania Health Bulletin* No. 104 (February 1920):117–121, and Albert P. Francine, "Local Responsibility for Community Welfare with Special Reference to Tuberculosis," *Pennsylvania Medical Journal* 25 (April 1922):469–471.

29. Arnold C. Klebs, *Tuberculosis* (New York: D. Appleton, 1909), pp. 105–112.

30. "Hamburg Highlights," *Spunk* 14 (September 1922):44.

31. *Pennsylvania Department of Health Yearbook 1929*, pp. 154, 163.

32. W. G. Turnbull, "County Tuberculosis Needs," *Listening Post* 4 (April 1929):6.

33. Lewis Thomas, *The Lives of a Cell: Notes of a Biology Watcher* (New York: Viking Press, 1974), pp. 31–36.

34. Maurice B. Strauss, ed., *Familiar Medical Quotations* (Boston: Little Brown, 1968), p. 410a, and Edward L. Trudeau, *An Autobiography* (Philadelphia: Lea and Febiger, 1916), p. 243.

35. Hans O. Mauksch, "The Organizational Context of Nursing Practice," in *The Nursing Profession: Five Sociological Essays,* ed. F. Davis (New York: Wiley, 1966), pp. 109–137, and Renée C. Fox, Linda H. Aiken, and Carla M. Messikomer, "The Culture of Caring: AIDS and the Nursing Profession," *Milbank Quarterly* 68, suppl. 2 (1990):226–256.

12 STORIES OF EPILEPSY, 1880–1930

ELLEN DWYER

Epilepsy has had a long, complex, and illuminating history. Termed the "falling sickness" and "sacred disease" in antiquity, epilepsy's characteristic seizures have always been surrounded with an aura of specialness—even of the supernatural. Its symptoms could not well be ignored, or easily construed as diffuse and idiosyncratic. Even though modern neurologists may assume a diversity of causes for such alarming seizures, physicians and people in general have for thousands of years agreed in seeing them as constituting a discrete disease entity. The falling sickness has been a "disease" since at least classical antiquity. The Hippocratic text "On the Sacred Disease" (often considered the first monograph on a single disease) begins with a much-quoted dismissal of the widespread belief that epilepsy was a supernatural event; it was, the Hippocratic author argued, like all other ills, a natural outcome of natural processes.

Epilepsy's characteristic seizures provided it with a distinct social profile; in constructing responses to this symptom complex, societies have necessarily incorporated their time- and culture-specific reactions to such a frightening, biologically determined pathology. As has been suggested, and as common sense underlines, the history of epilepsy illustrates how the biological character of a disease can constrain the variety of social responses. It also illustrates how moral and physical elements are inextricably related in the process by which society frames disease. Epileptics have always been seen as having peculiar psychological and emotional qualities.

Epilepsy has been explained as long as it has been perceived. Explanations have varied from generation to generation and place to place, but have always been provided by medical authority. As Ellen Dwyer shows, for example, late-nineteenth-century physicians used such building blocks as evolutionism (in the inverted form of devolution) and hereditarianism to explain this condition. But as she shows with equal clarity, the total social picture of late-nineteenth- and early-twentieth-century epilepsy was far more complicated; lay people and

physicians did not necessarily agree, and the aggressive new specialty of neurology did not necessarily speak with the same interests
and perceptions as general practitioners.

As in the case of tuberculosis, another chronic disease implying
long-term care and treatment, late-nineteenth-century America created institutions to deal with epilepsy. Dwyer underlines how such
institutions—in this instance New York's Craig Colony—mediated
between the intellectual and professional world of medicine and the
lay world of patients and their families. As the colony's surviving correspondence makes clear, the arrangement was hardly symmetrical.
Society's response to epilepsy was neither unified nor coherent, but
an aggregate of social perceptions, family assumptions, and professional needs and explanatory schemes. Like the grain of sand in
which the universe is reproduced in microcosm, the history of
one institution's routine can exemplify the fine texture of such
relationships.

—C. E. R.

UNTIL THE 1930s, epilepsy remained a devastating and largely uncontrollable physical problem. As early as the 1870s, neurologists began to
understand something of the relationship between seizures and the
cerebral cortex of the brain, but most seizures remained in the "idiopathic" category. Their origins could be explained neither by the
medical profession nor by epileptics, and physicians were unable to
translate their limited new insights into efficacious therapies. In the
United States, as in western Europe, doctors responded to this impasse
in a variety of ways. They experimented with drugs, dosing and over-
dosing patients with everything from bromides, alone and in
combination with opium, to crotalin (an extract of snake venom, mercury, and iodides). They drilled holes in patients' heads, removed
women's ovaries, and applied blisters to necks as "counterirritants."
When such measures failed, many doctors turned to hereditarian explanations on the basis of which they called for the sterilization and
segregation of all epileptics. Particularly popular in the early twentieth
century were specialized state-funded colonies where epileptics, like
lepers, were expected to live their lives in tranquil isolation.

For American families during this period, the medical profession's

inability to control the symptoms of epilepsy was particularly discouraging. Parents struggled to meet the needs of brain-damaged
small children whose seizures were but one manifestation of multiple
neurological deficits. Adolescent epileptics found themselves cut off
from schools, churches, and peers. Adults experiencing their first seizures lost their jobs and were plunged with their families into penury
and social isolation. Many epileptics, especially adults, ended up in
poor houses or state psychiatric institutions where, along with syphili-
tics, they were segregated to prevent moral contamination of the rest of
the population. Families, too, experimented with drugs, especially patent medicines, and many greeted the development of state colonies for
epileptics with enthusiasm. Yet, although families and doctors embraced the same sorts of solutions, the ways in which they conceptualized epilepsy were very different.

Doctors shared the general social bias against epileptics, tending to
see them as moral as well as physical degenerates. To explain epilepsy,
they offered technical accounts of cerebral anemia, uremic poisoning,
and electrical discharges in the brain, along with highly charged
descriptions of flawed heredities and degeneration. Families, while
not immune to cultural stereotypes, feared epilepsy but seldom the
epileptic. The experiences of epileptics' families were more immediate than those of doctors. Families talked of problem pregnancies,
difficult births, accidents (especially involving the head), and bouts
with alcoholism. They sought the meaning of seizures in the events of
daily life, while doctors proposed general explanations of causality
and incidence based on aggregate clinical records, especially those
generated by specialized institutions for epileptics. Doctors and family
members thus created very different narratives to frame the epileptic's experience. Yet, despite their differences, both groups shared a
powerful sense, stated implicitly and explicitly, of the inadequacy of
their stories. Perhaps for that reason they told them over and over,
trying to make sense of the onset of a physical problem that remained
inexplicable.[1]

In the following pages, I look first for doctors' stories about epilepsy
in the American medical literature from the 1880s to the 1920s, and
then for families' stories in the clinical records of New York's Craig Colony, one of the first state-funded institutions for epileptics, which
opened in 1896. Writings on epilepsy appeared in a wide range of
sources, from *The Journal of Nervous and Mental Disease* and *Epilepsia*, to
general medical texts and manuals, to the publications of local and state
medical societies and the annual reports of institutions like Craig Colony. Families' views of epilepsy seldom appeared in print, but some can
be extracted from the Craig Colony records, particularly the parts of
patients' case histories filled in by family members and letters sent to
staff doctors. Unfortunately, the thoughts and feelings about seizures

of those afflicted by them has been lost. In contrast to mental patients, very few epileptics wrote first-hand accounts of their lives.

Late nineteenth- and early-twentieth-century doctors tended to view epilepsy with extraordinary hostility. For example, in an 1893 address before the International Medico-Legal Congress in Chicago, a California doctor proclaimed that "the epileptic is an individual of strange characteristics, and of a duality of personality which may quite outdo in viciousness and weakness, criminality and cunning, immorality and simulated innocence, the Jeckyl [sic] and Hyde creation of the novelist."[2] Such intense antipathy was not new. Oswei Temkin, in his study of attitudes toward "the falling sickness" from classical Greece and Rome to the 1880s, chronicles a long history of negative stereotypes and fear.[3] The expression of these attitudes in a particularly virulent form in the late nineteenth century was somewhat surprising, however, for during this period neurologists were increasingly interested in the way notions of cerebral localization helped to explain disorders such as epilepsy. The British neurologist John Hughlings Jackson was but the best known of many medical researchers then working toward a new neuroanatomical understanding of epilepsy.[4] Yet, instead of eradicating prejudice, this research, at least initially, helped to create a new justification for fear, built on allegedly objective, scientific data rather than superstition. Thus, scientific positivism gave new force to ancient stereotypes about the falling sickness.

How and why did this happen? The story is easier to tell than to explain. One clue can be found in the insistence of late-nineteenth- and early-twentieth-century neurologists on the frustrating and elusive nature of epilepsy. As one specialist observed, "The symptoms embraced under this name are as discordant as the pathological conditions which produce them." Another noted that epilepsy embraces a variety of clinical entities with little agreement as to how to subdivide them. A number of medical historians suggest that concepts of morbid heredity and degeneracy enabled psychiatrists to "gain intellectual legitimacy through identification with the more fashionable biological sciences, and accomodate themselves to a general pessimism" in late-nineteenth-century Western thought; the same can be said of neurologists.[5]

Most, but not all, doctors distinguished broadly between organic (or symptomatic) and idiopathic (or functional) epilepsy, and some argued that only the second constituted "true epilepsy." The organic category included seizures caused by head injuries, strokes, brain lesions, alcoholism, birth trauma, autointoxication, eclampsia, syphilis, and an amorphous syndrome known as "reflex irritation." Unfortunately, neurologists found that organic causes, despite their large number, accounted for only a small percentage of seizures. By default, the rest were relegated to the frustratingly large category known as "idiopathic." As a result, many general physicians, bewildered by epilepsy's clinical variety, fell back on the old dictum "fits are fits" and thereby condemned their patients to custodial care.[6]

According to Andrew Scull, in such situations of "therapeutic impotence and aeteological ignorance" physicians become more willing to turn to what he calls "desperate remedies." The medical literature on epilepsy supports his argument. Particularly during the late nineteenth century, doctors' ways of treating epilepsy were as diverse as their definitions of the disorder. In addition to dietetic and hygenic measures, probably the most popular strategy was the prescription of bromides. This had a somewhat curious origin. In 1857, the English physician Sir Charles Locock announced that, after reading that a German physician had produced temporary impotence in men by administering bromide of potassium, he had successfully used the drug to cure "hysterical epilepsy" in several young women in his care. Although Locock was primarily interested in the effect of bromide on "uterine epilepsy," other British doctors used it more widely. Bromides quickly became popular and, particularly in the late nineteenth century, were prescribed in high doses, despite serious side effects. Better to have abscesses and a weakened body than epilepsy, proclaimed the New York neurologist William Hammond. Doctors continued to experiment with a wide range of other drugs, from crotalin and solanum coralinense (horse-nettle berries) to borate of soda and sulfonal, but none proved particularly efficacious.[7]

Because drugs more often masked than cured the symptoms of epilepsy, doctors also experimented with surgical procedures. During the 1880s trephining was especially common. It was an old remedy, once used to release evil spirits or morbid secretions from the brain, that enjoyed a new popularity during this period. The term was applied to various procedures, from a simple drilling into the brain to relieve cranial pressure to the removal of large sections of bone from the top of the head. Trephining was felt to be particularly appropriate for epilepsy caused by head trauma, but was seldom used for the idiopathic kind. Because improvement was rare, whatever the cause of seizures, and many patients died or deteriorated, trephining of epileptics enjoyed only a brief vogue.[8]

Supporters of the reflex irritation theory argued that neurologists and general practitioners should not dismiss most seizures as idiopathic but should look for an exciting cause. They would then have a much firmer basis for determining appropriate treatment. For example, if gastrointestinal disturbances precipitated seizures, the diet should restrict meat and sugar and consist mainly of carbohydrates. If the cause seemed to be glandular malfunctioning, doctors should consider galvanizing the thyroid. Correction of vision problems sometimes also cured seizures. Counterirritants, including cauterization and blistering, were occasionally prescribed. A neurologist who attributed the seizures of many young boys to masturbatory habits induced by strictures in the urethra claimed that he could cure the seizures and stop the masturbation by introducing a mild galvanic current through a special urethral electrode he had developed, and urged others to follow his example.[9]

In a similar fashion, doctors who found anomalies in the reproductive organs or sexual habits of female epileptics occasionally resorted to corrective surgery. They were most likely to operate on adolescent girls whose seizures seemed to worsen just before or during menstruation. But, like trephining, ovariotomies produced few, if any, long-term cures and provoked much criticism. In a cautionary tale published in the *Alienist and Neurologist,* the New York neurologist William Hammond described a female epileptic whose ovaries had been removed in an unsuccessful effort to control her seizures. She later asked a gynecologist to examine her uterus, which she had been told also was diseased. Ignorant of her earlier surgery, he reported that her ovaries were much inflamed and recommended their removal. According to Hammond, the woman then showed the doctor her ovaries, suspended in a preservative liquid and "no longer capable of being charged with causing epilepsy or being guilty of other heinous offenses."[10]

The debate over surgical therapies was often more serious and heated than the humorous tone of Hammond's story suggests. When Maryland doctor J. Taber Johnson discussed at a medical conference in the 1880s the oophorectomy he had performed on an overmedicated nineteen-year-old, he characterized her case as "menstrual or hystero-epilepsy." He was then questioned closely. Did he know what a "normal" ovarian cyst looked like, one colleague inquired. Another noted that men were not castrated unless a serious disease of the testicles had been found, and surgeons never attributed nervous phenomena to the testicles. On the defensive, Dr. Johnson responded that he had performed the surgery only as a last resort for a desperate patient.[11] Yet, as late as 1931 Craig Colony continued to order occasional ovariotomies as a way of controlling seizures in adolescent females.[12] Johnson's comment (as well as the sheer volume of articles on trephining) suggests that, despite the intense late-nineteenth-century debate over the appropriateness of drastic surgical remedies, some frustrated doctors still considered them a legitimate therapeutic option well into the twentieth century.

Even when doctors successfully removed what they had identified as the exciting cause of patients' epilepsy, especially in cases involving brain lesions, seizures sometimes continued. They attributed such recalcitrance to the power of what they called the "epileptic habit." This notion shifted responsibility for therapeutic failures away from the doctors and to a fundamental constitutional quality of their patients. Those who had once suffered seizures were always susceptible to a recurrence. If patients maintained well-balanced digestive systems, good personal habits, a regular occupation, and strong self-discipline, they might be able to ward off subsequent seizures. Thus, neurologists made clear that the "epileptic habit" was a "bad habit" that might be controlled by self-discipline.[13]

From the notion of an epileptic habit, neurologists moved easily to one of an epileptic (or "epileptoid") personality. A text for general practitioners described epileptics as "always balancing on the verge, ready to topple over into a convulsion at the slightest provocation." The book went on to warn that "epileptics are self-willed, obstinate as a rule, easily angered, and especially need the controlling influence of a strong mind and a strong hand. . . . Whether young or old, as a rule, they require a master." At the 1916 annual meeting of the American Medical Association, doctors from the Craig Colony for Epileptics mounted a display that depicted epilepsy as made up of two "essential factors": the convulsion and the epileptic personality. The latter, they proclaimed, "exists from infancy independent of seizures" and encompasses a number of character defects, including exaggerated selfishness, "primitive" infantile behavior, and "deficient social interests."[14]

The view that epileptics were weak of will as well as of body continued to appear in medical writings for many years. Epileptics generally were considered to be unpleasant: careless in speech, prone to exaggeration, illogical in conversation, irritable in controversy, and erratic in ideas. The result of such personality defects, doctors argued, was social isolation (although they might easily have reversed the order of cause and effect). A grim picture emerged: "The confidant of no one, in close companionship with his own perverted imaginings, no wonder in brooding over his fate he sees the world all against him." In 1916 New York neurologist L. Pierce Clark added a Freudian twist, describing seizures as withdrawals from the pressures of daily life. He elaborated these views in an extraordinary flood of articles and even in Osler's *Modern Medicine,* where he argued that epilepsy is due not to a propensity to seizures but to an underlying weakness of character. "The epileptic character," he asserted, "is embraced in an undue development of the ego and an arrest in emotional development resulting in hypersensitiveness and emotional deficiency. . . . Under stress, such individuals respond by losing consciousness and convulsing in a fashion that constitutes a violent regression to a lower level of adaptation."[15]

As a consequence of their concern about the epileptic habit and the epileptic personality, doctors began to broaden their definition of the sorts of seizures requiring medical attention. Because only early treatment could prevent establishment of the "habit," they urged general practitioners to regard all possible symptoms with suspicion, to treat infantile convulsions, night terrors, vertigo, dizziness, and even faintness as potential epilepsy.[16] Although the public and general practitioners often equated epilepsy with grand mal seizures, neurologists were more interested in petit mal seizures, Jacksonian or focal seizures, and a somewhat amorphous phenomenon known variously as "psycho-epilepsy," "masked epilepsy," "cerebral epilepsy," or "non-convulsive epilepsy." While the most frightening to observe, grand mal seizures were considered the most curable form. It was the more subtle, hard-to-detect forms, often unaccompanied by motor disturbances, neurologists declared, that were the most frustrating to treat and the most dangerous. Because doctors felt that the loss of physical control during a seizure led to a loss of spiritual control, they viewed seizures that were difficult to detect as particularly ominous. The existence of such "hidden" epilepsy too often surfaced only in the wake of some horrible crime. Yet, while "masked epilepsy" was the most terrifying form of the disease, all epileptics were to be regarded with alarm. Most people knew, argued a Ohio physician, that "moral liberty and responsibility were entirely suspended" during a seizure, but few realized that, for an unpredictable period of time before and after a seizure, the patient's mind was "so sensitive, irritable, and liable to hallucinations as to make him even more dangerous to others than during his acute convulsions."[17]

Such fear of epileptics was not new, but late-nineteenth-century neurologists offered a novel scientific justification of it. They not only described in detail inter- and postictal outbreaks of violence but attributed them to the temporary paralysis of the highest centers of the brain. The influential British neurologist John Hughlings Jackson, for example, drawing on his rich clinical experience, described instances of what he called epileptic and postepileptic automatism, unconscious actions for which the individual involved was not responsible. They ranged from a patient who suddenly began to chew a packet of prescription papers, to the man who took a fellow worker's coat while in a postseizure state of temporary mental confusion, to a woman who seriously mutilated herself with a knife in a maniacal frenzy. These cases helped support Jackson's argument that epilepsy was an example of evolutionary dissolution.[18]

While Jackson's portrayal was complex and nuanced, many who were influenced by it translated his "dissolution" into a somewhat crude scenario of corruption within which epileptics became menacing degenerates. Even while conceding the accomplishments of such noted epileptics as Dostoevski and Napoleon, degenerationists felt that, for the most part, epilepsy was synonymous with imbecility, madness, and unpredictable violence. Such ideas were to have a powerful impact on political and social policy. Typically, one doctor claimed that even those who developed the ailments as adults subsequently suffered "the progressive deterioration of all faculties, mental, moral and physical" characteristic of epilepsy. As their intellects dulled, so did epileptics' physical features. They acquired a distinctive physiognomy: a "heavy, lost look," a "bloated and livid appearance," trembling of the limbs, a slow respiration.[19] Those advocating such a view of epileptics paid little attention to the minority of their colleagues who pointed out that many of the so-called stigmata of epilepsy could be attributed to bromide poisoning. High dosages of bromides had known side-effects easily misinterpreted as signs of degeneration, including the physical (acnelike rash, bad breath, slowed heartbeat and breathing, violent headaches, bronchial catarrh, staggering gait, and lusterless eyes) and the mental (intellectual apathy, irritability, memory failure, and a maniacal form of behavior called "bromania").[20] Bromism could and did transform even the liveliest patients into unattractive, withdrawn dements.

Epilepsy was even more commonly equated with madness and criminality than with imbecility (indeed, the three were often linked together). "Many extraordinary misdeeds, ascribed to crime," claimed M. Echeverria, "are originated by epilepsy." The law, he protested, paid too little attention to the frequency with which homicide and other violent acts resulted from the "perverted affection, or well-marked intellectual impairment, which irresistably compels the epileptic to obey his morbid impulses, and destroys the judgment." Another neurologist attributed "the most brutal and senseless crimes" to what he called "the epileptic alienation." Because of the likelihood that epileptics would commit criminal acts, all family doctors were urged to familiarize themselves with "this neurosis" and to search for and carefully observe its earliest mental manifestations. If they could not thus prevent misdeeds, at least they would be well prepared to testify in criminal or probate trials involving epileptics.[21] Far too often, some experts contended, the physical side of epilepsy had a psychological component variously called mental epilepsy, cerebral mania, or psychic epilepsy. Its characteristics included "aggressive violence without cause, an irresistible impulse, leading to suicide and murder; delusions and hallucinations; reproduction of identical insane ideas in each seizure; loss of consciousness of the cerebral regions involved and gradually increasing debility of the mind . . . eventuating in more or less complete dementia, and possibly in death."[22]

Such claims of epileptics' predeliction for violent behavior were lent credence by highly charged descriptions of incidents of homicidal epilepsy in medical and legal journals, as well as in the popular press. A few dramatic cases received a great deal of attention, one of the most notorious being that of Richard Barber. Barber, a twenty-seven-year-old farm laborer, brutally assaulted an elderly couple who had befriended him. The attack began unexpectedly during an evening visit. After chatting several hours with his host, Richard Mason, Barber suddenly began to beat him with a heavy stick. When the elderly man fell and managed to crawl under a piece of furniture, Barber went into the bedroom and beat Mason's wife so ferociously that she died. He then set the house on fire and walked away. Shortly thereafter he was picked up by a neighbor and arrested.[23]

At Barber's trial, his lawyers introduced his childhood medical history of violent seizures in support of an insanity plea. They noted that, although he had not suffered grand mal seizures as an adult, there was some evidence of nocturnal seizures. Because Barber lacked a discernible motive for the crime, several psychiatrists testified that his acts must be "the unconscious and uncontrollable result of epileptic mania." While they felt Barber should be confined in a psychiatric rather a penal institution, the expert witnesses clearly considered him (and, by implication, most epileptics) highly dangerous. Many of their professional colleagues agreed. Calling for the establishment of a specialized state institution for epileptics, the *American Journal of Insanity* warned that "the history of epilepsy is the history of violence, of crime, of homicide." A colony, it said, would provide not only hospital care and schooling for unfortunate epileptics but also a home for those with "no place in ordinary social life, their seizures depriving them of the privileges of companionship, family, congenial occupation, and life."[24]

John Ordronaux, a prominent late-nineteenth-century jurist, made similar points in his discussion of the case of Jacob Stauderman, a German immigrant who suffered from both epilepsy and imbecility. After months of being ridiculed by his friends, Stauderman killed a young woman with whom he had fallen in love, seemingly without realizing what he was doing. Such motiveless crimes, Ordronaux exclaimed, aroused concern about the extent to which states of latent or undiscovered epilepsy might "mold individual character and shape its destiny." They demonstrated the fragility of epileptics' "moral varnish," which, once worn through, left them "pregnant illustrations for the doctrines of Mr. Darwin, since the least shimmer of passion, revives in them the unexpired embers of an aboriginal ferocity." Ordronaux then progressed smoothly from themes of physical degeneration and moral deterioration to nativist and eugenicist rhetoric. Staudermann never should have been permitted to enter the United States, he argued, "America is becoming, in fact, a sort of international dust-bin, into which the old civilizations sweep their human refuse."[25]

Such sentiments helped to win support for early twentieth-century immigration laws that restricted the admission of epileptics into the United States until 1965. They also fueled eugenicist campaigns to control the sexuality and reproductive capacity of a wide range of so-called defectives, including epileptics. A number of neurologists applauded the 1895 Connecticut law making it a crime for epileptics, imbeciles, and the feebleminded to marry or live together as man and wife if the woman was under 45. Also forbidden was carnal knowledge of such females, whether inside or outside of marriage. The penalty for violation was at least three years' imprisonment.[26] Despite such efforts, a contributor to *Epilepsia* complained, the number of defectives continued to increase. In an age when nursery stock was protected and the reproduction of animals regulated so as to produce the strongest strains, why should the state not "regulate the forces tending to the proper propagation of the human race?" Many agreed and went even further, advocating the compulsory sterilization and segregation of "gross defectives" such as epileptics.[27]

In many ways, the response of the late-nineteenth- and early-twentieth-century biomedical community to epilepsy closely resembled its condemnation of the feebleminded, homosexuals, criminals, and others considered "defective." The culmination of that view in the formation of a eugenics movement has been ably told by several historians, most notably in the United States by Mark Haller and Charles Rosenberg,[28] but it merits retelling from the special perspective of epilepsy for several reasons. While attitudes toward socially deviant behavior like homosexuality and criminality certainly became medicalized during this period, epilepsy had always been seen as a medical as well as a social problem. Why and how did late-nineteenth- and early-twentieth-century neurologists, interested in cerebral localization, continue to think of epilepsy in terms of moral contagion? What made them so willing to try drastic surgical and drug therapies that more often weakened than cured? I can offer only preliminary answers. Even though doctors' knowledge of the brain and its functions was expanding, it did not translate into improved therapy. Epileptic seizures, devastating to experience and to witness, were seldom limited to a single episode or even to a period of months. They went on and on, often leaving the patient mentally enfeebled and physically debilitated. Again and again doctors endorsed what seemed to be promising therapies, only to find that their impact was shortlived; after some weeks or months, the seizures returned. By the late nineteenth century, discouraged and frustrated, the medical community began to think in terms of social solutions. Epileptics would be best off in isolated colonies, out of sight of the broader community, where they could receive the best possible care and treatment.[29]

Craig Colony for Epileptics was one of the first such institutions, opening in New York in 1896. Its establishment owed much to the lobbying efforts of Frederick Peterson, who had become interested in the epileptics under his charge while an assistant physician at New York's Hudson River State Hospital for the Insane. In 1886 Peterson visited the Germany colony for epileptics at Bielefeld and, on his return, wrote an influential description of its work for the New York *Medical Record.* Convinced of the necessity of a state institution for epileptics, he lobbied throughout New York State and eventually, with the help of the State Charities Aid Association, won legislative support. The founders chose an ideal site in southwestern New York, an isolated 1800-acre former Shaker colony in the fertile Genesee Valley. One of its many advantages was a deep gorge that divided the property and thus made it easier to separate patients by sex. According to its enabling legislation, Craig Colony was to offer "the humane, curative, scientific, and economical care and treatment of epilepsy."[30]

According to a survey conducted by the New York State Board of Charities, many epileptics were being cared for in poorhouses and state lunatic asylums, not because they had psychiatric problems but because they could not support themselves.[31] Craig Colony was to give first priority to such individuals, who constituted the majority of its early patients. It accepted only those without serious psychological problems, however, because it was not able to provide the intensive supervision required to deal with outbursts of psychotic violence or attempted suicide. Difficult-to-control patients were usually transferred to state hospitals. The decision to establish a colony rather than a large institution was made because most of the colonists (as they were called) were expected to spend their lives at Craig Colony. They were to live in cottages and learn to support themselves through trade, industry, and agriculture.

Peterson was initially optimistic about the benefits of colony life. While approximately 90 percent of epileptics differed little from the general population except during their short seizures, social prejudice isolated them and led to their exclusion from schools and churches. It was not surprising, Peterson observed, that, under such conditions, men and women who might otherwise have developed great talents grew up feebleminded, ignorant, and "an easy prey to all the degenerative tendencies which are prone to show themselves when a mind is left . . . unguided and uncared for."[32] Despite his sensitivity to the barriers faced by epileptics, Peterson, who became the first president of the Craig Colony Board of Managers, agreed that they were best off spending their lives in isolation, which thus would prevent the spread of epilepsy to subsequent generations.

Like most physicians of this period, the Craig Colony medical staff strongly supported the notions of hereditary degeneration and the eugenicist legislation that developed out of them. Thus, every year until 1917 the Colony published aggregate data about patients' physical anomalies—referred to by its first superintendent, William P. Spratling, as "the so-called stigmata of degeneration." It also compiled charts showing the frequency with which other degenerative diseases, such as insanity, alcoholism, and sometimes even rheumatism, ran in patients' families, and cooperated with several state-funded generational studies involving Craig Colony patients. In 1911 William T. Shanahan, a Craig Colony physician, called for sterilization and segregation of all epileptics and the legally mandated reporting of all such "defectives" to a central state office. Thirteen years later, as superintendent, Shanahan asked the legislature for money to build new staff housing far away from the patients, so that employees' children would not be "contaminated" by their negative moral influence.[33] Yet the staff was not uniformly unsympathetic to patients, A female teacher, for example, movingly described her students' (especially those from large cities) ignorance of the most basic sorts of information as a result of their having been closely confined for many years.[34]

The Craig Colony staff performed many studies of how families passed epilepsy and other defects from one generation to the next but were little interested in the stories of individual families. They also admitted (at least in some of their writings) that the population of institutions like theirs was not representative of epileptics in general because it included disproportionate numbers of severely brain-damaged children. As one Craig Colony superintendent noted, "many who have the disease fail to recognize it; others have it in so inconspicuous a form that they feel no alarm and seek no treatment for it; others still have attacks only at night and nocturnal attacks may occur for years before the disease is finally recognized."[35] Nonetheless, Craig Colony patient records (especially family case histories collected at the time of commitment and records of subsequent interactions of staff and patients with families) offer a rare glimpse of how the experience of epilepsy affected both patients and those who cared for them. Especially before the development of effective drugs in the late 1930s, the adequate care of epileptics by families and institutions was not easy. And the burden was increased by medical and legal writings that reinforced negative stereotypes and intensified stigmatization. Despite this, families who made the painful decision to institutionalize a child, parent, or spouse often did so in search of a cure, not out of a desire to hide family shame. Many families retained strong ties to institutionalized members and, in some cases, brought them home again against medical advice when they did not seem to be benefiting from institutional care.

The Craig Colony records are a particularly useful source of information on families' views of epilepsy. From the late nineteenth century on, its doctors collected information about patients' heredity, earliest experiences, family histories, illnesses, and seizures, in addition to their age, sex, race, nativity, education, and occupation. Although the heads of nineteenth-century psychiatric institutions had questioned the reliability of information provided by patients' families, Craig Colony doctors initially were less skeptical. At the top of their earliest forms, they proclaimed that "Facts are vital; details valuable." If the "facts" could not be obtained, they added without hesitation, "give hearsay evidence," although the records did not distinguish between the two. Particularly crucial, they felt, were details about early life, but, since "knowledge of a thing apparently of no importance may be of vital importance to a patient," no facts were too trivial to include. Concerned to gather information about heredity, Superintendent William Spratling claimed to have formulated his questions so as to elicit specific information on points that "the patient or his friends under direct questioning might evade or deny." Yet, as his successor, William Shanahan, candidly admitted in 1917, the case histories were much better as sources for "facts upon facts" than for diagnosis.[36]

The heart of the bulky Craig Colony files are the lengthy question-naires about "family history," "history of the patient," and "history of the epilepsy," filled out by patients, families, friends, and doctors at the time of admission. How accurate were the entries? In 1911, Superin-tendent William Shanahan warned, "The difficulty in obtaining correct information from relatives or friends in regard to alcoholism, syphilis, insanity, and similar conditions is marked and oftentimes well nigh impossible. They feel that some of these matters should be con-cealed from everyone, even the physician. In many cases, there is . . . a complete or practically complete ignorance of facts concerning the ex-act state of health of living members and the cause of death of those not living. [Even] the family doctor . . . is sometimes afraid to offend the relatives by asking for such information." Shanahan was equally skepti-cal of technical information such as descriptions of the epileptic aura when provided by "mental defectives."[37]

For a variety of reasons, in additional to lay naivete, parents and doctors differed in their analyses of patients' case histories. Corre-spondence between families and Colony doctors suggests that even well-informed parents were reluctant to accept the implications of their children's seizures. As the Craig Colony pathologist scathingly noted in 1910, family members' assessment of the mental grade of epileptics was useless. "Bright" to them often means "feeble-minded" or "imbecile" to us, he scoffed. The same doctor who wrote eloquently of his colonists' search for education warned of the futility of trying "to force a $5000 education into a $500.00 brain." "We recognize that the epileptic is largely a defective being even though many could acquire sufficient vocational skills as to support themselves."[38]

Despite the negative views of epilepsy articulated with such force in the general culture, these parents refused to abandon their hopes and expectations for epileptic children. For example, the schoolteacher mother of a twelve-year-old boy described his severe convulsions in great detail but although he attended school for only a short time in first grade, his mother listed his occupation as "student," and sent him to Craig Colony in the hope that he would finally get an education. When the institution's doctors decided that his low intelligence made him uneducable, she remonstrated angrily, but to no effect. Less than two months after his admission, she took the child home "on parole" and never returned him.[39]

Other parents similarly contested the institution's assessment of their children, although with less force. Most frequent were disagreements over the causes and implications of seizures in the children's case histo-ries, almost always the most detailed of the records. For example, a father attributed the severe seizures of his twelve-year-old to a fall down a flight of eighteen stairs because the first convulsion started six hours later. Craig Colony doctors decided that the fall was incidental to cerebrospinal meningitis, for which the patient was treated at roughly the same age. But both father and doctors agreed that, whatever the cause of the seizures, the child required custodial care.[40] More wrenching is a father's account of his daughter's epilepsy blaming himself for her seizures. A slow developer, the child still was not toilet trained at six. Asked about the onset of seizures at six, the father reported that Abby was "allways a nervous child." Frustrated by her bedwetting, he told her one night he "would leave her and get another abbie [sic]." Her first convulsions came the next night. When, later in the case history form, the father came to the question "cause of first attack," he repeated sadly, "Well . . . it was caused I think by my threat to leave her and get another abbie." By contrast, the Craig Colony doctors, upon learning that the child's mother was in a state psychiatric institution, attributed her epilepsy to heredity. A sad postscript is a newspaper story included in the child's files. Depressed by his inability to care for his child and out of work because of ill health, the father committed suicide in his boarding house shortly after the commitment. The child lived on at Craig Colony for another 47 years.[41]

Even when acutely aware of their children's limitations, parents often refused to abandon hope of a cure. For example, when a New York City couple decided to send their only daughter, a retarded ten-year-old, to Craig Colony, they did so on the advice of their local doctor. At admission, Colony physicians described Elizabeth as a drooling moron; the child neither responded to taste and smell tests nor answered any questions. Even though her retardation had preceded the onset of seizures by five years, her parents clearly hoped that institutional care might cure her. Despite having been told several times that Elizabeth's mental age was only 2 or 3, the mother continued to send letters to the doctors for many years, asking when her daughter might begin school.[42]

The puzzle of epilepsy's origins, so difficult for the medical profession to unravel, created even more anguish and confusion for parents. When a three-year-old boy was sent to Craig Colony in March 1935, Colony doctors under the category "cause of epilepsy" wrote "idiopathic" and "hereditary." (The child's father had had seizures from 22 to 35 years of age; a paternal uncle was also afflicted.) Whatever the cause of his seizures, the child was severely disabled and needed a great deal of care. After his commitment, the boy's mother flooded the institution with letters of inquiry. Again and again she asked about his health and speculated about the cause of his condition. Had she done something during pregnancy? While her husband had had fits, he had recovered. Once Bobby fell and hit his head. Could that be the reason? None of these possibilities satisfied her. Each was too ordinary to account for an event as terrifying as a grand mal seizure, for the illness that had turned her beautiful firstborn into a bedridden burden, about whose continued existence she felt much ambiguity. Her letters suggest that, no matter how inadequate the doctors found the label of "idiopathic epilepsy," it left families even more dissatisfied.[43]

A wide range of possible causes was also offered by the mother of a ten-year-old girl sent to Craig Colony in January 1924. Perhaps the epilepsy had resulted from an automobile accident six months before the first attack, the mother suggested. Maybe it had been inherited from her alcoholic father, who had tuberculosis. Before the child's birth, the mother had been distressed by her husband's drinking and recent leg amputation. At the age of five, the child fell off a veranda and lost consciousness for about one minute. Her second attack came one month later, since which time she had suffered almost monthly seizures. The mother's scattered list of possible explanations underscored her feelings of desperation and helplessness.[44]

Different issues appear in the case histories of the Colony's adolescent patients. From the perspective of the irritated medical staff, too many of these youngsters suffered primarily from delinquency and only secondarily from epilepsy. Teenagers with "fixed asocial tendencies" needed more restrictive environments, doctors asserted.[45] While they transferred the most difficult to schools for juvenile delinquents, they still had to restrain runaways and to control physical violence and sexual experimentation. In many cases, parents who needed the labor of their teenage children took them home after relatively short stays. Craig Colony aftercare records from the 1930s and 1940s show that some of those with mild seizures subsequently found employment but that most ended up reinstitutionalized, often in a state psychiatric institution.

Tensions between parents and their adolescent children appeared most obviously in parents' responses to checklists of personality traits near the end of case-history forms. The mother of one severely retarded twelve-year-old, who could neither speak nor care for himself, simply wrote "no" at the end of the list, but the parents of other teenagers stressed their children's selfishness, stubbornness, overanxiety, and resistance to advice. A widowed mother saw her fourteen-year-old as "quick-tempered like his father," whom he also resembled physically. Another youngster, described as resembling his recently deceased father, an epileptic alcoholic, was cross after seizures, showed little concern for others, and was quarrelsome, indifferent, and critical. Perhaps not surprisingly, a fourteen-year-old who annoyed his mother by wandering away from home did the same at Craig Colony. He was eventually confined to a locked ward.[46]

As sources for the personal accounts of adult patients about their epilepsy, the Craig Colony records are disappointingly thin. Few adults whose epilepsy dated back to childbirth could remember the age of onset. Others were too disoriented to offer many details about their seizures. Most had long since lost the family networks that usually buffered children and adolescents; their case histories make clear the devastating social disorganization and poverty that a highly stigmatized and disabling disease like epilepsy could create. This was particularly true for those whose epilepsy was a by-product of alcoholism, a group whose presence Colony doctors resented. Their case files were slim, containing few references to visits by family members or gifts from friends.

Despite their limitations and biases, the Craig Colony patient records illustrate the extent to which chronic diseases, like epilepsy, involve an entire family, not just the individual patient. In the case of a retarded patient, the last of seven children in a close-knit Italian family, who was sent to Craig Colony at the age of 12, family members wrote monthly for thirty years to inquire about his health and to thank the superintendent for "your kind attention to this matter." Only the handwriting changed, as first the mother took over for her husband, and then a sister took over for her mother. A medical crisis in 1943 interrupted the seemingly endless steam of identical letters with an exchange of telegrams. That March, some 27 years after their brother had been sent to Craig Colony, the patient's sisters and brother-in-law set out from Brooklyn to visit but were stopped by a snowstorm in Peekskill. Occasionally, when the sister who did the writing became ill, the flow of letters faltered; she always apologized. After the patient died in September 1946, the sister remembered to send Christmas cookies to all "the boys" on his ward.[47]

Family comments reflect a perspective on epilepsy far different from the one that prevailed in the medical and legal literature well into the 1930s, with its references to degeneracy, violence, imbecility, dementia, and mania. The two worlds obviously touched, if only in the institution and the clinic, but families' views seemed to have had little impact on the medical profession. In his book *Illness Narratives*, Arthur Kleinman maintains that patients and their families bring illness problems to a practitioner, who then reconfigures them as narrow technical issues, disease problems.[48] Glimpses of this process can be seen in the Craig Colony patient files. Families brought children, siblings, and spouses to the institution in search not just of treatments but of explanations adequate to explain the source of their pain and disappointment. When doctors responded with questions about exciting and predisposing causes, with talk of cranial irregularities, Simon-Binet intelligence tests, and idiopathic epilepsy, some families retreated in bewilderment, never to return. Others immediately began efforts to remove family members from the institution that could do so little for them. A third group used letters to continue their exchanges with the medical staff, to try (with little success) to bridge the gulf between the medical discourse and the realities of disease.

Eventually, by the late 1920s, the strongly negative attitude toward epilepsy that had dominated the medical, legal, and social welfare literature since the 1880s began to recede. With the advent of phenobarbital (in the early 1920s) and dilantin (in the late 1930s), which controlled seizures more effectively and with fewer negative side effects than did bromides, the number of articles on epilepsy in both specialized and general interest journals diminished, suggesting that the medical profession perceived the disease as less problematic, more manageable, than in the past. Even institutions like Craig Colony began to feel the impact of new drug therapies. As fewer families sought institutional care for members with relatively mild cases of epilepsy, the Colony closed first its school then its more challenging occupational therapy programs. Eventually, in 1967, Craig Colony for Epileptics was converted into the Craig Development Center.

Yet, negative stereotypes of epilepsy never completely died. They live into the present, although they no longer fit into a tightly woven medical-legal-social ideology. For example, in his sympathetic introduction in 1974, to the autobiography of a nonconvulsive epileptic, the popular writer and physician Walter C. Alvarez still portrays unpredictable violence as an important and dangerous characteristic of such individuals. And Carl Sagan, in his best-selling *Dragons of Eden* (1977), describes grand mal seizures as "temporarily regressing the victim back several hundreds of millions of years."[49] Such persistent scientific biases have their counterpart in popular culture; two sociologists recently remarked that many Americans continue to regard seizures as evidence of "a more fundamental, essential disreputability." As a result, families and the medical profession continue to be concerned with the effect of biological and social pressures on epileptics.[50]

Epilepsy thus remains a disease that is both physiological and, due to its social reception, psychological. It is not communicable and yet profoundly distressing, to both those suffering and those observing its ravages. The neurological literature no longer speaks of homicidal maniacs or epileptoid personalities; it is less concerned with issues of heredity and degeneracy than with subtle brain malfunctions. But it does continue to dwell more on aggregate patterns and technicalities than with what it means to have epilepsy.[51] For the full drama of epilepsy, we still must look to the stories of individuals and families as well as of doctors and try to reconstruct the important, if sometimes elusive, interactions between the two worlds.

NOTES

1. The narrative quality of families' accounts, whether looked at one by one or as a group, is easier to see than that of doctors' published essays. But the devices of story telling help to bring some coherence to what otherwise seems

ELLEN DWYER

an almost absurdly broad range of theories about epilepsy, many tinged with intense hostility, and often drastic prescribed therapies. Viewed as attempts to tell the "story" of epilepsy over and over, the doctors' willingness to entertain many different explanations and to experiment with almost any conceivable therapy begins to make sense. For some of the popular theories, see G. C. Bolton, "Researches on the Pathogenesis of Genuine Epilepsy," *Epilepsia* 5 (1914–1915):300–309. S. Weir Mitchell's frustration was widely shared; in "College of Physicians and the Public Health," *Journal of the American Medical Association* (hereafter referred to as *JAMA*) 58 (March 30, 1912):966, he wrote that "I must frankly admit that none of the remedial measures we apply for the general treatment of idiopathic epilepsy has any firm foundation on reason or trustworthy experimentation on the lower animals." Seventeen years later the sentiment reappeared in "Treatment of Epilepsy," *JAMA* 93(July 6, 1929):78, phrased in very similar terms. I do not distinguish here between the writings of neurologists, general physicians, and the medical staff of institutions like Craig Colony because I do not find major differences in their positions. Few reports of laboratory work appeared in American medical journals during this period.

    2. A. E. Osborne, "Responsibility of Epileptics," *Medico-Legal Journal* 11 (1893):210–220.

    3. Oswei Temkin, *The Falling Sickness: A History of Epilepsy from the Greeks to the Beginnings of Modern Neurology* (Baltimore: Johns Hopkins University Press, 1945, 1971).

    4. Anne Harrington, *Medicine, Mind, and the Double-Brain: A Study in Nineteenth-Century Thought* (Princeton: Princeton University Press, 1987), pp. 234, 270; Robert H. Wilkins and Irwin A. Brody, "Jacksonian Epilepsy," *Archives of Neurology* 22 (February 1970):183–188; James Taylor, ed., *Selected Writings of John Hughlings Jackson* (London: Hodder and Stoughton, 1931, 1932), 2 vols.; Temkin, pp. 303–388.

    5. For some of the many articles on the definition of epilepsy, see: "The Diagnosis of Epilepsy and Hystero-Epilepsy," *Journal of Nervous and Mental Disease* (hereafter referred to as *JNMD*) 3, new series, (January 1878):602–603; Reuben A. Vance, *On Syphilitic Epilepsy* (New York: F. W. Christern, 1871), p. 3; Frederick Peterson, "The Treatment of Epilepsy," *Boston Medical and Surgical Journal* 32 (August 1892):12–16; William C. Krauss, "Reflex Disturbances in the Causation of Epilepsy," *Transactions of the Medical Society of the State of New York* (1893):58–163; John Ferguson, "Some Remarks on Epilepsy," *Alienist and Neurologist* 4 (August 1893):235–262; Everett Flood, "Epilepsy as a Symptom," *Proceedings of the American Medico-Psychological Association* (1905):251–258; T. P. Staunton, "The Treatment of Epilepsy," *Medicine* 1 (1895):156–160; Abner Post, "Syphilitic Epilepsy," *Transactions of the National Association for the Study of Epilepsy and the Care and Treatment of Epileptics* (hereafter referred to as *NASECTE*) 3 (November 1905):156–166; Charles Emerson Ingberg, "A Case of Reflex Epilepsy," *Alienist and Neurologist* 27 (May 1906):170–188; Smith Ely Jeliffe, "A Contribution to the Pathogenesis of Some Epilepsies: A Preliminary Contribution," *Transactions of the American Neurological Association* 33 (1907):304–316; "Psychoasthenic Attacks Simulating Epilepsy," *Medical Review of Reviews* 13 (April 24/5, 1907):24; "Abstract: 'A Contribution to the Study of Epilepsy and Borderline Cases'," *Archives of Neurology and Psychiatry* 1 (March 1919):339–342; Beverly Tucker, "Consideration of the Classification of Recurrent Convulsions," *Transactions of the American Neurological Association* (1921):321–327; L.J.J. Musckens, "Note-taking in the Case of Epilepsy," *Epilep-*

sia 1 (1909–1910):387. For the views of historians of psychiatry, see Ian Dowbiggin, "Degeneration and Hereditarianism in French Mental Medicine 1840–1890: Psychiatric Theory as Ideological Adaptation," in W. Bynum, R. Porter, and M. Shepherd, eds., *The Anatomy of Madness* (London: Tavistock, 1985), vol. 2 pp. 188–232; L. S. Jacyna, "Somatic Theories of Mind and the Interests of Medicine in Britain, 1850–1879," *Medical History* 26 (1982):233–258.

6. A. L. Shaw, "The Story of Epilepsy for the Clinician," *Epilepsia* 5 (1914–1915):179, 182. H. Allen Starr, "Is Epilepsy a Functional Disease?" *JNMD* 31 (March 1904):104–112, 145–156; Staunton, pp. 156–160; Pepper, pp. 141–152; L. Pierce Clark, "A Digest of Recent Work on Epilepsy," *JNMD* 27 (July 1900):387–404; "The Pathology and Treatment of Epilepsy," *JAMA* 8 (March 19, 1887):321–322; William Aldren Turner, "Remarks upon the Outlook in Epilepsy," *Epilepsia* 5 (1914–1915):40–47.

7. Andrew Scull, "Desperate Remedies," *Psychological Medicine* 17 (1987):561–577. Articles on antiepileptic drugs include: Allan McLane Hamilton, "The Therapeutics of Epilepsy," *Chicago Medical Journal and Examiner* (December 1876):1–15; M. Benedikt, "The Nature and Treatment of Epilepsy," *JNMD* new series, 1, (January 1876):395–403; "The Treatment of Epilepsy," *American Journal of Insanity* (hereafter referred to as *AJI*) 49 (July 1892):8; "Sulfonal in Epilepsy," *AJI* 49 (July 1892):92; "Borax in Epilepsy," *AJI* 48 (January 1892):376–377; "Case of Epilepsy Cured by Antipyrine," *AJI* 48 (July 1889):24; "Amylene Hydrate in Epilepsy," *AJI* 48, no. 3, (January 1892):388; "Bromides in Epilepsy," *Alienist and Neurologist* 22 (April 1901):361–362; William P. Spratling, "The Treatment of Epilepsy by the General Practitioner," *American Medicine* 5 (January 10, 1903):53–55; "Borax in Epilepsy," *Current Medical Literature* (October 1904):740–743; Ralph H. Spangler, "The Crotalin Treatment of Epilepsy," *Epilepsia* 4 (1912–1914):307–318; Douglas A. Thom, "Crotalin and Its Value in the Treatment of Epilepsy," *Epilepsia* 5 (1914–1915):291–299; "Case of Epilepsy Cured by Antipyrene," *AJI* 48 (July 1891):25. For secondary literature, see R. H. Balme, "Early Medicinal Use of Bromides," *Journal of the Royal College of Physicians of London* 10 (January 1976):205–208; Evart A. Swingard, "History of the Antiepileptic Drugs," in G. H. Glaser et al., eds., *Antiepileptic Drugs: Mechanisms of Action* (New York: Raven Press, 1980), pp. 1–9; Dennis B. Smith et al., "Historical Perspectives on the Choice of Antiepileptic Drugs for the Treatment of Seizures in Adults," *Neurology* 33, suppl. 1, (March 1983):2–7; Walter J. Friedlander, "Putnam, Merritt, and the Discovery of Dilantin," *Epilepsia* 27, suppl. 3, (196):S1–S21. For Locock's discovery, see Temkin, pp. 298–299.

8. E. S. Cooper, "Fracture with Depression of the Skull," *American Medical Times* 3 (June 7, 1862):319; "Trephining in Epilepsy," *American Medical Times* 4 (January 11, 1862):31; Trephining in Epilepsy," *JAMA* 9 (November 12, 1887):626; Pepper, p. 147; B. Merrill Ricketts, "Brain Surgery for Epilepsy," *Cincinnati Lance-Clinic* (November 9, 1895):1–4; "Epilepsy Operations a Failure," *Buffalo Medical Journal* 30–40 (October 1899):188; Charles L. Dana, "A Case of Cortical Sclerosis, Hemiplegia and Epilepsy, with Autopsy," *JNMD* 28 (February 1901):67–73. As late as 1915, *Practical Therapeutics* still advocated a range of measures including venesection, trephining, and lumbar puncture to "reduce intracranial tension." "The Treatment of Epilepsy," *Practical Therapeutics* 28 (August 1918):491–492. For an even later, if more limited endorsement, see "Indications for Surgical Intervention in Traumatic Epilepsy," *JAMA* 93 (August 3, 1929):420.

9. Notions of reflex epilepsy persisted well into the twentieth century; for examples, see Edward A. Tract, "The Treatment of Epilepsy, Incipient and Chronic," *American Journal of Clinical Medicine* 29 (July 1922):643–646; "What Epilepsy Is," *American Journal of Clinical Medicine* 29 (July 1922). In "Idiopathic Epilepsy," *American Journal of Clinical Medicine* 29 (November 1922):808–811, B. M. Abramson discussed autointoxication as a cause of epilepsy. He noted that, because the large intestine was the "chief seat" of autointoxication, some doctors advocated cutting off large sections of it. Abramson disapproved of such surgery as too heroic. For vision problems, see J. Elliott Colburn, "A Case of Epilepsy Cured (Apparently) by the Correction of an Error of Refraction," *JAMA* 10 (February 18, 1888):189–191, 207–210; W. P. Spratling, "The Non-operative Relief of Eyestrain for the Possible Cure of Epilepsy as Tested in Sixty-Eight Cases at the Craig Colony," *American Medicine* 7 (1904):585–589; "Is Eye Strain an Aetiological Factor in Epilepsy?" *Medical Review of Reviews* 13 (March 25, 1907):195–197. For discussions of counterirritants, see Robert Far-tholow, "Epilepsy-Torticollis-Irritable Heart," *JAMA* 6 (February 13, 1886): 169; "The Treatment of Partial Epilepsy by Encircling Blisters with Transfer of the Aura," *JAMA* 2 (April 1884):409; "The Treatment of Epilepsy," *JAMA* 5 (August 8, 1885):151. For treatment of problems related to the thyroid, see William Browning, "Pseudo-Epilepsies, and the Relief of Some Forms by Thyroid," *JNMD* 29 (October 1902):610–619; "Galvanization of the Thyroid in Epilepsy," *JAMA* 10 (February 18, 1888):204; "Galvanization of the Thyroid," *JAMA* 11 (August 18, 1888):235. The discussion of urethral disturbances is in Krauss, 1893, pp. 162–163. Attribution of epilepsy to masturbation appeared in the twentieth century as well; see L. Pierce Clark, "Clinical Studies in Epilepsy," *Psychiatric Bulletin* 1 (1916):133–134; H. C. Kehoe, "What is Epilepsy," *Alienist and Neurologist* 36 (February 1915):10–11.

10. "Oophorectomy and Epilepsy," *Alienist and Neurologist* 17 (July 1896):394–395.

11. J. Taber Johnson, "Specimen from an Oophorectomy," *JAMA* 7 (September 18, 1886):326–332. For support of such surgery, see C. D. Palmer, "Ovarian Epilepsy; Four Cases," *JAMA* 6 (May 1886):480–482.

12. William Spratling, "Description of the Work of the Craig Colony for Epileptics," *Transactions of the College of Physicians* 28, 3d series (1906):10; Craig Colony, *Eighth Annual Report* (1901):45. For cases of patients who had ovaries removed, see Craig Colony Clinical Records, patients numbers 01801, 04288, 08438. For the case of a four-year-old girl whose clitoris was removed to control her seizures, see Craig Colony Clinical Records, patient number 08436, New York State Archives, Albany. (All patient records can be found at the Archives.)

13. In a typical statement, Boswell Park, "On the Surgical Treatment of Epilepsy," in *The Retrospect of Practical Medicine and Surgery* 107 (July 1885):102–105, commented that "a mere removal of the lesion is not necessarily or always enough to break up the well-formed habit." See also "Nancrede," *Buffalo Medical Journal* 36 (November 1896):304–305; A. W. Dunning, "Clinical Observations upon the Treatment of Idiopathic Epilepsy," *Northwestern Lancet* 17 (1897):21–23. In a discussion of reflex causes of epilepsy, especially eyestrain, Ambrose L. Ranney, "The Rational and Scientific Investigation and Treatment of Epilepsy," *Medical Bulletin* 22 (March 1900):81–91, gave a less-value-laden view of the epileptic habit. In situations of extreme nervousness or disturbance, he believed, chronic epileptics were likely to have seizures, whereas others would get headaches or something less. See also William Spratling, "The Treatment of Epilepsy in Its Incipience," *Buffalo Medical Journal* 39–40 (June 1900):799–808; Spratling, "The Value of an Occasional Convulsion in Certain Cases," *Albany Medical Annals* 23 (May 1902):263–267.

14. A. N. Williamson, "The Supplementary Treatment of Epilepsy," in *The Retrospect of Practical Medicine and Surgery* 113 (July 1896):45; Craig Colony, *Twenty-Third Annual Report* (1916), p. 128. For examples of patients characterized as having an "epileptoid personality makeup," see Craig Colony Clinical Records, patients numbers 00966, 02240, 02515, 04539.

15. J. F. Munson, "Public Care for the Epileptic," *Epilepsia* 3 (1911–1912):38–40; L. Pierce Clark, "The Epileptic Psyche," *State Hospital Quarterly* 11 (May 1926):335–362; L. Pierce Clark, "Epilepsy," *Modern Medicine* 6 (1928):596–597, ed. William Osler, re-edited by Thomas McCrae.

16. Alfred Clum, "Epilepsy and Its Relation to Insanity and Crime," *Cleveland Medical Gazette* 10 (September 1895):513–526; William Spratling, "On Epilepsy in Early Life, with Especial Reference to the Colony System in the Care and Treatment of Epileptics," pamphlet, New York Academy of Medicine, reprinted from *New York Medical News* (September 15, 1894); Spratling, "The Treatment of Epilepsy in Its Incipience," pp. 39–40.

17. W. A. Hunt, "The Relation of the State to the Epileptic," *Northwestern Lancet* 17 (1897):2; M. Echeverria, "Violence and Unconscious States of Epilepsies, in Their Relations to Medical Jurisprudence," *AJI* 29 (April 1873):508–556; Maurice D. Lynch, "Social Aspects of Epilepsy," *Transactions of NASECTE* (1914):39–40;

18. John Hughlings Jackson, "On Temporary Paralysis after Epileptiform and Epileptic Seizures: A Contribution to the Study of the Dissolution of the Nervous System," *Brain* 3 (1880–1881):433–451; Jackson, "On Epilepsy and Epileptiform Convulsions," in James Taylor, ed., *Selected Writings of John Hughlings Jackson* (London: Hoder and Stoughton, 1931), vol. 1, pp. 119–134, first published as "Temporary Mental Disorders after Epileptic Paroxysms," *West Riding Lunatic Asylum Reports* 5 (1875).

19. J. B. Stonehouse, "The Psychoses of Epilepsy," *Albany Medical Annals* 14 (August 1893):224, 235; see also A. Ferree Witmer, "Stigmata of Degeneration in Epilepsy," *Pediatrics* 4 (1897):295–299; "Stigmata of Degeneration" (a comment on Witmer), *Pediatrics* 4 (1897):282–283; J. F. Munson, "The Role of Heredity and Other Factors in the Production of Traumatic Epilepsy," *Epilepsia* 2 (1910–1911):343–347; Turner, pp. 23–24. For other discussions of epilepsy as involving dissolution, see John T. McCurdy, "Epileptic Dementia," *Psychiatric Bulletin* 1 (1916):341–352; Alfred Gordon, "Epileptic Dementia," *Proceedings of the American Medico-Psychological Association* 21 (1914):513–519.

20. Harriet C. B. Alexander, "Abuse of the Bromides," *Alienist and Neurologist* 17 (July 1896):279–294; William P. Spratling, "The Treatment of Medicine by the G.P.," *American Medicine* 5 (January 10, 1903):53–55; Ben Karpman, "Bromide Delirium—Part I," *State Hospital Quarterly* 7 (November 1921):66–114.

21. M. Echeverria, "Violence and Unconscious States of Epilepsies"; Echeverria, "Criminal Responsibility of Epileptics, as Illustrated by the Case of David Montgomery," *AJI* 29 (January 1873):341–425; J. M. Mosher, "Mental Epilepsy," *Transactions of the Medical Society of New York State* (1893):169–179; M.

Benedikt, "The Nature and Treatment of Epilepsy," *JNMD* 1, new series, (January 1876):394–403; William Spratling, "On Some of the More Unusual Forms of Epilepsy," *St. Louis Medical Review* 52 (September 23, 1905):253–254; Everett Flood, "Epilepsy as an Inheritance and Other Factors," *Transactions of NASECTE* (1911):73–87. The durability of such views, and the ease with which they continued to be linked to degeneration and dissolution, can be seen in Hubert W. Smith, "Medico-Legal Facets of Epilepsy," *Texas Law Review* 31 (1953):765–793. Smith discussed how epileptics sometimes commit criminal acts during psychomotor seizures. "Such episodes," he wrote, "involve a discontinuity, or break in the stream of consciousness, with a resultant loss of part of all of the past knowledge and conditioning which help the ordinary man to inhibit primitive drives."

22. J. M. Mosher, "Mental Epilepsy," *Transactions of the Medical Society of the State of New York* (1893):169–179; James G. Kiernam, "Epileptic Insanity," *AJI* 52 (April 1896):516–529. See also W. R. Gowers, "Relation of Epilepsy to Insanity," *Retrospect of Practical Medicine and Surgery* 112 (January 1896):49–53; Clark, "Clinical Studies," pp. 133–134; William P. Spratling, "Epilepsy in Its Relation to Crime," *JNMD* 29 (August 1902):481–496; D. J. McCarthy, "Epileptic Ambulatory Automatism," *JNMD* 27 (March 1900):143; Charles Cary, "Psychical Form of Epileptic Equivalent," *JNMD* 28 (May 1901):280–284.

23. Accounts of the case include P. M. Wise, "The Barber Case: The Legal Responsibility of Epileptics," *AJI* 45 (January 1888):360–373; "The People versus Barber," *AJI* 46 (January 1890):375–379.

24. "The People versus Barber," p. 375.

25. John Ordronaux, "Case of Jacob Staudermann," *AJI* 32 (April 1876):451–74.

26. Roscoe L. Barrow and Howard D. Fabing, *Epilepsy and the Law,* 2d ed. (Harper & Row, 1966), p. *x;* "Intermarriage of Epileptics," *Medico-Legal Journal* 14 (1896):133–135.

27. H. M. Carey, "Compulsory Segregation and Sterilization of the Feeble-Minded Epileptic," *Epilepsia* 4 (1912–1914):86–118; Charles B. Davenport, "State Laws Limiting Marriage Selection Examined in the Light of Eugenics," Eugenics Record Office, Bulletin no. 19 (1913).

28. Mark Haller, *Hereditarian Attitudes in American Thought* (New Brunswick: Rutgers University Press, 1963); Charles Rosenberg, *No Other Gods: On Science and American Social Thought* (Baltimore: Johns Hopkins University Press, 1976, 1978), pp. 25–53. See also Daniel Kevles, *In the Name of Eugenics: Genetics and the Uses of Human Heredity* (New York: Knopf, 1985), pp. 70–112. For neurology's eventual rejection of eugenics, see Abraham Myerson et al., *Eugenical Sterilization: A Reorientation of the Problem* (New York: Macmillan, 1936).

29. "Special Provision for Epileptics," *AJI* 48 (January 1892):409–410; "Special Provision for Epileptics," *AJI* 49 (July 1892):124–125. "Craig Colony for Epileptics," *Alienist and Neurologist* 15 (July 1894):388–389. Everett Flood, "President's Address," *Transactions of NASACTE* 5 (1907):21. A major goal of the NASACTE was to help states build specialized institutions for epileptics. See also W. Marion Bevis, "Colony Care for the Epileptic and Feeble-Minded in Florida," *Journal of the Florida Medical Association* 4 (May 1919):322; William Pryor Letchworth, "Provision for Epileptics," *Buffalo Medical and Surgical Journal* 34 (August 1894):18–19. For examples of "model" sterilization laws, see

Harry H. Laughlin, "The Legal, Legislative, and Administrative Aspects of Sterilization," Eugenics Record Office, Bulletin no. 10B (1914).

30. Joseph N. Larner, *The Life and Work of William Pryor Letchworth* (Boston: Houghton Mifflin, 1912), pp. 333–338; William P. Spratling, "Two and a Half Years' Work at the Craig Colony, with Notes on Future Development," *AJI* 55 (October 1898):241–251; William T. Shanahan, "History of the Development of Specialized Institutions for Epileptics in the United States," *Psychiatric Quarterly* 2 (1928):422–434; Shanahan, "The Problem of Epilepsy in New York State," *Psychiatric Quarterly* 1 (1927):160–183; Munson, p. 41.

31. New York (State) Board of Charities, "Report on Institutionalized Epileptics, 1885," New York State Archives, Albany.

32. Frederick Peterson, "Colonies for Epileptics," *Medical Record* (September 19, 1896):404–408; Peterson, "On the Care of Epileptics," *AJI* 50 (January 1894):362–371; Peterson, "The Craig Colony for Epileptics," *Pediatrics* 1 (1896):157–162.

33. For the most extensive statement of Spratling's views, see his 522-page *Epilepsy and Its Treatment* (New York: Saunders, 1904). For a typical reference to "stigmata of degeneration," see Edward Livingston Hunt, "Epilepsy and the Epileptic Temperament," *Medical Record* (Aug. 5, 1911):261–264. For studies of Craig Colony patients and their families, commissioned by the State Board of Charities, see "Nine Family Histories of Epileptics in One Rural County" (1916) and "Nineteen Epileptic Families: A Study" (1917), (New York) State Board of Charities, National Library of Medicine, Albany; and David F. Weeks, "Field Work in the Study of Epilepsy: The Value and Importance of Systematic Work and Tabulation of Findings," *New York State Journal of Medicine* 14 (1914):370–376. See also William T. Shanahan, "The Care and Treatment of Epileptics," *New York State Journal of Medicine* 11 (November 1911):18–24; Shanahan, "Why the Marriage of Defectives Should Be Prevented when Possible," *Epilepsia* 15 (1914–1915):94–100; Shanahan, "Custodial Power over Inmates of State Institutions for Defectives," *Medical Record* (July 5, 1913); "Hygiene of the Epileptic School Child," *Epilepsia* 5 (1914–1915):337; Craig Colony, *Fifteenth Annual Report* (1908), p. 13. (Copies of the Craig Colony annual reports produced for the New York State legislature are at the New York State Library, Albany.)

34. Shanahan, "Hygiene of the Epileptic School Child," pp. 340–341.

35. William P. Spratling, "Two and a Half Years' Work at the Craig Colony, with Notes on Future Development," *AJI* 55 (October 1898):244–251. Craig Colony, *Third Annual Report* (1896), p. 36.

36. Spratling, "The Treatment of Epilepsy by the G.P.," *American Medicine* (January 10, 1903):53–54; Shanahan, "The Care and Treatment of Epileptics," pp. 18–24.

37. Shanahan, "The Care and Treatment of Epileptics," p. 19.

38. Munson, p. 347; Craig Colony, *Seventh Annual Report* (1900), p. 67.

39. Craig Colony Clinical Records, patient number 11431.

40. Ibid., patient number 6413.

41. Ibid., patient number 321.

42. Ibid., patient number 7830.

43. Ibid., patient number 9707.

44. Ibid., patient number 6257.

45. Craig Colony, *Thirty-first Annual Report* (1925), p. 10.

46. Craig Colony Clinical Records, patients numbers 6413, 6258, 6259.

47. Ibid., patient number 6413.

48. Arthur Kleinman, *The Illness Narratives: Suffering, Healing, and the Human Condition* (New York: Basic Books, 1988), p. 5.

49. Ruth C. Adam, in Walter C. Alvarez, ed., *Living with Mysterious Epilepsy: My Forty-Eight Year Victory over Fear* (New York: Exposition Press, 1974), pp. 7–11; Carl Sagan, *The Dragons of Eden: Speculations on the Evolution of Human Intelligence* (New York: Ballantine, 1977), p. 58.

50. Joseph W. Schneider and Peter Conrad, "In the Closet with Illness: Epilepsy, Stigma Potential, and Information Control," *Social Problems* 28 (October 1890):36; Steven Whitman and Bruce P. Hermann, eds., *Psychopathology in Epilepsy: Social Dimensions* (New York: Oxford University Press, 1986).

51. For a rare exception, see J. W. Schneider and P. Conrad, *Having Epilepsy: The Experience and Control of Illness* (Philadelphia: Temple University Press, 1983).

DISEASE AS SOCIAL DIAGNOSIS

13 THE SICK POOR AND THE STATE: Arthur Newsholme on Poverty, Disease, and Responsibility

JOHN M. EYLER

The rich historical tradition of social medicine as potential social criticism was built logically—and emotionally—around the response to disease incidence. But as John Eyler demonstrates in his discussion of English public health authority Arthur Newsholme's thought, this style of social analysis was complex and often inconsistent. One could easily agree that excessive levels of disease could and must be ameliorated, but the means for accomplishing this were neither obvious nor conflict-free. Were individuals responsible for the behavior that placed them at risk—or were they passive victims of inimical social circumstances? Were the poor dirty as a matter of choice or did their circumstances make cleanliness impossible? Did poverty "cause" disease, or did endemic and localized epidemic disease cause poverty? Few late-nineteenth-century physicians could, in fact, ignore either kind of causation—and were well aware that poverty and disease could and did interact, creating "vicious cycles" of environmental deprivation, immorality, and ultimate illness.

The debate often turned on the question of individual responsibility. How were Newsholme and his contemporaries to balance the dangers of working-class alcohol abuse against the even more alarming, if less morally culpable, specter of hereditary weakness and degeneracy? How could one formulate a humane and viable state policy that reflected a due consideration of all such factors? Disease in this context provided an occasion and an agenda for a generation-long debate about the interrelation of state policy, physician responsibility, and individual culpability. It is still a very live debate, as the recent history of AIDS so forcefully demonstrates.

—C. E. R.

Between 1880 and 1920 the diseases of greatest concern to public authorities were dramatically reframed as a consequence of both scientific and political changes. On the one hand the status of such diseases as pathological entities was altered by bacteriological research. What had been known as zymotic disease processes, because of their presumed resemblance to fermentation, were reframed as infectious processes and viewed as the body's response to the growth and reproduction of specific microbes. Related research on the transmission of such specific living agents provided a biological explanation for the results of epidemiological studies, which, since mid-century, had suggested that epidemic diseases were contagious and had suggested vehicles, such as food and water, for their transmission.

On the other hand, the public response to such diseases also changed. In the last quarter of the nineteenth century civil authorities in Britain tried much more actively to prevent disease and premature death. While the classic documents of the sanitary movement may date from the 1840s,[1] a comprehensive network of local health authorities was created only in 1872, with the passage of legislation dividing the nation into sanitary districts and requiring each sanitary authority to appoint a Medical Officer of Health (M.O.H.).[2] This legislation, and the financial incentives that the central government provided in the form of low-cost public loans, stimulated local initiative and caused the majority of local authorities to do what only a minority had done previously: construct or improve public water supplies and sewers, build public abattoirs, condemn and tear down slum housing, and appoint a corps of sanitary inspectors to visit markets and shops, common lodging houses, and private homes.[3]

The reframing of epidemic and endemic diseases soon redirected the eyes of local health authorities from environmental sanitation to the control of infectious patients. In 1889 notification, or case reporting, of about a dozen of the most common acute infectious diseases became mandatory in London and a local option in the provinces, and a decade later notification was universally required.[4] Visiting homes following notification, tracing contacts and antecedent cases, isolating cases, and building isolation hospitals all followed, as people joined things as the focus of health departments' attention. In the process, local authorities began to provide treatment and other personal services to the diseased and their families, in addition to the older police and regulatory functions implicit in sanitary reform. By the second decade of the new century, beginning with infants, local authorities were offering personal services to both the sick and the well in the hope of preventing future disease or disability.

Finally, and of greatest interest in the present context, is the fact that the relation of public authority to the sick poor, the most frequent recipient of these services, dramatically changed during these four decades. The clearest indicators of the altered relationship between the sick or physically helpless and the state were the new social programs the Liberal Party introduced between 1906 and 1914: old age pensions, national health insurance, unemployment insurance, free meals in school for malnourished children, a school medical service, infant welfare services, and national systems of gratuitous treatment for tuberculosis and venereal disease.[5] These new programs were a deliberate repudiation of the strategy of public assistance codified by the Poor Law Amendment Act of 1834.

The New Poor Law had established deterrents and penalties intended to make the receipt of relief so unpleasant that it would be the last resort of all who applied.[6] The workhouse test, which required that the able-bodied receive assistance only in the workhouse, and the principle of least eligibility, which insured that the life of the pauper in the workhouse was less attractive than that of the most poorly paid laborer on the outside, were attempts by the Poor Law reformers of the 1830s to separate the pauper from the laboring poor and to deter the latter from sliding downward into destitution and dependency. It has often been pointed out that the New Poor Law of 1834 both reflected and fostered the view that destitution was a sign of personal failing and character defect.[7] To receive relief was not only to lose some of one's civil rights and sometimes one's freedom, but to be morally stigmatized.

By contrast, the new welfare initiatives of the first two decades of the twentieth century provided services as entitlements: personal services on the basis of need, often without prior contribution, and without penalty or civil disability. The appearance of such programs is ample proof that poverty as well as disease was being understood in new terms. If poverty were being reframed, then attitudes toward the sick poor also might be expected to have changed. As we shall see, changes in conviction among those providing the new public services were less swift and conclusive than the change in policy.

We can conveniently explore this problem by considering the works of a leading British public health figure whose official career almost exactly corresponds to this forty-year period. Sir Arthur Newsholme began his public career in 1883 as part-time Medical Officer of Health to the London vestry of Clapham. In 1888 he became the first full-time M.O.H. of Brighton and from 1908 to 1919 served as Medical Officer of the Local Government Board, the effective head of the British public health service.[8] His public career reflects the reorientations in public health programs necessitated by both scientific discovery and political change. His prominence and his interest in the social components of disease make Newsholme's thought revealing.

Public health activists who, like Newsholme, began their careers in the early eighties were heirs to a well-established Victorian belief that associated disease with poverty and destitution. At midcentury the journalist and investigator of London's poor Henry Mayhew noted:

> Indeed, so well known are the localities of fever and disease, that London would almost admit of being mapped out pathologically, and divided into its morbid districts and deadly cantons. We might lay our fingers on the Ordnance map, and say here is the typhoid parish, and there the ward of cholera; for truly as the West-end rejoices in the title of Belgravia, might the southern shores of the Thames be christened Pestilentia. As season follows season, so does disease follow disease in the quarters that may be more literally than metaphorically styled the plague-spots of London.[9]

In such a passage Mayhew was doing little more than echoing investigators like William Farr of the General Register Office, who over the last decade had compared the vital statistics of town and countryside, industrial town and cathedral town, working-class district and middle-class district to show that the urban laboring poor had much higher mortality rates, lower life expectancy, and presumably more frequent bouts with disease than other groups in the population:[10]

> The annual rate of mortality in some districts will be found to be 4 per cent., in others 2 per cent.; in other words, the people in one set of circumstances live 50 years, while in another set of circumstances . . . they do not live more than 25 years. In these wretched districts, nearly 8 per cent. are constantly sick, and the energy of the whole population is withered to the roots. Their arms are weak, their bodies wasted, and their sensations embittered by privation and suffering. Half the life is passed in infancy, sickness, and dependent helplessness. . . . Diseases are the iron index of misery, which recedes before strength, health, and happiness, as the mortality declines.[11]

For such commentators disease and consequent mortality were measures of relative well-being, indices of civilization. The excess sickness and premature deaths of the poor were evidence of social defects demanding remedy.

Even the Poor Law reformers of the 1830s, who argued most forcefully for a system of deterrents in the granting of relief, acknowledged that the sick poor were a special case. It was not the intention of either Edwin Chadwick or Naussau Senior, chief architects of the New Poor Law, that the sick poor be subject to conditions intended to force the lazy, the imprudent, or the drunken into productive labor. Both conceded that sickness, disability, and old age rendered some of the laboring poor unable to support themselves and their families. Such misfortune was not blamed on the destitute themselves, who were held entitled to assistance in their own homes and were to be taken into special facilities only at their own request.[12] This recognition may have been small consolation to the destitute sick who, in subsequent decades and contrary to Chadwick's and Senior's intent, often had no choice but to enter a general workhouse where they endured conditions that, when made public, shocked mid-Victorians. Nor did it alter the fact that in the Metropolis, prior to 1885, receiving any relief, even medical relief, pauperized the recipients and imposed the same civil penalties the law laid on other paupers.[13] But it does show that early Victorians, even Utilitarian bureaucrats like Chadwick, were capable of recognizing a distinction between destitution caused by sickness or infirmity and that caused by the alleged character flaws blamed for most destitution.

Chadwick in fact recognized that sickness was a major cause of pauperism and destitution. He repeatedly made this point in his 1842 *Report on the Sanitary Condition of the Labouring Population*, where he gave several estimates of the annual loss of human life from disease and the numbers of widows and orphans thrown on the relief rolls by the premature death of wage earners.[14] This reading of the relationship between poverty and disease further removed the destitute sick from the stigma of pauperism. Much pauperism stemmed from disease, which, Chadwick's medical advisers assured him, was due to environmental defects that could be prevented by sanitation.

Viewing the problem in this way permitted Chadwick to broaden the appeal of public health reform and to make intervention more acceptable to an age that preferred laissez-faire economics and government retrenchment. The removal of sanitary hazards, often beyond the control of individual laboring families, would economically prevent disease and thus reduce pauperism and the resulting burden on the rates without compromising individual responsibility or personal liberty. The sanitary movement was founded on the conviction that squalor caused disease; disease in turn caused destitution. Destitution threatened the social and moral order. Viewed in these terms, sanitation was the complement of the deterrent policy of the New Poor Law. The latter would insure that the giving of relief did not undermine initiative or encourage malingering. Sanitation would reduce the number thrown into destitution through no fault of their own.

Newsholme fell squarely in this tradition of viewing disease as a cause of poverty. In a 1909 paper intended for fellow M.O.H.s, written in response to the deliberations of a Royal Commission appointed to consider revising the poor-law policy Chadwick had helped design nearly three quarters of a century earlier, Newsholme summarized the Commission's evidence linking poverty and disease. Thirty percent of English paupers were sick, and 50 percent of the funds spent in their behalf went to the relief of sickness. In interpreting these results, Newsholme revived the past:

> We need to learn again the lessons taught to our parents by Southwood Smith, Chadwick, and their co-workers, that one of the chief causes of poverty is disease, and that extended public health administration must continue to be a chief means of removing destitution from our midst.[15]

Pulmonary tuberculosis was a case in point. Newsholme estimated that, between 1907 and 1916, fatal cases of tuberculosis in men alone would cost the nation some 58.3 million pounds, excluding wages lost due to prolonged illness.[16] What better investment in national efficiency could there be than preventive measures against the disease? Tuberculosis was an especially important example because it was popularly associated with poverty, but statistics showed that mortality from this dread ailment had been falling for several decades. Had improvements in the standard of living caused this mortality decline? In studies of the epidemiology of tuberculosis Newsholme found a strong correlation between the decline of pauperism and the decline in the mortality from pulmonary tuberculosis in the United Kingdom.[17] He cautioned, however, that pauperism was not poverty but rather poverty relieved at state expense, and that closer analysis showed that a decline in actual privation was not the crucial factor in the decline of phthisis.

Using census and trade figures for various nations and capital cities, Newsholme attempted to demonstrate that falling mortality from phthisis appeared unrelated to several relevant standard-of-living indicators: improvements in nutrition, measured by lower wheat prices or food costs to feed a working-class family; reduced total cost of living; and better housing. The most important factor in tuberculosis's decline, Newsholme found, was institutional segregation of the sick. Poor Law records showed that tuberculosis deaths fell as the ratio of indoor to outdoor relief rose (i.e., when the proportion of institutionalized sick paupers grew), which had occurred in London and overall in England and Wales, and Scotland. But where the ratio of indoor to outdoor relief had fallen, as in Ireland, tuberculosis mortality, and presumably morbidity, increased. These findings supported Newsholme's claim that isolation was the best preventive measure against tuberculosis and added plausibility to his claim that specific administrative action rather than general social amelioration was the best solution to the health problems of the poor.

Like Chadwick and the early Victorian sanitary reformers, Newsholme justified expenditures on public health programs in part as community investments; even economic calculations of the cost of sickness indicated that

by combined efforts on the part of the community the total mass of poverty and sickness can be decreased, and that such efforts are not mere unproductive expressions of altruism. They represent, in fact, the best investment that the community can make. . . . Health is always cheaper than disease, and without an efficient state of health progress in other social concerns is being constantly impeded.[18]

It was, of course, entirely possible to turn Chadwick's equation around and blame economic deprivation for excessive disease and premature deaths among the poor. But this interpretation was less congenial to Victorians because it suggested the need for intervention in the economic system to solve the problems of both pauperism and preventable disease. While Newsholme's predecessors had considered the possibility that low income fostered disease, most of them had judged its contribution at best secondary.[19] Newsholme's four major epidemiological studies of infant mortality also reached this conclusion.[20] High infant mortality, he easily verified, was a fact of urban working-class life, but low income per se did not explain its distribution. Poor nations such as Ireland and Norway had lower infant mortality rates than richer Britain, and within Britain high wages did not necessarily diminish mortality. Jews living in great poverty in London's East End had remarkably low infant mortality rates, while the rates for children of miners, among the best-paid English workers, were very high. Also, if greater income reduced infant mortality, why hadn't mortality declined for infants as it did for all other age groups when real wages rose during the last twenty years of the nineteenth century? While domestic overcrowding seemed to have an important influence, it was not an invariable determinant of high infant mortality. High infant mortality was tied to the life of the urban poor, Newsholme concluded, but some feature of industrial working-class life, not low income per se, was its major preventable cause.

But Newsholme was under no illusions about health hazards in the lives of the poor. Few other observers had such regular contact with the domestic life of the poor as the M.O.H. Even at the beginning of his public career he recognized a barrier to the efficacy of public health work:

> I refer to the extreme poverty among certain sections of the population, which checkmates efforts made to prevent overcrowding and ensure cleanliness. A low rate of mortality among children is difficult to attain when they are insufficiently clad and fed, and live under conditions of poverty which by some strange fatality appear to render more rapid the multiplication of the population.[21]

Over the next few years Newsholme's thinking about poverty responded to the more systematic social investigations that marked the last two decades of the century.[22] Investigators like Charles Booth and B. Seebohm Rowntree not only replaced Mayhew's anecdotal accounts of destitute individuals and his vivid portraits of the more colorful of London's street people with aggregate measurements of the extent and depth of destitution, but, by computing a poverty line, they showed in hardheaded terms the meager life that even the most sober and prudent of unskilled laborers could hope to attain.

Within months of the appearance of Rowntree's influential study of poverty in York, Newsholme published a careful summary of its evidence and conclusions for the medical profession.[23] He explained

Rowntree's division of the town's population by social class, the meaning and use of the poverty line, a typical working-class family's odyssey above and below the poverty line as earning capacity and number of dependents changed year by year, and the fact that at the time of the survey, a period of relative prosperity, 28 percent of York's population lived below the poverty line. He painted a sensitive picture of the plight of unskilled laborers who lived just above the poverty line. They could have no luxuries. "The family must never spend a penny on 'bus or railway, or on newspapers; they must write no letters; they can join no sick club." The calculation of the poverty line made no allowance for pleasures like tobacco or beer and certainly none for sickness. Any deepening of poverty could be met only by reducing a diet that was barely able to sustain physical efficiency. "To give the father sufficient food, wife and children go short."[24] The nature and consequences of such privation were missed by more casual observers.

The growing knowledge of the life of the poor encouraged Liberal economists to discuss poverty as a problem of an economic system rather than as a personal problem of individuals. John A. Hobson, for example, attacked the complacent assumption that the poor had only themselves to blame and that education and self-help were the primary remedies for destitution. Hobson instead taught that regardless of any personal shortcomings of the poor, poverty resulted from economic and legal systems that denied opportunities and cheapened the value of labor.[25] In these years Medical Officers of Health, more haltingly to be sure, also began to reexamine their assumptions about the role of privation in the health of a community. James Niven, the M.O.H. of Manchester, for one, regarded the casual labor system and trade cycles as major causes of both poverty and disease in industrial cities, adding ignorance and irresponsibility as secondary factors.[26] While poverty per se need not cause disease, Niven argued, as long as the casual laborer lived in a precarious hand-to-mouth existence poverty could fairly be said to cause disease. Low or irregular wages forced the poor to live in conditions that exposed them to infection and lowered their resistance.

Newsholme also was learning to view the issue in economic terms. Influenced by the investigations of the Royal Commission on the Poor Laws, he labeled the meager outdoor relief given to widows with small children as "extravagant parsimony."[27] Not only would such inadequate support cause chronic malnutrition, lost "efficiency" and disease, but the desperate efforts of the widows to supplement their relief by entering the labor market at the lowest level would help depress the wages of other workers. In certain contexts Newsholme was willing to depart from the traditional Victorian position and to argue that destitution caused disease, but typically he assigned poverty an indirect role.[28] It necessitated overcrowding in working-class dwellings, discouraged cleanliness, and it promoted irresponsible behavior. But he sometimes indicted economic privation more directly. In his study of the history of typhus in Ireland, for example, he concluded that extreme poverty had fostered typhus mortality by promoting disease transmission and raising case fatality.[29] And in a context different from his writing on the epidemiology of tuberculosis, he could even advance a conclusion he would soon repudiate: that falling wheat prices had significantly contributed to the improved health of the English people and even to the decline of tuberculosis.[30]

What then was his position? Did poverty cause disease, or did disease cause poverty? He tried to explain with a simile:

> The conditions of poverty in a community exposed to typhus as to phthisis, may be compared with the dryness of timber exposed to the onset of fire. The poorer and the more over-crowded the population, the drier and the more densely aggregated the timber, the more extensive will be the epidemic or the conflagration produced by infection or flame.[31]

In a town free from fire the most protective action might be to fireproof buildings, but the presence of fire allows no time to treat timbers. The best strategy then is to protect buildings from the spread of the fire. Analogously, measures to increase the resistance of a population to infection, with the exception of smallpox vaccinations, work too slowly and uncertainly to be used in a crisis.

It seems then that Newsholme's position was partly a matter of expediency. Poverty, he came to realize, was a complex phenomenon with economic, behavioral, and biological components. He was fond of explaining that poverty and disease, like many social evils, formed a vicious circle, each contributing to the other.[32] This fact gave grounds for optimism, not discouragement, because a circle can be broken at any point. Given the state of knowledge at that time, the most efficient means of attacking the problem of poverty was to keep people from getting sick, which in practice meant breaking the chain of contagion.[33] The knowledge of social problems, he explained in 1909, was at the stage once occupied by the understanding of disease. Using available sociological and economic knowledge, and the crude solutions they suggest, one could remove some of the symptoms of poverty. But as knowledge improved, more exact means would become available.[34] The existing means of dealing with poverty, like medical treatment in the past, showed "the mischief and the hindrance to real progress which are caused by adopting an empirical treatment of symptoms instead of a scientific treatment of disease."[35]

That answer coming from an M.O.H. was hardly surprising and was in sympathy with the Minority Report of the Royal Commission on the Poor Laws with its insistence on attacking the causes rather than treating the symptoms of poverty. Newsholme had thus reached a reasonable synthesis, one that came to terms with new understandings of poverty but that continued to justify public health work as a force for general social amelioration. In the first decade of the twentieth century, however, the issue was not so simple. Poverty forced the nation's leaders to face urgent new political problems.

At home there was chronic worry, especially in the Liberal party, about how to channel working-class aspirations and recently acquired votes into the programs of the traditional parties.

By 1890 the old problem of pauperism had become a problem of poverty and the essentially economic dilemma was political and social: What, it was asked, can the governors of the nation do to prevent the poor from using their franchise to overturn a society based on capitalist wealth? As it turned out, the defence against socialism was social legislation.[36]

This worry was accompanied by even more immediate racial and imperial anxieties. The nation was, in fact, facing a crisis of confidence. Foreign competition in industry, trade, and agriculture, the growth of German military might, and the disastrous showing of British forces in the South African War shattered Britain's complacency about its place in the world and provoked in reaction a movement for "national efficiency" that cut across party lines and aimed at halting the nation's decline.[37] Especially troubling was the news from urban recruiting stations during the Boer War. Disease and poor physical condition in potential recruits meant that many, it seemed far too many, urban working-class males were unfit to help defend the Empire. This revelation gave urgency to the troubling information about the lot of the urban poor collected by social investigators since the middle eighties. Perhaps the race had grown too puny to rule a great empire. Were the British, or at least the urban working class, no longer an imperial race? Was it possible that both poverty and disease were due to defects inherent in the poor?

Nothing reveals the strength of hereditarian thought in the Edwardian period more clearly than the use of biological arguments at the turn of the century by the New Liberals, the economists and political philosophers on the left flank of liberalism, to justify a more active, less deterrent welfare policy. The New Liberals refashioned Liberalism to make it a political philosophy capable of dealing with the social problems of industrial society, while preserving the party's traditional commitment to individual liberty.[38] This intellectual transformation was bold and far-reaching. The New Liberals dissociated themselves from laissez-faire economics and embraced collectivism. They also moved away from their antipathy to the state, which they now saw as the dominant agent for the creation of a just society. But this change of opinion depended heavily on the example provided by biology. Evolutionary biology gave assurance that the principles governing human progress were open to human understanding, and encouraged the view of society as an organic entity in which collective choice, that is, state action, was not only justified but essential.[39]

J. A. Hobson argued mightily against blaming the poor for their own misery:

> How shall a child of the slums, ill-fed in body and mind, brought up in the industrial and moral degradation of low city life, without a chance of learning how to use hands or head, and to acquire habits of steady industry, become an efficient workman? . . . It is the bitterest portion of the lot of the poor that they are deprived of the opportunity of learning to work well. To taunt them with their incapacity, and to regard it as the cause of poverty, is nothing else than a piece of blind insolence.[40]

But even Hobson shared the common suspicion that at least some of the poor were biologically inferior. In a passage much like the one just quoted about the disadvantages slum children face he concluded that "bad seed sown in poor earth will not grow into flourishing and fruitful plants, even if carefully watered, pruned, and protected as it grows."[41] He could embrace eugenics in the same spirit in which he criticized the old economic system:

> Selection of the fittest, or at least, rejection of the unfittest, is essential to all progress in life and character. . . . To abandon the production of children to unrestricted private enterprise is the most dangerous abnegation of its functions which any Government can practice.[42]

Newsholme also was keenly interested in applying the principles of biological evolution to social problems.[43] This preoccupation is most evident in his concern for changes in human fertility patterns. He, in fact, went to greater lengths than most other observers to measure precisely the much-publicized fall in birth rates and wondered about its economic and political consequences.[44] Scattered throughout his writings one finds evidence of the then-common anxieties about national power, interracial competition within the Empire, and the future of Anglo-Saxon cultural dominance. Such worries were created by the realization that at home the birthrates of the middle and upper classes were falling while those of the working classes were not, and that abroad other nations and races had higher fertility rates than Britain.[45] But Newsholme recognized much sooner than many New Liberals the dangers that lay in hereditarian or eugenic solutions to social problems. His caution was first alerted in the early 1890s. The Boer War rekindled it a decade later during the investigations of alleged physical degeneration among the poor, and it was sustained by a lengthy disagreement with Karl Pearson and his associates. Pearson's group saw both pauperism and the health problems of the poor as evidence of adverse heredity and questioned the utility of preventive measures against diseases such as tuberculosis.[46]

Newsholme challenged the hereditarian assumption that current socioeconomic standing reflected inherent differences in ability.[47] Success also depended on opportunities that were very unequally distributed.

> Even the greatest ability may fail through lack of favoring circumstances; and it is impossible to say how many mute inglorious Miltons may have failed to be discovered. The fact that the poorest are lowest in the social scale cannot be used as a completely satisfactory argument that—as proved by selection—they are the poorest stock.[48]

Borrowing heavily on evidence provided to the Royal Commission on Physical Training in Scotland and the Interdepartmental Committee on Physical Deterioration, he argued that although much physical debility could be found among the poor, that debility was better explained by poor nurture than by hereditary degeneration.[49] Adopting the environmentalist stance he maintained through his entire career, Newsholme wrote that the troublesome thing about the differing upper- and lower-class birthrates was not that the inherently inferior were outbreeding the inherently superior, but that those least able to nurture children adequately were having the most children.[50] He also cited empirical evidence that natural selection was not working as predicted by hereditarians. For example, if high infant mortality served to weed out weak, inferior stock, then older people in populations so affected should enjoy greater health. To the contrary, his studies on infant mortality showed that districts having high mortality rates in the first year of life also had high rates at later ages.[51]

But there was yet another possibility one that also had preoccupied the early Victorians. Might not poverty and diseases of the poor both be attributed to moral inferiority? Did the slums' much-publicized idleness, drunkenness, gambling, vice, child abuse, and congenital syphilis prove what the respectable classes had long suspected: The poor were ignorant, irresponsible, and cruel? Early Victorian investigations had tended to confirm the view that the poorest classes were morally degraded. Both Chadwick and Mayhew, for example, linked the destitute and miserable with the filth alleged to cause their ill health: "It is significant that the same words—'residuum,' 'refuse,' 'offal'—were used to denote the sewage waste that constituted the sanitary problem and the human waste that constituted the sanitary problem and the human waste that constituted the social problem."[52] Chadwick quoted one of his witnesses to the effect that

> The bone-pickers are the dirtiest of all the inmates of our workhouse; I have seen them take a bone from a dung-heap, and gnaw it while reeking hot with the fermentation of decay. Bones, from which the meat had been cut raw and which had still thin strips of flesh adhering to them, they scraped carefully with their knives, and put the bits, no matter how befouled with dirt, into a

wallet or pocket appropriated to the purpose. They have told me, that whether in broth or grilled, they were the most savoury dish that could be imagined. I have not observed these creatures were savage, but they were thoroughly debased. Often hardly human in appearance, they had neither human tastes or sympathies, not even human sensations, for they revelled in the filth which is grateful to dogs, and other lower animals, and which to our apprehensions is redolent only of nausea and abomination.[53]

Suspicion as to the moral sensitivity of the poor if not their very humanity, survived the discovery that poverty was an economic phenomenon. Social investigations of the last two decades of the nineteenth century left room to blame part of the plight of the poor on their own behavior. Rowntree, for example, found that while 28 percent of the population of York lived below the poverty line, 18 percent (almost 65 percent of the poor) lived in what he called secondary poverty; that is, their earnings would have sufficed for basic needs if not for wasteful expense, including drinking and gambling.[54] According to some troubling estimates, such people spent 25 percent of their income on drink.[55] Certainly this self-imposed destitution should be remedied by moral reform.

New Liberal economists such as Hobson demurred. He argued that temperance might be prudent behavior for individual laborers, but it offered no solution to the problems of destitution. He predicted that if all laborers gave up their beer, the most likely result would be a reduction in their wages by the same amount they now spent in the pub unless some other essential expense replaced that of beer.[56] The state might have moral purposes, Hobson argued, but moralizing was not the way to realize them. Self-help and moral reform could not be expected to civilize the poor so long as the struggle to survive absorbed all their energy:

We cannot go to the lowest of our slum population and teach them to be clean, thrifty, industrious, steady, moral, intellectual, and religious, until we have first taught them how to secure for themselves the industrial conditions of healthy physical life. . . . What these people do want is better food, and more of it; warmer clothes; better and surer shelter; and greater security of permanent employment on decent wages. Until we can assist them to gratify these "lower" desires, we shall try in vain to awaken "higher" ones.[57]

Newsholme's differences with the New Liberals were greatest on this point. He agreed that the basic problems of health and human welfare needed a broader range of state initiatives but did not follow those like Hobson who demanded little of individuals. Newsholme believed that only changes in human behavior, that is, moral reform, could adequately address some of the most difficult health problems of his age. Consider the problem of alcoholism, part of the vicious circles that intensified other health and social problems. Newsholme was troubled by the amount the working class spent on drink. As a temperance advocate he blamed the consumption of alcoholic beverages for diminished national efficiency and for crime, poverty, disease, and high infant mortality.[58] Expenditure on drink, he feared, had absorbed much of the gain in human welfare that might have resulted from higher wages.[59] But as an advocate of environmental reform he blamed the living conditions of the poor for their irresponsible or destructive behavior. Poor housing, a monotonous diet, fatigue, chronic pain, and social custom all encouraged drinking.[60] A lack of knowledge and domestic skills made many working-class families poor managers of their small resources. The poor needed to be taught to choose and prepare nourishing foods, and to protect their health.[61]

Ignorance, of course, was not a monopoly of the poor. Newsholme also had harsh words for those who would explain away high infant-mortality rates in urban working-class districts as due to the ignorance and irresponsibility of mothers. "It is a comfortable doctrine for the well-to-do person to adopt; and it goes far to relieve his conscience in the contemplation of excessive suffering and mortality among the poor."[62] He felt that what made the poor mother's ignorance or carelessness so tragic was her helplessness and economic vulnerability. Poor housing and sanitation, a lack of domestic and medical help, overwork, and a pinched budget all conspired to make the consequences of her mistakes graver than those of more prosperous mothers.[63]

But environment could not be blamed for all human shortcomings. After all, some slum dwellings were "an oasis of cleanliness and sweetness in a desert of dirt and neglect. The structural condition of the house is not better than that of the neighboring houses but the character of the inmates has impressed itself upon the house."[64] Not all poor parents squandered their resources, neglected their children, or led drunken lives. Certainly part of the condition of the poor was the result of decisions made by individuals. Conquering poverty and disease required more than public initiative; it also demanded responsible individual behavior.

Newsholme encouraged the sanitary inspector to think of himself as a "home missionary to the poor and helpless" who, by promoting "cleanliness and self-respect" and the "decencies of family life," made "moral and social improvement" possible.[65] Where individual responsibility faltered, public authority might have to add incentive. Local authorities might have to intervene to see that alcoholic parents met their responsibilities to their children, even by drastic measures such as garnishing wages.[66]

Especially in his writings about alcoholism and venereal disease published after 1920, Newsholme blamed character faults and could be harsh in his condemnation of individuals. Syphilis, he held, was spread almost exclusively by "sexually immoral persons," and the disease could not be controlled until public opinion viewed "the sexually immoral man as an enemy of society, who cannot be tolerated" and promiscuity as "the chief enemy of the social order," a form of barbarism.[67]

Newsholme's emphasis on the reformation of character as a solution for problems of human welfare may at first seem anomalous, a remnant of his strict Methodist upbringing. It was undoubtedly that. It was also a reflection of conventional Victorian middle-class views of human sexuality, family life, and public decorum. He was offended by drunkenness, prostitution, contraception, and abortion, as well as by child abuse, and responded with an evangelical zeal well-nurtured in his childhood. However, it would be wrong to pass off his pronouncements so simply; moral reform played a more reasoned part in his outlook. It allowed him to reconcile his politics with his science, his liberal faith in progress, amelioration, and liberty with his belief in the fundamental truth of Darwinian evolution.

Following the cue of Thomas Henry Huxley's Romanes Lectures, Newsholme argued that the evolution of intelligence and cooperation sheltered humans from the brutal and blind force of the competition for survival.[68] Fitness, after all, was relative, a ratio of individual strength to environmental strain. Hence, fitness to survive could be enhanced by either increasing strength or reducing strain. By acting collectively and with forethought humans had learned to transfer much of the pressure of the struggle for existence from the individual to the group. Newsholme held, in other words, that a process of moral evolution paralleled biological and social evolution. The self-denial that accompanied the evolution of morality created "a new moral environment which alters very largely the results of evolution."[69] As a result, public opinion and collective action began to protect the weak and helpless from the crushing force of competition.

Newsholme held further that the progress of civilization, recently evidenced by the abolition of slavery and the launching of campaigns to end cruelty to children and animals and to abolish the double standard of sexual conduct, reflected a grand evolutionary process in which selfishness gives way to altruism and competition to cooperation.[70] Private philanthropy, public assistance, and medical charity were all indications of the altruism that moral evolution created. In Newsholme's view, crime, vice, and other antisocial activity were survivals of earlier less-civilized ages, and the problems of human conduct in contemporary society were reduced to a tension between coexisting human impulses: the primitive and selfish versus the more evolved and altruistic.[71]

The same moral evolution had created new ethical standards that rejected eugenic remedies for disease and disability. In the new moral environment a civilized nation simply could not allow a disease like tuberculosis to run unchecked:

> The logical alternative [to preventive work] is to kill off the susceptible stock or, as has been suggested, to allow them to infect their susceptible brethren and together with them perish of their disease. Such proposals have only to be stated in their crude terms in order to be apprehended and reprehended as an unsocial negation of civilization.[72]

A decade later Newsholme argued that reverting to natural selection in human affairs would destroy civilization. "To think otherwise is the secret behind German aggression; to act otherwise is to revert to barbarism. Man has definitely replaced natural by rational selection, and will, I have no doubt, to a steadily increasing extent replace competition by cooperation."[73]

Humans could escape the tyranny of Darwinian evolution because they possessed the ability to act purposefully, collectively, and altruistically. In a liberal state much of the motive for right conduct had to come, Newsholme believed, from within. For this reason he found the final solution to some of the most intractable health and social problems in the reformation of human character. The state could do some things: see that assistance did not undermine individual responsibility, and insist that help was contingent on responsible behavior.[74] Compulsion might sometimes be needed for "those who do not evolve in response to the advancing tide of morality."[75] But ultimately humans had to learn to behave responsibly. He was aware that he was asking a great deal of the poor, more in fact than of the more fortunate: "Sometimes I think that those to whom a farthing is a rare coin do not realize effectively the demands on character which are made by a life in which every farthing counts." But those oases of "cleanliness and sweetness" in deserts of "dirt and neglect" proved, to his satisfaction at least, that

> the necessary sacrifices can be made and made cheerfully. The constant self-denial which they entail cannot conceivably be improvised. It must result from the possession of habits both of conduct and of mind. Life without such habits in conditions which hourly call for them is not only miserable for the individual but also mischievous to the community. And yet how small a part of our national educational system is devoted to the training of character, which is, I take it, the resultant of habits of mind and conduct.[76]

On February 3, 1908, the day before he took up his duties in Whitehall as Medical Officer of the Local Government Board, Newsholme appeared as a witness before the Royal Commission on the Poor Laws.[77] The Commissioners were trying to decide whether the New Poor Law Chadwick had helped design should remain the nation's primary response to destitution and dependency. Newsholme's testimony indicated that much had obviously changed since Chadwick's day. Newsholme condemned the Poor Law medical services, maintaining that efficient, humane care of the sick poor could be provided only by agencies completely removed from the Poor Law. His own proposals revealed not only a vastly different understanding of disease causation but the acceptance of a much expanded sphere of public initiative. Environmental sanitation was now only one routine function of many municipal health services. In addition, a network of publicly financed diagnostic and therapeutic services had evolved. Newsholme proposed to expand and unify this ad hoc system into nothing less than a comprehensive municipal medical service under the supervision of the local M.O.H., which would offer gratuitous preventive, diagnostic, and therapeutic services. This municipal system would in effect take over all medical services previously rendered the working class except those of the voluntary hospitals.

In other ways Newsholme's testimony strikingly resembled the views of the Poor Law reformers of the 1830s. He, as Chadwick had, pointed to the waste of life and money caused by preventable disease. Public health work was a good investment after all. Newsholme, more surprisingly, was also concerned about the moral effect of medical services on the recipient; such assistance must not undermine character. He argued that free medical care would not be demoralizing because cooperation and personal sacrifices would be required of the recipient:

> The form in which medical aid would be given would be such as constantly to enforce on the minds of the patients their duty to the community and to themselves in matters of health. Though they would pay nothing, they would not be merely passive recipients of advice and attention. The influence of the doctor would demand from them habits of life and even sacrifices of personal taste in the interest of the health of the community, their families and themselves, which would leave them conscious of a sensible discharge of duty in return for the attention which they received. The discipline of responsibility into which the system would educate them should, in my judgment, suffice to avoid the loss of self-respect liable to arise from the merely passive receipt of gifts; and it would introduce into the national life an attitude towards matters of personal health that would have an indirect influence upon conduct, while directly restricting disease.[78]

So firm was Newsholme's commitment to the need for character reform that it survived the scrapping of the Poor Law and the advent of the welfare state. He was most insistent on this point in discussions of infant welfare and venereal disease. Both were new additions to the public medical service in which the alternatives to voluntary changes in human behavior were unacceptable invasions of privacy and breaches of personal liberty. But Newsholme's insistence on the necessity of high standards of personal conduct was also founded on his conviction that

JOHN M. EYLER

only ethical evolution guaranteed continued human progress. Ultimately, moral, not economic, imperatives must drive social and medical
policy.

ACKNOWLEDGMENT

The research for this piece was supported in part by NIH Grant LM03765
from the National Library of Medicine, which the author gratefully acknowledges.

NOTES

1. Edwin Chadwick, *Report on the Sanitary Condition of the Labouring Popula-
tion of Gt. Britain* [1842], ed. M. W. Flinn (Edinburgh: Edinburgh University
Press, 1965); & the Report of the Metropolitan Sanitary Commission, more
precisely, *Report of the Commissioners Appointed to Inquire whether Any and What
Special Means May Be Resquite for the Improvement of the Health of the Metropolis,*
B.P.P. 1847–1848,XXIII.

2. 35 & 36 Vict., C. 79, Public Health Act, 1872.

3. See for example Anthony S. Wohl, *Endangered Lives: Public Health in Victorian Britain* (Cambridge, MA: Harvard University Press, 1983), pp. 80–116,
166–199.

4. 52 & 53 Vict., C. 72, Infectious Disease (Notification) Act, 1889; and 62
& 63 Vict., C. 8, Infectious Disease (Notification) Extension Act, 1899.

5. For descriptions of the new welfare services see Bentley B. Gilbert, *The
Evolution of National Insurance in Great Britain: The Origins of the Welfare State*
(London: Michael Joseph, 1966); Deborah Dwork, *War Is Good for Babies and
Other Young Children: A History of the Infant and Child Welfare Movement in England
1898–1918* (London & New York: Tavistock, 1987).

6. Perhaps the best introduction to Victorian attitudes toward poverty and
pauperism is Gertrude Himmelfarb, *Idea of Poverty: England in the Early Industrial Age* (New York: Knopf, 1984).

7. Gilbert, *Evolution of National Insurance,* p. 21.

8. There is no biography of Newsholme. He did leave two accounts of public health work that are autobiographical in character: Arthur Newsholme, *Fifty
Years in Public Health: A Personal Narrative with Comments* (London: George Allen

& Unwin, 1935); *The Last Thirty Years in Public Health: Recollections and Reflections on my Official and Post-Official Life* (London: George Allen & Unwin, 1936). For the more useful obituaries see also *British Medical Journal* no. 1, (1943):680–681; *Lancet*, no. 1, (1943):696; *Nature* 151 (1943):635–636.

9. Henry Mayhew, "A Visit to the Cholera Districts of Bermondsey," *Morning Chronicle*, Sept. 24, 1849, cited in Himmelfarb, *The Idea of Poverty*, p. 315.

10. For a brief discussion of Farr's studies of urban mortality see John M. Eyler, *Victorian Social Medicine: The Ideas and Methods of William Farr* (Baltimore: Johns Hopkins University Press, 1979), pp. 123–154.

11. William Farr, "Letter to the Registrar-General," *Annual Report of the Registrar-General of Births, Deaths, and Marriages in England*, 1 (1838):88–89.

12. Himmelfarb, pp. 158, 160–161, 164.

13. For a brief description of the 1866 investigations of the condition of the sick poor in Metropolitan workhouses, see Brian Able-Smith, *The Hospitals 1800–1948: A Study in Social Administration in England and Wales* (Cambridge, MA: Harvard University Press, 1964), pp. 50–64. See also ibid., pp. 124–127 for the ending of pauperization of patients in the Metropolitan Asylums Board.

14. Chadwick, *Sanitary Condition*, pp. 76–79, 254–276. See also S. E. Finer, *The Life and Times of Sir Edwin Chadwick* (London: Methuen, 1952), pp. 147–149, 154–163, 209–229.

15. Arthur Newsholme, "Some Conditions of Social Efficiency in Relation to Local Public Administration," *Public Health* 22 (1909–1910):404–405.

16. Ibid., p. 408.

17. For the following comments see Arthur Newsholme, "The Relative Importance of the Constituent Factors Involved in the Control of Pulmonary Tuberculosis," *Transactions of the Epidemiological Society, London* ns 25 (1905–1906):49–65; "An Inquiry into the Principal Causes of the Reduction in the Death-Rate from Phthisis during the Last Forty Years, with Special Reference to the Segregation of Phthisical Patients in General Institutions," *Journal of Hygiene* 6 (1906):324–350; *The Prevention of Tuberculosis* (New York: E. P. Dutton, 1908), pp. 224–251; "The Causes of the Past Decline in Tuberculosis," *Charities* 21 (1908–1909):217–222.

18. Newsholme, "Conditions of Social Efficiency," p. 408.

19. For Farr's consideration of this problem see Eyler, *Victorian Social Medicine*, pp. 125–126.

20. Arthur Newsholme, *Report on Infant and Child Mortality: Supplement to the Thirty-ninth Annual Report of the Medical Officer of the Local Government Board* (London: His Majesty's Stationery Office, 1910), pp. 54–56, 60–63, 68–69; *Second Report on Infant and Child Mortality: Supplement to the Forty-second Annual Report of the Medical Officer of the Local Government Board* (London: His Majesty's Stationery Office, 1913), pp. 73–78; "Preliminary Report by the Medical Officer," in *Third Report on Infant Mortality Dealing with Infant Mortality in Lancashire: Supplement to the Forty-third Annual Report of the Local Government Board* (London: His Majesty's Stationery Office, 1914), pp. 14, 21; *Report on Child Mortality at Ages 0–5 in England and Wales: Supplement to the Forty-fifth Annual Report of the Local Government Board, 1915–1916* (London: His Majesty's Stationery Office, 1916), pp. 68–71.

21. Arthur Newsholme, *Report of the Medical Officer of Health for the Third Quarter of 1889* (Brighton, 1889), p. 1.

22. For a brief discussion of the variety of investigations and related proposals for scientific social reform, see Gilbert, *Evolution of National Insurance*, pp. 40–58.

23. Arthur Newsholme, "Poverty in Town Life," *Practitioner* 69 (1902):682–694.

24. Ibid., p. 690.

25. J. A. Hobson, *The Crisis of Liberalism: New Issues of Democracy* (London: P. S. King & Sons, 1909), pp. 159–175, and *Problems of Poverty: An Inquiry into the Industrial Condition of the Poor*, 8th ed. (London: Methuen, 1913), pp. 171–182.

26. James Niven, "Poverty and Disease," *Proceedings of the Royal Society of Medicine* 3, part 2, Epidemiological Section, 3 (1910):4–11.

27. Newsholme, "Conditions of Social Efficiency," p. 406.

28. Arthur Newsholme, "A Discussion on the Co-ordination of the Public Medical Services," *British Medical Journal*, no. 2 (1907):656–657.

29. Arthur Newsholme, "Poverty and Disease, as Illustrated by the Course of Typhus Fever and Phthisis in Ireland," *Proceedings of the Royal Society of Medicine* Part 1, Epidemiological Section, 1 (1908):2–3, 10–14.

30. Arthur Newsholme, "Alleged Physical Degeneration in Towns," *Public Health* 17 (1904–1905):299.

31. Newsholme, "Poverty and Disease," p. 4.

32. Newsholme, "Conditions of Social Efficiency," p. 406; "Causes of the Past Decline," p. 222; *Public Health and Insurance: American Addresses* (Baltimore: Johns Hopkins University Press, 1920), p. 148.

33. Newsholme, "Co-ordination of the Public Medical Services," p. 657.

34. Newsholme, "Conditions of Social Efficiency," p. 409.

35. Newsholme, "An Address on Social Evolution and Public Health," *Lancet*, no. 2 (1904):1334.

36. Gilbert, *Evolution of National Insurance*, p. 19.

37. For good discussions of the political response to the postwar crisis, see G. R. Searle, *The Quest for National Efficiency: A Study in British Politics and Political Thought, 1899–1914* (Oxford: Basil Blackwell, 1971); Gilbert, *Evolution of National*, pp. 59–101; Bernard Semmel, *Imperialism and Social Reform: English Social-Imperial Thought 1895–1914* (Cambridge, MA: Harvard University Press, 1960), pp. 53–82.

38. For an excellent study of this problem see Michael Freeden, *New Liberalism: An Ideology of Social Reform* (Oxford: Clarendon Press, 1978).

39. Ibid., pp. 76–116.

40. Hobson, *Problems of Poverty*, p. 177.

41. Hobson, *Crisis of Liberalism*, p. 165.

42. J. A. Hobson, *The Social Problem* (London, 1901), p. 214, and in Freeden, *New Liberalism*, p. 178.

43. Arthur Newsholme, "The Influence of Civilization upon the Survival of the Fittest," *Annual Report of the Brighton and Sussex Natural History Society*, 1893–1894, pp. 5–12, and "Some Aspects of Heredity," ibid., 1894–1895, pp. 6–8.

44. Arthur Newsholme and T.H.C. Stevenson, "The Decline of Human Fertility in the United Kingdom and Other Countries as Shown by Corrected Birth-Rates," *Journal of the Royal Statistical Society* 69 (1906):34–87; Arthur Newsholme, *The Declining Birth Rate: Its National and International Significance* (London, New York, etc.: Cassell, 1911).

45. See, for example, Newsholme, "Influence of Civilization," pp. 10–12; *Declining Birth Rate*, pp. 57–58; "The Control of Conception," in *Sexual Prob-*

lems of To-Day, ed. Mary Scharlieb (London: Williams and Norgate, 1924), pp. 152, 153; "Some Public Health Aspects of 'Birth Control,'" in *Medical Views on Birth Control*, ed. James Marchant (London: Martin Hopkinson, 1926), p. 132.

46. Newsholme, *Prevention of Tuberculosis*, p. 187; Karl Pearson, *The Fight against Tuberculosis and the Death-rate from Phthisis* (London: Dulau, 1911); Karl Pearson, *Tuberculosis, Hereditary and Environment* (London: Dulau, 1912); Newsholme, *Second Report on Infant Mortality*, pp. 46–48; Karl Pearson, "The Check to the Fall in the Phthisis Death-rate since the Discovery of Tuberculosis and the Adoption of Modern Treatment," *Biometrika* 12 (1918–1919):374–376.

47. Newsholme, *Declining Birth Rate*, pp. 44, 49–52; "Public Health Aspects of 'Birth Control,'" pp. 145–146.

48. Newsholme, *Declining Birth Rate*, pp. 51–52.

49. Newsholme, "Alleged Physical Degeneration," pp. 293–294; "Physical Inspection," *Journal of the Royal Sanitary Institute* 26 (1905):67; *Declining Birth Rate*, pp. 49–50.

50. Newsholme, "Control of Conception," p. 154.

51. Newsholme, *Report on Infant & Child Mortality*, pp. 9–18, 78–82; *Second Report on Infant and Child Mortality*, pp. 43–53.

52. Himmelfarb, *Idea of Poverty*, p. 358.

53. Chadwick, *Sanitary Condition*, pp. 164–165. This passage is reproduced and discussed in Himmelfarb, *Idea of Poverty*, pp. 358–359.

54. Newsholme, "Poverty in Town Life," pp. 685, 688.

55. Newsholme, "Social Evolution," p. 1336.

56. Hobson, *Problems of Poverty*, pp. 178–179.

57. Ibid., p. 175.

58. Newsholme, "Alleged Physical Degeneration," p. 300; "Alcohol and Public Health," in *The Drink Problem in its Medico-Sociological Aspects*, ed. T. N. Kelynack (London: Methuen, 1907), pp. 122–151; *Public Health and Insurance*, pp. 123–124; *Second Report on Infant and Child Mortality*, pp. 78–82.

59. Newsholme, "Alleged Physical Degeneration, p. 300.

60. Newsholme, "Conditions of Social Efficiency," p. 406; *Public Health and Insurance*, pp. 149–150.

61. Newsholme, "Social Evolution," p. 1333; "On the Study of Hygiene in Elementary Schools," *Public Health* 3 (1890–1891):134–136.

62. Newsholme, *Report on Child Mortality*, p. 64.

63. Newsholme, *Report on Infant and Child Mortality*, pp. 70–74; *Report on Child Mortality*, pp. 64–66.

64. Newsholme, "Social Evolution," p. 1333.

65. Arthur Newsholme, "The Duties and Difficulties of Sanitary Inspectors," *Sanitary Record* ns 10 (1888–1889):412.

66. Newsholme, "Social Evolution," p. 1336; "Conditions of Social Efficiency," p. 405.

67. Arthur Newsholme, *Health Problems in Organized Society: Studies in the Social Aspects of Public Health* (London: P. S. King & Son, 1927), pp. 107, 177–179, 188.

68. Newsholme, "Influence of Civilization," pp. 6–8; "Social Evolution," pp. 1331, 1332. See also Thomas Henry Huxley, "Evolution and Ethics" (the Romanes Lecture, 1893), in *Evolution and Ethics and Other Essays* (New York: D. Appleton, 1899), pp. 46–86.

69. Newsholme, "Influence of Civilization," p. 7.

70. Newsholme, *Health Problems*, pp. 98–99, 118–119.

71. Ibid., pp. 118–19, 182–183.

72. Newsholme, *Prevention of Tuberculosis*, p. 189.

73. Newsholme, *Public Health and Insurance*, p. 165.

74. Newsholme, "Social Evolution," pp. 1334–1335.

75. Newsholme, *Health Problems*, p. 119.

76. Newsholme, "Social Evolution," p. 1333.

77. For Newsholme's testimony see *Report of the Royal Commission on the Poor Laws and Relief of Distress*, appendix, vol. 9, *B.P.P.*, 1910, *XLIX* (Cd. 5068), pp. 155–182.

78. Ibid., pp. 164–165.

14 HENRY E. SIGERIST:
His Interpretations of the History of Disease and the Future of Medicine

ELIZABETH FEE

Henry E. Sigerist was America's most widely read, influential—and controversial—historian of medicine during the middle third of this century. A fluent and productive writer, he wrote for an audience far broader than only professional historians, or even physicians. And he wrote in terms that seemed radical to many Americans, including a good number of his Johns Hopkins colleagues.

Sigerist in the 1930s, for example, like many of his generation, saw disease incidence as in part a consequence of—and comment on—capitalist social relations; in his younger years, as Elizabeth Fee emphasizes, he had tended to see it more broadly, as a reflection of culture in general. What both of Sigerist's positions had in common was a centuries-old emphasis on the relation between particular incidences of disease and particular social realities.

Like many of his radical and reform-minded generational peers, Sigerist maintained a strong faith in the social promise of science and medicine. He had no doubt that real diseases existed in the bodies of men and women—in a biological sense unmediated by the constructions placed on those phenomena by society. But he was also much concerned about the contingent aspect of society's response to disease, just as he was with the way economic and ecological factors affected morbidity and mortality. His writings illustrate both the strength and longevity of the social medicine tradition and of the centrality of comparative disease incidence in that tradition. As we have also seen in the case of Arthur Newsholme, Sigerist's changing views reflect the sobering truth that the social policy implications of this emotional and analytical stance are neither obvious nor predictable.

—C. E. R.

HENRY E. SIGERIST DID more than any other individual to establish, promote, and popularize the history of medicine in America.[1] He made the history of medicine relevant to contemporary concerns and greatly broadened its appeal beyond the small company of scholars, collectors, and amateur "gentlemen" physicians who had earlier been interested in the field.[2] On arriving from Germany to become professor of the history of medicine at Johns Hopkins in 1932, he immediately began to make the history of medicine "more comprehensive, more comprehensible, more significant in human and social terms."[3]

Here, I will be concerned with Sigerist's analysis of the social production of disease and its relation to his proposals for the reform of medical care. I will argue that his work on the history of disease has more radical implications than much of his more overtly political writing on the sociology of medicine and medical care organization. Indeed, Sigerist articulated two, largely distinct, positions in relation to the politics of health and disease. On the one hand, his work on the history of disease suggested that the incidence of disease was generated by social and economic conditions and therefore had to be addressed by social and economic reorganization, promoted in part by a "people's war" for health. On the other hand, his active political and sociological work concentrated on the more limited (if still ambitious) goal of changing the organization, delivery, and financing of health services. Sigerist's dual vision of the physician's role—as participant in or leader of the "people's war" for health, and as the provider of individualized preventive and curative care—was mediated by the politics of the possible. These positions were connected for Sigerist by his view of the politics of science and scientific medicine: Socialism represented the form of society in which the benefits of science would be distributed to all.

Some of the apparent inconsistencies in Sigerist's writings stem from his simultaneous involvement in theoretical work and in the practical political struggles of his day; while he permitted himself revolutionary visions, he committed his energy to the more limited goals of the liberal reform movements of the 1930s and 1940s. The first represented a long-term or ideal future, the second, the goals that seemed politically feasible in the United States at that time. His engagement in medical reform has already been discussed by a number of the major participants, who trace their own involvement at least in part to his influence.[4]

Sigerist's views and politics also changed over time. His earliest papers, in Germany in the 1920s, were somewhat abstract explorations of the relation between disease and culture; as the Depression underlined the problems of poverty and unemployment in the United States in the 1930s, he addressed more specific connections between disease and social conditions. His travels in the late 1930s and early 1940s—to South Africa, India, and especially the Soviet Union—also clearly influenced him. In Sigerist's final work in the history of medicine, he attempted to synthesize his knowledge and ideas into a single, comprehensive account of the historical relation of medicine to civilization.[5] He died in Switzerland while writing the second book of his projected eight-volume history of medicine, so this ambitious effort was never completed.

Contemporary readers of Sigerist's work may have difficulty bridging the apparent gap between his analysis of disease as caused by social conditions and his advocacy of universal access to medical care as the solution. One major question in the politics of medical care is whether the provision of medical services has much impact on people's health, at least in comparison to such factors as nutrition, housing, sanitation, education, and employment. The controversy, while not new, has recently been provoked by the work of McKeown and others who have asserted that medical care has been only one, and probably not the most important, factor in improving health.[6] Sigerist's views on the social causation of disease certainly pointed in this direction, while his sociology of medicine retained a central role for medical care and, by extension, for the medical profession. As Charles Rosenberg has noted, many of Sigerist's generation saw no conflict between the call for radical social change and the plea for a more equitable distribution of the benefits of scientific medicine.[7]

Sigerist and many of his contemporaries perceived science and medicine as positive and liberating forces; science was constrained under capitalism, perverted under fascism, and would be fully developed only under socialism.[8] They believed that scientific and medical knowledge were themselves value-free, but that the *uses* of science were determined by the structures of social and economic power. Capitalism thus thwarted science's socially beneficial role by forcing it to serve the ends of private profit and military production rather than human needs. Science and technology—the forces of production—were the main motor of history; the contradiction between the development of these productive forces and outdated social relations would eventually lead to a new mode of production under socialism.

Most contemporary radical and Marxist analysts, more critical than Sigerist of science, technology, and medicine, have abandoned this view of the neutrality of scientific and medical knowledge, and with it the belief that the contradiction between technological developments and capitalist social relations will lead inevitably to a socialist future.[9] Instead, medical knowledge is seen as itself socially constructed, its content reproducing the ideological power relations within any given society, and reflecting struggles over class, gender, and racial divisions. In this view, the class struggle ultimately determines both the form of medical knowledge and larger social transformations; medicine itself is not neutral but performs both liberating and repressive functions. Different positions on the politics and epistemology of science may in part account for the surprising range of opinion about Sigerist's politics: He has been variously described as a liberal, a radical reformer, a Fabian, a socialist, a Marxist, a communist, and a "twentieth-century *philosophe*."[10]

Sigerist's Analysis of the History of Disease

Sigerist repeatedly stated that the history and geography of disease were the foundation of all medicohistorical work and that historians could not fully understand medical theory and practice unless they were familiar with the common disease problems of the relevant period and place.[11] Take one example: To understand the significant role given to the spleen and black bile in the theory of the four humors, one needs to know that the theory was elaborated in a region of endemic malaria, where people's enlarged spleens could easily be felt through the skin. Sigerist argued that the history of disease was also an essential part of the history of civilization. Any history of civilization that failed to investigate disease problems would, at best, be incomplete; health and disease were intimately related to wars, famines, and the fates of nations, and also to art, culture, religion, and philosophy. Studies in the history of disease should therefore aim to illuminate the relation between disease and civilization, between biological and social existence.

Sigerist repeatedly returned to the problems of civilization and disease throughout his scholarly career. He wrote his first papers at the Karl-Sudhoff-Institut für Geschichte der Medizin in Leipzig, where he succeeded Karl Sudhoff as professor of the history of medicine in 1925. At Leipzig, he had already declared that the history of medicine should play the role of mediator between scientific medicine and the humanistic tradition: "The medical historian should help prepare the ground for a new humanism which will harmoniously unify the old humanism with modern science."[12] Members of the *Kyklos* group who worked in the institute saw the history of medicine as a means to clarify current problems of medicine; their interests ranged from the history of disease, to philosophical problems of medical theory, to ethical issues of medical practice.[13]

Disease as a Cultural Expression

Sigerist wanted to tie the history of medicine to larger patterns of culture and cultural transformation. In this search, he was influenced by the great cultural historian Jacob Burckhardt, by the art historian Heinrich Wölfflin, and by other German historians interested in the relation of disease to economic, political, philosophical, and religious influences.[14] He was attracted to Spengler's concept of cultural morphology—the idea that all aspects of a culture reproduce the same structural themes—but later adopted a more flexible relativism in viewing all cultural manifestations of a period as expressions of its "style."[15] Sigerist's essay on William Harvey, for example, showed that both Harvey's work on the circulation of the blood and the baroque artist's contemporary style displayed a common preoccupation with movement.[16] Sigerist also argued, less successfully in Oswei Temkin's view, that the forms of disease prevalent in any period were culturally determined and reflected the "style" of that period.[17]

Many of Sigerist's early associations between forms of civilization and disease now seem abstract and metaphoric. Although he clearly intended the connections between disease and civilization to be more than metaphors, he did not specify the mechanisms relating specific disease problems to cultural forms: The nature of the connection remained unclear. Thus, he described the plague of Justinian as a symptom of the crisis affecting the Mediterranean world in the sixth century and as an expression of the struggle between a dying culture and one striving to emerge.[18] He saw tuberculosis as a pathological expression of the romantic period (an idea that René and Jean Dubos and Susan Sontag would later elaborate in considerable literary detail).[19] Industrial diseases, nervousness, and neuroses to him were the pathological expression of nineteenth-century industrialization—again, a familiar and plausible notion at the metaphorical level.[20]

Throughout his later work, Sigerist would return to the idea that diseases reflected the cultural style of a period. In *Civilization and Disease* (1943), for example, he said that:

> It is interesting to see that there is a certain relation between the prevailing diseases of a given period and their general character and style. The Middle Ages was a period of collectivism and the dominating diseases were such collective diseases as leprosy, plague, or dancing mania that befell entire groups. In the highly individualistic Renaissance, syphilis was in the foreground, a disease that does not attack just anybody, but is acquired through a highly individualistic act. The Baroque period was one of tremendous contrasts and contradictions. . . . The diseases most frequently pictured were deficiency diseases such as hunger-typhus and ergotism, and luxury diseases such as gout and dropsy.[21]

His first effort to link disease to civilization was thus abstract, symbolic, and idealist; gradually, it would be supplanted by a more materialist conception of disease causation.

Disease as an Environmental Response

In *Man and Medicine* (1932), developed from his introductory lectures for medical students in Leipzig, Sigerist made two arguments, one physiological, and one epidemiological, that would continue to structure his future work on the history of disease. From a physiological point of view, he stressed the basic permanence and biological identity of disease processes. Disease, he said, is as old as life itself, and is manifested in the same basic forms at all times: "For disease is after all nothing more than life, life under altered circumstances. Disease occurs as the result of the effect upon the organism of stimuli which exceed the limits of its adaptability."[22]

By contrast, Sigerist's epidemiological view emphasized the widely varying incidence of particular diseases over time and space: "Nothing could be more false than to assume that the diseases that we observe in our society today existed universally and at all times in the same intensity and the same distribution."[23] Because disease as a biological phenomenon reflected an organism's inability to adapt to elements in its environment, the actual occurrence of disease in time and place was caused by existing social conditions. Disease is a real pathological phenomenon, but is socially induced. It thus represents an indictment of the physical and social environment, an expression of the organism's distress. Studies of the history and geographical distribution of specific disease, therefore, provide tools for social criticism by showing where the dangers of the social and physical environment have exceeded the limits of health—the physiological adaptive capacity of the human organism.

Sigerist tried to promote historical and geographical studies of disease soon after his arrival in the United States where he was to succeed William Henry Welch as professor of the history of medicine at the Johns Hopkins Institute of the History of Medicine. In "Problems of Historical-Geographical Pathology" (1933), he suggested a dialectical relationship between civilization and disease: Civilization had solved certain disease problems, and thus promoted health, but had also created new hazards, and thus the emergence of new diseases.[24] With his usual ambitious scope, Sigerist now proposed an international journal for historical-geographical pathology, a series of monographs on the history and geography of different diseases, and an atlas to show the distribution of disease in time and space. This grand design for a research program was, however, never realized.

Disease as a Social Product

In Sigerist's early work, the idea that disease was created by social and economic conditions was present, but submerged. In the United States in the 1930s, this theme began to assume a more central importance in his writing. Economic collapse, poverty, and unemployment were devastating experiences for the United States after the optimism and prosperity of the 1920s; for Sigerist and others aware of the rise of fascism and the growing threat of a world war provoked by imperial ambitions, the future of civilization was at stake. He now began to pay closer attention to the social and economic organization of society and its specific impacts on people's health, and to stress the significance of poverty and associated malnutrition as a major cause of disease, indeed, as *the* major cause. He was struck by the discrepancy in the United States between medicine's highly developed technical capacities and the very limited access to health care available to the working class.[25]

Sigerist began to pay more attention to the effect of working conditions, occupational diseases, and industrial accidents on the health of the working population in Europe and the United States. In 1936 he surveyed the "Historical Background of Industrial and Occupational Diseases" and stated that, while civilizations are valued according to their artistic achievements, the "blood and tears of thousands of human beings" who labored to build those monuments were too often forgotten.[26] From the point of view of human health, the working conditions of each period and country should be an important criterion for judging its civilization. In that same year, Sigerist wrote several times to George Rosen, encouraging him to study the history of occupational diseases and suggesting a monograph on the diseases of miners.[27] Seven years later, Rosen published *The History of Miners' Diseases: A Medical and Social Interpretation*, for which Sigerist wrote the introduction.[28] From the mid-1930s, while engaged in a multitude of scholarly activities, Sigerist also became increasingly active in antifascist movements, such as the Medical Bureau to Aid Spanish Democracy, the Federation of Faculty Committees for Aid to the Spanish People, and the American Committee for Democracy and Intellectual Freedom; he was national sponsor of the American Association of Scientific Workers, an organization of radical and left-wing scientists.[29]

Soon after coming to the United States, Sigerist published *American Medicine* (1934). In its epilogue he declared that the United States and the Soviet Union would determine the future of medical care. He used the contrast between their political and economic structures to relate their different forms of medical care to differences in their social structures and political philosophies.[30] He had first become interested in Soviet medicine at Leipzig through personal acquaintance with the Soviet historian of medicine Ilya Davidovič Strašun.[31] He now began to discuss the organization of capitalist and socialist societies and to prepare for a projected trip to the Soviet Union by extensive background reading and language studies.

In 1935, 1936, and 1938, Sigerist spent his summers traveling in the Soviet Union, returning highly enthusiastic about the Soviet reorganization of medical care and efforts to integrate preventive and curative medicine.[32] These experiences undoubtedly strengthened his vision of the future of medical care, the need for national organization and financing of health care services, and the extension of services to the whole population. They also strengthened his conviction that social and economic conditions were largely responsible for ill health and disease. In *Socialized Medicine in the Soviet Union* (1937), for example, he expanded his definition of the "diseases" caused by economic conditions. In addition to stressing the need for protecting the health of industrial workers, he outlined a broad new category of "social diseases" caused by poor economic conditions.[33] He had earlier termed tuberculosis a "disease of romanticism," but now redefined it as a disease generated by unhealthy living and working conditions. Once having depicted venereal diseases as an expression of Renaissance individualism, he now linked them to prostitution, in turn caused by poverty, unemployment, and lack of economic opportunities for women.[34] He now defined alcoholism as a disease of misery: Poor living conditions, a sense of oppression, and a lack of educational and recreational facilities drove men to drink. He defined crime, too, as a social disease, caused by poverty, unemployment, and frustration. The "treatment" for crime was full employment and reeducating criminals for work in labor communes. (Sigerist reported favorably on one visit to a labor camp in the Soviet Union where 5,000 former thieves and their families were manufacturing skis, tennis rackets, and footballs.)[35]

The Soviet Union seemed to offer a vital alternative to the stagnation of a depression economy in the United States. It demonstrated an unprecedented rate of economic growth, generating more jobs than the number of workers to fill them; at least to sympathetic observers, centralized planning was a rational alternative to the economic disorganization of capitalism in crisis. Sigerist was not the only enthusiastic observer of the "great social experiment" of the Soviet Union; Karl Compton, an American physicist and chairman of President Roosevelt's Science Advisory Board, praised Soviet scientific achievements.[36] Arthur Newsholme, former head of the public health service in England, and John Adams Kingsbury, secretary of the Milbank Memorial Fund, also wrote favorably about Soviet medicine.[37] And Beatrice and Sidney Webb, the British Fabians, declared that the Soviet Union had ushered in a "new civilization."[38] Sigerist's work on the Soviet Union aroused considerable public interest, and he was deluged with invitations to address both medical and popular audiences in the late 1930s.

In "The History of Medical History" (1938), Sigerist again argued that the incidence of particular diseases depended on social, economic, and geographic factors. Since the changing social environment had so profoundly altered health conditions, the historian of medicine must first acquire a thorough knowledge of social and economic history:

When we study the history of disease we will soon find that its incidence is determined primarily by the *economic and social conditions* of a society. . . . The mode of production and the working conditions are largely responsible for whether a man's life will be healthy or not. . . . In other words: we must be thoroughly familiar with economic and social history before we can approach the history of disease.[39]

In *Civilization and Disease* (1943), Sigerist showed how disease was related to such influences as hunger and diet, clothing, housing, water supplies, sanitation, and working conditions. Drawing on his observations in the Soviet Union, he advocated adequate rest and recreation for the working population, paid annual vacations, wages sufficient for a good basic standard of living, periodic health examinations, and comprehensive facilities for treating minor ailments as well as serious diseases: "Steady employment under the best possible hygienic conditions, the correct balance between work, rest and recreation, and wages that permit a decent standard of living—these are basic and significant factors of public health."[40]

While civilization had, in general, led to rising standards of living and improvements in health conditions, this was not universally the case. Sigerist's travels in South Africa persuaded him that, for those subjected to colonial exploitation, civilization had meant a decline in standards of living, especially in nutritional standards relative to "primitive" society:

Under the most different climates primitives devised a balanced diet. . . . The great deal of malnutrition among them today is the result of prevailing social and economic conditions, the consequence of colonial exploitation. As long as the Bantu were in possession of their homeland, they had a balanced diet consisting chiefly of milk, mealy meal (ground African corn), and indigenous herbs. Once the white man took their land away and they were reduced to living on small overstocked farms, the cows had not enough milk, the land produced not enough corn, and the people in contact with the white man forgot the use of herbs.[41]

Similarly, Sigerist noted that, while Herodotus had considered the ancient Egyptians among the healthiest of men, Egypt in 1938 had one of the highest recorded death rates in the world. Had Herodotus been wrong? Sigerist argued that health conditions had objectively deteriorated because the population had doubled, methods of cultivation had remained primitive, and much fertile land had been turned over to the production of such export cash crops as cotton. The result was nutritional impoverishment for the majority and lowered standards of health.[42] "Civilization" could thus mean immiserization, poverty, and ill health, just as it could lead to health, wealth, and happiness.

The Physician's Task: The "People's War" for Health

If, as Sigerist often stated, the problems of disease were primarily caused by social and economic conditions, what were the implications for the future of medicine and for the task of the physician? Were physicians simply to treat the symptoms consequent upon social problems or were they to deal with the prevention of disease at its source? Sigerist was not content solely to analyze the history of disease; his aim was to transform and improve the present. But his program for dealing with disease problems and his agenda for the "new physician" seem to have combined two largely distinct and possibly incompatible strategies. If disease incidence was due to social and economic conditions, then the solution must involve social, economic, and political changes. Here, Sigerist took inspiration from Rudolf Virchow and the German health movement of 1848, although he was critical of their failure to involve the mass of the population in the struggle for health.

Sigerist frequently cited two declarations from Virchow's journal *Die medizinische Reform:* "The physicians are the natural attorneys of the poor, and social problems fall to a large extent within their jurisdiction," and, "Medicine is a social science, and politics is nothing else but medicine on a large scale."[43] The physician who is close to the people knows social conditions better than anyone else and must therefore be committed to social reform:

The social causes of illness are just as important as the physical ones. Etiological therapy means more than killing a few bugs. The medical officer of health and the practitioners of a distressed area are the natural advocates of the people. They well know the factors that paralyze all their efforts. They are not only scientists but also responsible citizens, and if they did not raise their voice, who else should?[44]

Similarly, the physician's role in the case of occupational health was not simply to treat sick workers but to fight for the reduction of working hours, the provision of adequate rest and recreation, and the improvement of wages and working conditions: "The physician . . . who is familiar with [working] conditions and knows the evil effects of such work on the people's health must assume leadership in the struggle for the improvement of conditions: His concern is not whether an enterprise is profitable or not. His place is with the workers, whose protector he is."[45]

Internationally, the people of the poorer countries of the world also had to organize and fight for better social and economic conditions. These countries were plagued by preventable diseases, long since vanished from Europe and North America, and their populations would continue to suffer until economic conditions were transformed. The solution to their problems was more political than medical.

The Physician's Task: A New System of Medicine

In the United States in the 1930s, the medical world was deeply embroiled in controversies over the organization and financing of medical care. Liberal and progressive reformers agreed that medical care was a poorly distributed social good that should be made available to all the American people at a price they could afford. The majority final report of the Committee on the Costs of Medical Care (CCMC) in 1932 recommended that medical care be provided by groups of physicians organized around hospitals and health centers.[46] Its costs were to be covered by group payment, financed either by insurance or taxation, thus removing economic barriers to care. This report, supported by major foundations, placed medical care organization on the national political agenda, but provoked a storm of controversy within the medical profession.[47]

Sigerist became heavily involved in the subsequent debates and acted as a spokesperson for those advocating national health insurance.[48] He saw national health insurance as a relatively conservative measure: "The idea of social insurance is by no means new but has a history of over sixty years. It is not a revolutionary but on the contrary a conservative issue. It does not tend to overthrow the existing economic order but provides a corrective mechanism that mitigates its hardships."[49] He supported national health insurance as the best that could be hoped for under a Roosevelt administration but personally favored a national health service, financed through taxation, with physicians on salary, and he enthusiastically embraced the concept of a national health program as developed by the National Health Conference of 1938.

In speaking about the need to reorganize medical care, Sigerist redefined the tasks of the physician: the promotion of health, prevention of illness, restoration of health, and rehabilitation of the patient. But even this broad vision of medical aims failed to incorporate the social goals of full employment and the abolition of poverty that he had previously argued were the real basis of health.

In common with other progressive reformers, Sigerist placed more emphasis on access to medical care than on access to political power. And even when he broadened the concept of medical care to include health promotion and disease prevention, he tended to focus on education as the route to health promotion, and individualized medical attention as the route to disease prevention. The physician was to see people before they became sick and to advise them how best to maintain their health.[50] Similarly, the statesman was to ask the physician for advice on such questions as nutrition and housing.[51] The shift in emphasis is significant: The physician, participant in the "people's war" for health, becomes instead an "adviser to the state."[52]

The Chronic Diseases of "Old Age"

When Sigerist spoke about chronic diseases, he did not seem to perceive them as a consequence of environmental or industrial toxins, intense and highly pressured pace of work, people's lack of control over their work and their lives, poor nutrition, or unhealthy habits linked to social stress—all explanations that might have been compatible with his general orientation to the social causes of disease. Instead, he defined chronic diseases as diseases of "old age." His model for their control was that of clinical medicine, with the clinical gaze focused on the individual patient. The physician was to monitor the person rather than the social or physical environment: "Acute diseases are no longer in the foreground, but the chronic diseases of mature and old age, those diseases that require close and steady supervision by the physician."[53] In the case of the chronic diseases, Sigerist thus reduced the problem of health to the problem of medical care: "We no longer accept the Greek view that health is a privilege of the rich, but agree with the medieval idea that everybody, rich and poor, should have all the medical care that science can give."[54]

Sigerist consistently argued that chronic illness required individual medical care rather than the community-wide public health measures appropriate to the infectious diseases of the past. "As long as acute infectious and communicable diseases dominate the scene, they will be opposed by general public health measures, quarantine, sanitation, immunizations, and similar measures; while the chronic diseases of wear and tear call for individual services of the general practitioner and specialist, preventive and curative, and the major organizational task at such a stage is to make all such services easily available to all the people."[55]

Sigerist's thinking on chronic illness resembled that of Charles-

Edward A. Winslow, who celebrated the shift from a social or collective public health orientation to a focus on individual health.[56] In Sigerist's understanding of public health history, the first stage of public health was concerned with removing the environmental causes of disease: providing clean water and improving the urban environment. The second stage, exemplified by the tuberculosis movement, involved both the care of patients and supervision of their immediate environment. The third stage, the child hygiene movement, brought individual health examinations of schoolchildren. The next stage would extend periodic health examinations to the entire adult population. Attention had thus moved from community public health to individual preventive medicine; dealing with the chronic diseases was "a question of getting hold of the individual."[57]

Clinical Preventive Medicine

When public health focused on the individual, it became clinical preventive medicine: a personal relationship between a physician and patient. Sigerist claimed that this emphasis on individual health represented not a constriction but an expansion of vision: "But today hygiene has much wider possibilities of action. In approaching the individual, in supervising him, in determining by periodic examinations his constitution, his hereditary defects, the dangers which menace him, the doctor can certainly prevent a great many diseases and can give effective care before it is too late."[58]

In the field of the chronic diseases, preventive and curative medicine merged in the person of the general practitioner. Sigerist's "new physician" would apply a social understanding of disease causation to the care of patients in a clinical setting: "Clinical medicine must be taught differently than heretofore. Every case must be analyzed medically and socially as to the factors that have made it possible, and conclusions must be drawn how to prevent similar cases in the future."[59] The new physician would be aided by a medical curriculum that integrated epidemiology and preventive medicine with traditional biomedical and clinical studies. Departments of preventive medicine in every medical school were to demonstrate that "preventive medicine is no longer the prerogative of public health officers but the concern of every practitioner of medicine."[60]

The general practitioner would extend medical care from a base in a clinic or hospital out to patients' homes and workplaces, and return to consult a group of specialists at a health center for any needed help or advice. Doctors would cooperate in teams, to prevent illness before it struck. The new physician would be "scientist and social worker, ready to cooperate in teamwork, in close touch with the people he disinterestedly serves, a friend and leader, he directs all his efforts toward the prevention of disease and becomes a therapist where prevention has broken down—the social physician protecting the people and guiding them toward a healthier and happier life."[61] Promotion of health, however, was an individual responsibility: "The state can protect society very effectively against a great many dangers, but the cultivation of health, which requires a definite mode of living, remains to a large extent an individual matter and is the result of education."[62]

In discussing his personal experience of chronic disease, Sigerist noted his own unhealthy mode of living: He was overweight, a heavy smoker, did practically no exercise, and often worked seven days a week and slept five hours a night. His personal therapy included "a few weeks of complete rest and relaxation away from everybody, with light exercise, walks in the enchanting landscape of New York State, combined with a strict reducing diet, mineral baths, massage, nasal inhalations . . . solitude and meditation."[63] He failed to notice that his own restful therapeutic regimen was largely determined by his class position.

We have seen that, in his analyses of the history of disease, Sigerist advocated social and economic reforms to adapt the environment to man's needs.[64] But in discussing medical reforms, he tended to speak instead of adjusting the individual to the environment: "Medicine, usually regarded as a natural science, actually is a social science because its goal is social. Its primary target must be to keep individuals adjusted to their environment as useful members of society, or to readjust them when they have dropped out as a result of illness."[65]

The U.S.S.R.: The Future of Medical Care

Sigerist's proposals for the future of medicine were strikingly similar to his descriptions of the organization of medical care in the Soviet Union. He claimed that the Soviet health care system had abolished the distinction between preventive and curative medicine and was built entirely around the idea of prevention; every medical worker tried to prevent disease.[66] And prevention was carried out through close surveillance of each individual's health: "The general idea is to supervise the human being medically, in a discrete and unobtrusive way, from the moment of conception to the moment of death. . . . Medical supervision begins with the pregnant woman and the woman in childbirth, proceeds to the infant, the pre-school and school child, the adolescent, and finally the man and woman at work."[67] He now applauded the view that disease was simply a biological process that must be dealt with scientifically: "In a society that is based on scientific principles and whose philosophy is rational, disease has lost its magical implications and is considered for what it is, a biological process that has to be faced openly without fussing and has to be treated scientifically."[68] Especially in the case of chronic diseases, this scientific approach meant clinical preventive medicine and a focus on the individual.

Science as Social Progress

Sigerist's dual vision of the physician's role—as participant in or leader of the "people's war" for health, and as the provider of individualized preventive and curative care—was mediated in part by the politics of the possible. If the existing political and economic situation precluded the basic conditions for healthy living—good food, housing, and a safe workplace—then the physician could at least help the individual patient "adjust" to the prevailing social and physical environment. But Sigerist's views were also influenced by his understanding of the chronic diseases as diseases of "old age" requiring individual medical care more than social or environmental reforms. He further bridged the apparent disjunction between the need for fundamental social and economic reform and the need for access to medical care by placing a high value on science and technology, especially on scientific medicine.

Sigerist displayed a highly positive, even uncritical view of the progress and achievements of medical science. The problem with medicine, as he saw it, was not the nature of medical knowledge but the failure to distribute the fruits of that knowledge equitably: The infinite technological possibilities of medicine were constrained by market forces.[69] The United States exemplified both this abundance of medical knowledge and the social failure to make its benefits fully available in the form of medical care.

In *Civilization and Disease* (1943), Sigerist tried to integrate his analysis of the history of disease—disease generated by social and economic conditions—and the progress of medicine—generated by advances in scientific knowledge. He admitted that medical science could not take all the credit for improved health conditions since the seventeenth century; they were due mainly to rising standards of living. But he tacitly contradicted himself by reasserting the importance of medicine: "Civilization fights disease in many ways, but medicine nevertheless is its most powerful weapon."[70]

The conclusion of *Civilization and Disease* restated this longstanding ambivalence about the relative importance of economic conditions and medical care. Sigerist argued that over half the world's population lives in such atrocious health conditions that medical care is essentially irrelevant: "To immunize colonial people against disease with one hand and exploit them into starvation with the other is a grim joke."[71] When poverty is the chief cause of disease, the remedy must be an adequate basic standard of living, with access to the necessities of life.

For Sigerist, however, the answer to poverty lay in the application of scientific knowledge to agricultural production. Again, he believed that the Soviet Union could serve as a model and that the reorganization of agriculture after the Revolution had ended famines and crop failures.[72] Just as science could show how to improve the fertility of the soil and the quality and quantity of crops, it could also show how to distribute food and avoid starvation in the midst of plenty. In the United States, he said, food could be abundantly produced but not rationally distributed; an irrational economic system had thus resulted in the slaughtering of millions of animals in the midst of the Depression.[73]

Sigerist believed that scientific knowledge could also help solve the social problems of war, poverty, and crime: "War is a social disease, like poverty or crime. When it breaks out it reminds us that we are still in the initial stages, in the prehistory of civilization, not far removed from savagery. It reminds us that although we like to play with science and kill with scientific weapons, we have not yet learned to approach the basic problems of social life—production, distribution, and consumption—scientifically."[74]

A scientifically organized and rational society would in turn increase social interdependence; individuals would give up some liberties in favor of increased social responsibility. Here, Sigerist introduced a new disease analogy: Within a rationally organized society, excessive individualism assumes a pathological form. Individuals who insist on their freedom at the expense of the social good are like cancer cells that multiply at the expense of the rest of the organism: "As soon as a cell-group begins to lead an independent life without regard to the rest of the cells, a malignant tumor develops which will destroy the whole organism. This applies to society as well. The more specialized, the more differentiated a society becomes, the more the individual members have to give up liberties and assume duties toward society."[75]

Sigerist argued further that, in a rationally organized society, the application of scientific knowledge and principles may appropriately replace much of the traditional realm of politics. The physician becomes a scientific manager whose job is to readjust patients to their social and physical environment. Physicians are also responsible for managing criminals, who are simply sick people in need of scientific care and attention.[76]

The idea of scientific progress gave Sigerist a coherent framework for his analysis of history and his politics. His plan to write the whole history of medicine as a single connected account would hardly have been possible without some unifying progressive theme. Oswei Temkin had objected to Sigerist's beginning the *History of Medicine* (1951) with a section describing medicine among contemporary "primitive peoples," on the grounds that this could not substitute for an account of the origins of medicine—about which little could be known.[77] But starting with "primitive medicine" gave Sigerist the framework for an overall argument that was historical, progressive, and political: a study of uneven progress, to be sure, but one that moved gradually onward and upward from magic and mysticism, through philosophy, to science. The story of the scientific revolution was one of liberation from the "bonds of magic and religion";[78] from the time of Andreas Vesalius to the twentieth century, scientific modes of thought and understanding had penetrated ever further into medical knowledge and practice.[79] Science was still young, and our knowledge incomplete, but optimism about the future was justified by the prospects of continued developments.[80]

Sigerist's historiographical framework closely identified scientific and social progress; scientific knowledge was a motor of human progress. For him, the application of science to society was also synonymous with the application of Marxist philosophy: "The philosophy of Marxism is erected upon the foundation of the natural sciences and the science of economics. It is rational. Where such a philosophy prevails, scientific research has the best possible chances of development. The two characteristic features of Soviet science are the disappearance of the distinction between theory and practice, or between pure and applied science, and the planning of scientific research on a nation-wide scale."[81] Because the most effective organization of medical care was essentially a rational process whereby a given level of medical technology was scientifically distributed, the rational reorganization of medical care in the United States should result in a system similar to that of the Soviet Union: "Once we resolve to bring health to all the people in town and country, irrespective of race, creed or economic status, I feel that the methods we develop to do so will resemble those of the USSR, despite our different social and economic structures, because, after all, the technology we use is the same."[82] In this reading, the level of science and technology, rather than the form of economic and political organization, determines the structure of medical care.

It will be evident from this discussion that I believe Sigerist, in common with many intellectuals of the 1930s and 1940s, overvalued the inherent progressive force of scientific and technological development and failed to see the ways in which science itself is culturally determined. His studies of the history of disease were remarkable in linking disease to broader social, economic, cultural, and political forces. I would argue, however, that his view of the history of disease was culturally richer and more complex than his view of science. His admiration for scientific achievements led him to a form of technological determinism, to a view of the physician's role as scientific and social manager, and to a particularly enthusiastic and uncritical approval of social and scientific developments in the Soviet Union in the 1930s.

As historians of science have gained a more complex understanding of the social and cultural determinants of scientific knowledge and practice, it has been—at least in part—at the cost of the political optimism shared by Sigerist and many of his contemporaries, that scientific knowledge would lead us into a new world of reason and social equality. If we have abandoned the scientific optimism of that era, and also the uncritical faith that the Soviet Union represents perfected social justice, we may still respect and endorse Sigerist's conviction that studies in the history of medicine can help us address the broader philosophical, ethical, and political issues confronting contemporary medicine and health care.

NOTES

1. For a complete bibliography of Sigerist's work, see Genevieve Miller, ed., *A Bibliography of the Writings of Henry E. Sigerist* (Montreal: McGill University Press, 1966). Some of his more influential articles have been republished in F. Marti-Ibañez, ed., *Henry E. Sigerist on the History of Medicine* (New York: MD Publications, 1960) and Milton I. Roemer, ed., *Henry E. Sigerist on the Sociology of Medicine* (New York: MD Publications, 1960).

2. G. Miller, "Medical History," in Ronald L. Numbers, ed., *The Education of American Physicians* (Berkeley: University of California Press, 1980), pp. 290–308.

3. L. G. Stevenson, "A Tribute to the Influence of Henry Sigerist," *Journal of the History of Medicine and Allied Sciences* 13 (1958):212–213.

4. L. A. Falk, "Medical Sociology: The Contributions of Dr. Henry E. Sigerist," *Journal of the History of Medicine and Allied Sciences* 13 (1958):214–228; M. I. Roemer, "Henry Ernest Sigerist: Internationalist of Social Medicine," *Journal of the History of Medicine and Allied Sciences* 13 (1958):229–243; M. I. Roemer, "Medical Care Programs in Other Countries: Henry Sigerist and International Medicine," *American Journal of Public Health* 48 (1958):425–427; M. Terris, "The Contributions of Henry E. Sigerist to Health Service Organization," *Milbank Memorial Fund Quarterly* 53 (1975):489–530; G. Silver, "Social Medicine and Social Policy," *Yale Journal of Biology and Medicine* 57 (1984):851–864.

5. H. E. Sigerist, *A History of Medicine. I. Primitive and Archaic Medicine* (New York: Oxford University Press, 1951); *A History of Medicine: II. Early Greek, Hindu and Persian Medicine* (New York: Oxford University Press, 1961).

6. T. McKeown, *The Modern Rise of Population* (New York: Academic Press, 1976).

7. C. Rosenberg, "Disease and Social Order in America: Perceptions and Expectations," *Milbank Quarterly* 64, suppl. 1 (1986):34–35.

8. P. J. Kuznick, *Beyond the Laboratory: Scientists as Political Activists in 1930s America* (Chicago: University of Chicago Press, 1987); G. Werskey, *The Visible College: The Collective Biography of British Scientific Socialists of the 1930s* (New York: Holt, Rinehart and Winston, 1978).

9. V. Navarro, *Crisis, Health, and Medicine: A Social Critique* (New York and London: Tavistock, 1986).

10. See, for example, R. Frankenberg, "Functionalism and After? Theory and Developments in Social Science Applied to the Health Field," *International Journal of Health Services* 4 (1974):411–427; M. Terris, "The Contributions of Henry E. Sigerist to Health Service Organization," *Milbank Memorial Fund Quarterly* 53 (1975):489–530; F. G. Vescia, "Henry E. Sigerist: The Years in America," *Medizinhistorisches Journal* 14 (1979):218–232; G. Rosen, "Toward a Historical Sociology of Medicine: The Endeavor of Henry E. Sigerist," *Bulletin of the History of Medicine* 32 (1958):500–516.

11. H. E. Sigerist, "The History of Medical History," in New York Academy of Medicine, *Milestones in Medicine* (New York: D. Appleton-Century, 1938), pp. 163–184.

12. H. E. Sigerist, "Aufgaben und Ziele der Medizingeschichte," *Schweizerische medizinische Wochenschrift* 3 (1922):12.

13. A. Thom and K.-H. Karbe, *Henry Earnest Sigerist (1891–1957): Ausgewählte Texte* (Leipzig: Johann Ambrosius Barth, 1981), p. 18.

14. G. Rosen, "Toward a Historical Sociology of Medicine: The Endeavor of Henry E. Sigerist," *Bulletin of the History of Medicine* 32 (1958):500–516; "Critical Levels in Historical Process: A Theoretical Exploration Dedicated to Henry Ernest Sigerist," *Journal of the History of Medicine and Allied Sciences* 13 (1958):179–185.

15. O. Temkin, "Henry E. Sigerist and Aspects of Medical Historiography," *Bulletin of the History of Medicine* 32 (1958):485–499.

16. H. E. Sigerist, "William Harvey's Stellung in der europäischen Geistesgeschichte," *Archiv für Kulturgeschichte* 19 (1928):158–168.

17. H. E. Sigerist, "Kultur und Krankheit," *Kyklos* 1 (1928):147–156.

18. G. Rosen, "Toward a Historical Sociology of Medicine: The Endeavor of Henry E. Sigerist," *Bulletin of the History of Medicine* 32 (1958):508.

19. R. Dubos and J. Dubos, *The White Plague: Tuberculosis, Man, and Society* (Boston: Little, Brown, 1952); S. Sontag, *Illness as Metaphor* (New York: Farrar, Straus and Giroux, 1977).

20. H. E. Sigerist, *Man and Medicine: An Introduction to Medical Knowledge* (New York: Norton, 1932), p. 180.

21. H. E. Sigerist, *Civilization and Disease* (Chicago: University of Chicago Press, 1943), p. 186.

22. H. E. Sigerist, *Man and Medicine: An Introduction to Medical Knowledge* (New York: Norton, 1932), p. 172.

23. Ibid., p. 173.

24. H. E. Sigerist, "Problems of Historical-Geographical Pathology," *Bulletin of the Institute of the History of Medicine* 1 (1933):10–18.

25. A. Thom and K.-H. Karbe, *Henry Ernest Sigerist (1891–1957): Ausgewählte Texte* (Leipzig: Johann Ambrosius Barth, 1981), p. 25.

26. H. E. Sigerist, "Historical Background of Industrial and Occupational Diseases," *Bulletin of the New York Academy of Medicine* 12 (1936):597–609.

27. A. J. Viseltear, "The George Rosen–Henry E. Sigerist Correspondence," *Journal of the History of Medicine and Allied Sciences* 33 (1978):281–313.

28. G. Rosen, *The History of Miners' Diseases: A Medical and Social Interpretation* (New York: Schuman, 1943).

29. N. S. Beeson, ed., *Henry E. Sigerist: Autobiographical Writings* (Montreal: McGill University Press, 1966); H. E. Sigerist, "A Summer of Research in European Libraries," *Bulletin of the History of Medicine* 2 (1934):402–409.

30. H. E. Sigerist, *American Medicine* (New York: Norton, 1934).

31. A. Thom and K.-H. Karbe, *Henry E. Sigerist (1891–1957)*, p. 26.

32. H. E. Sigerist, *Socialized Medicine in the Soviet Union* (New York: Norton, 1937); *Medicine and Health in the Soviet Union* (New York: Citadel Press, 1947).

33. H. E. Sigerist, *Socialized Medicine in the Soviet Union*, p. 217.

34. Ibid., p. 228.

35. Ibid., pp. 235–236.

36. P. J. Kuznick, *Beyond the Laboratory: Scientists as Political Activists in 1930s America* (Chicago: University of Chicago Press, 1987), p. 126; K. T. Compton, "Put Science to Work: The National Welfare Demands a National Scientific Program," *Technology Review* 37 (1935):133–158.

37. A. Newsholme and J. A. Kingsbury, *Red Medicine: Socialized Health in Soviet Russia* (Garden City, NJ: Doubleday, Doran, 1933).

38. B. Webb and S. Webb, *Soviet Communism: A New Civilization?* (New York: Scribner's Sons, 1936).

39. H. E. Sigerist, "The History of Medical History," in New York Academy of Medicine, *Milestones in Medicine* (New York: D. Appleton-Century, 1938), pp. 179–180.

40. H. E. Sigerist, *Civilization and Disease* (Chicago: University of Chicago Press, 1943), p. 55.

41. H. E. Sigerist, *A History of Medicine. I. Primitive and Archaic Medicine* (New York: Oxford University Press, 1951), p. 147.

42. Ibid., p. 223.

43. H. E. Sigerist, *Medicine and Human Welfare* (New Haven: Yale University Press, 1941), p. 93.

44. Ibid., p. 134.

45. Ibid., p. 133.

46. Committee on the Costs of Medical Care, *Medical Care for the American People: The Final Report of the Committee on the Costs of Medical Care* (Chicago: University of Chicago Press, 1932).

47. F. A. Walker, "Americanism versus Sovietism: A Study of the Reaction to the Committee on the Costs of Medical Care," *Bulletin of the History of Medicine* 53 (1979):489–504.

48. N. S. Beeson, ed., *Henry E. Sigerist: Autobiographical Writings* (Montreal: McGill University Press, 1966).

49. H. E. Sigerist, "Medical Care for All the People," *Canadian Journal of Public Health* 35 (1944):232; *The University at the Crossroads: Addresses and Essays* (New York: Henry Schuman, 1946), p. 74.

50. H. E. Sigerist, "The Place of the Physician in Modern Society," *Proceedings of the American Philosophical Society* 90 (1946):73.

51. Ibid., p. 72.

52. H. E. Sigerist, "Remarks on Social Medicine in Medical Education," in Milton L. Roemer, ed., *On the Sociology of Medicine* (New York: MD Publications, 1952), p. 362.

53. Ibid., p. 361.

54. H. E. Sigerist, *Medicine and Human Welfare*, p. 139.

55. H. E. Sigerist, *A History of Medicine. I. Primitive and Archaic Medicine*, pp. 76–77.

56. C.-E. A. Winslow, *The Evolution and Significance of the Modern Public Health Campaign* (New Haven: Yale University Press, 1923).

57. H. E. Sigerist, *American Medicine* (New York: Norton, 1934), p. 264.

58. H. E. Sigerist, "L'inquiétude actuelle dans le monde médical," *Schweizerische medizinische Wochenschrift* 65 (1935):1007–1010.

59. H. E. Sigerist, *The University at the Crossroads: Addresses and Essays* (New York: Henry Schuman, 1946), p. 114.

60. Ibid., p. 131.

61. Ibid., p. 114.

62. H. E. Sigerist, *Medicine and Human Welfare* (New Haven: Yale University Press, 1941), p. 103.

63. H. E. Sigerist, "Living Under the Shadow," *Atlantic Monthly* 189 (1952):29–30.

64. H. E. Sigerist, *Landmarks in the History of Hygiene* (London: Oxford University Press, 1956).

65. H. E. Sigerist, "The Place of the Physician in Modern Society," *Proceedings of the American Philosophical Society* 90 (1946):275–279.

66. H. E. Sigerist, *Socialized Medicine in the Soviet Union* (New York: Norton, 1937), pp. 95–96.

67. Ibid., p. 96.

68. Ibid., pp. 97–98.

69. H. E. Sigerist, "Medical Care for All the People," *Canadian Journal of Public Health* 35 (1944):253–267.

70. H. E. Sigerist, *Civilization and Disease* (Chicago: University of Chicago Press, 1943), p. 234. See also H. E. Sigerist, "What Medicine Has Contributed to the Progress of Civilization," *International Record of Medicine* 168 (1955):383–391.

71. H. E. Sigerist, *Civilization and Disease*, p. 236.

72. Ibid., pp. 9–10.

73. J. Poppendieck, *Breadlines Knee-Deep in Wheat: Food Assistance in the Great Depression* (New Brunswick, NJ: Rutgers University Press, 1986).

74. H. E. Sigerist, *Medicine and Human Welfare* (New Haven: Yale University Press, 1941), p. 135.

75. H. E. Sigerist, *Socialized Medicine in the Soviet Union* (New York: Norton, 1937), p. 20.

76. H. E. Sigerist, *Medicine and Human Welfare*, pp. 100–101.

77. O. Temkin, "Henry E. Sigerist and Aspects of Medical Historiography," *Bulletin of the History of Medicine* 32 (1958):485–489.

78. H. E. Sigerist, *Civilization and Disease*, p. 133.

79. H. E. Sigerist, "The Foundation of Human Anatomy in the Renaissance," *Sigma Xi Quarterly* 22 (1934):8–12.

80. H. E. Sigerist, *Civilization and Disease*, p. 178.

81. H. E. Sigerist, *Socialized Medicine in the Soviet Union*, p. 292.

82. H. E. Sigerist, *Medicine and Health in the Soviet Union* (New York: Citadel Press, 1947), p. 10.

LIST OF CONTRIBUTORS

Robert A. Aronowitz is assistant professor of medicine in the University of Medicine and Dentistry of New Jersey/Robert Wood Johnson Medical School. His forthcoming book is a collection of essays on the social construction of chronic disease.

Barbara Bates is clinical professor in the School of Nursing, University of Pennsylvania, and in the Department of Medicine, Medical College of Pennsylvania and Hahnemann University. An authority on physical diagnosis, she recently published a historical study of the American experience with tuberculosis.

Joan Jacobs Brumberg is professor in the Department of Human Development and Family Studies at Cornell University. She has published in the area of nineteenth-century religious history as well as the history of American women. Author of a recent book on the history of anorexia nervosa, she recently completed a book on the changing experience of female sexual maturation in the United States entitled *Body Projects: An Intimate History of America's Adolescent Girls*.

Ellen Dwyer is associate professor in the Departments of Criminal Justice and History at Indiana University. Her previous research concerned the history of institutional psychiatry, and she is now completing a study of the treatment of epilepsy in twentieth-century America.

Peter C. English is associate professor of history and pediatrics at Duke University. His interests focus on the development of pediatrics and the history of medicine in general. Dr. English is completing a history of rheumatic fever from the eighteenth through the twentieth centuries.

John M. Eyler is associate professor in the Department of the History of Medicine at the University of Minnesota, Twin Cities. His previous publications dealt with the history of Anglo-American medicine, public health, and epidemiology. He is the author of a forth-coming biography of Sir Arthur Newsholme which focuses on his career in public health.

John Farley   retired as professor of biology at Dalhousie University, Halifax-Nova Scotia. His work in the history of biological sciences and tropical medicine includes parasitology and the analyses of marine invertebrates. He recently published a history of the study and treatment of bilharzia and is continuing his research on international health in the twentieth century.

Elizabeth Fee   is chief of the History of Medicine Division of the National Library of Medicine and adjunct professor of history and health policy at the Johns Hopkins University. She is contributing editor for history for the *American Journal of Public Health* and author of various books on the history of public health and health policy. Her forthcoming book, co-edited with Theodore M. Brown, is *Making Medical History: The Life and Times of Henry E. Sigerist.*

Janet Golden   is an assistant professor in the Department of History at Rutgers University, Camden. Author of *A Social History of Wet Nursing in America: From Breast to Bottle*, she is currently writing a cultural history of Fetal Alcohol Syndrome.

Bert Hansen   has written numerous articles on medical education, sexuality, and relations between science and other aspects of culture. His career encompasses administration at New York University and the City University of New York, as well as scholarship. He teaches in the history department of Baruch College in New York City.

Christopher Lawrence   is reader in the Wellcome Institute for the History of Medicine and in University College, London. He has written on the history of technology in medicine and on diagnosis, photography, and the Scottish Enlightenment. He is continuing research on the history of cardiology.

Michael MacDonald   is a professor of history at the University of Michigan. A specialist in English medical, social, and cultural history, he has focused on the phenomena of suicide and processes of healing in secular and religious perspectives. Author of a study of mental illness in early modern England, he has also published a monograph on suicide in that period (with co-author Terence R. Murphy).

Gerald Markowitz   is professor of history at John Jay College of Criminal Justice and the City University of New York Graduate Center. He has published work on the social and cultural history of the New Deal and on health care and occupational medicine. He is the co-author (with David Rosner) of a forthcoming book entitled *Children Apart: Kenneth and Mamie Clark's Northside Center and the Struggle Against Racism in New York.*

Steven J. Peitzman   is professor of medicine and associate professor of community and preventive medicine at the Medical College of Pennsylvania. His work in the history of medicine has centered on medical education, women in medicine, and his own clinical field, nephrology. He is the author of a forthcoming history of nephrology in the United States.

David Rosner   is distinguished professor of history at Baruch College and the City University of New York Graduate Center. He has published widely on the history of health care and occupational health. He is the co-author (with Gerald Markowitz) of a forthcoming book entitled *Children Apart: Kenneth and Mamie Clark's Northside Center and the Struggle Against Racism in New York.*

Charles E. Rosenberg   is Janice and Julian Bers Professor of the History of Science at the University of Pennsylvania. He has published widely on the history of health care and concepts of disease. His most recent publication is *Explaining Epidemics and Other Studies in the History of Medicine.* He is currently at work on a history of changing conceptions of disease from 1800 to 1985.

Janet A. Tighe   teaches in the Department of the History and Sociology of Science and in the American Civilization Program at the University of Pennsylvania. She has written about American psychiatry's efforts to shape the insanity defense and is currently completing a book about the history of medical education.

INDEX

Addis, Thomas, 9, 11, 12, 13, 14
Addison, Joseph, 94
AIDS, xxii, 134, 164, 166, 168, 243–244, 275
alcoholism, xiv, xvi, xviii, 104, 250, 251, 287–288
Allbutt, Clifford, 59, 64, 138
Andrews, Elmer, 197–199
Andrews, John B., 199
anorexia nervosa, 134–154
asbestosis, xxi, 186
Association of Casualty and Insurance Executives, 187, 194

Baillie, Matthew, 23–24
Bain, C. W. Curtis, 51–52
Barber, Richard, 256–257
Barnes, A. R., 72–73
Beard, George M., 116–117
Beccaria, Cesare Bonesana, marchese di, 93
Bedford, D. Evan, 51, 65, 72
Bell, Clark, 216–217, 219–220, 221
Bichat, Xavier, 8
Bigler, Mary, 165
Bloch, Marcus, 35
Bloor, David, 54, 62–63, 75
Blumer, G. Alder, 108–109, 116
Boone, Cherry, 143–144
Booth, Charles, 281
Bouillaud, Jean-Baptiste, 25–26
Boyd, Stanley, 55
Boyd, William, 56
Bradley, Stanley, 9
Bremser, Johann, 35–36
Bright, Richard, 3, 5–11
Bright's Disease, 3–19
Broadbent, William, 58
Bruch, Hilde, 138, 141–143

bulimia, 139–141, 144, 146
Burckhardt, Jacob, 301
Bury, George William Fleetwood, 26

cancer, xix, xx, 169–170, 194
Carpenter, Karen, 144
Carter, Edward Perkins, 70
Cassidy, Sir Maurice, 52
Cato, 87, 88, 93–94
Centers for Disease Control, 161–164, 168, 170, 173
Chadwick, Edwin, 278–280, 286, 290–291
Chamberlain, Joseph, 43
Chatterton, Thomas, 94–95
Cheadle, Walter Butler, 21, 22, 27
chlorosis, xv, 155
cholera, xv, xx, xxi, 35, 229
Christian, Henry, 9, 14, 65, 70
chronic disease, 185–205, 229–272, 308–310
Chronic Fatigue and Immune Disfunction Syndrome Association, 165, 167, 170
chronic fatigue syndrome, xiv, 155–181
Clark, L. Pierce, 254
Clifton, Martha, 24
Cocke, Walter, 213–214
Cohn, Alfred, 13
Committee on the Costs of Medical Care, 307
Compton, Karl, 304
consumption, see tuberculosis
coroners, 86, 90, 96, 98; juries, 86, 88–93, 95–98
Corvisart des Marets, Jean Nicolas, 22–23

Craig Colony (Craig Development Center), 249–250, 258–265
Cullen, William, 5, 11, 22, 24, 29

Darwin, Charles, 257, 290
De Armand, J. A., 120–121
Dixon, Samuel Gibson, 241
Douglas, Mary, 63
Draper, George, 165
dropsy, xix, 3–9, 16, 301
Drudget, Thomas, 10
Drury, A. N., 70–71
Dundas, David, 23–25, 28
Durkin, Martin, 192

East, C. F. Terence, 51, 52
Echeverria, M., 256
epilepsy, 248–272
Epstein-Barr syndrome, 155, 161–164, 166, 168–171, 172
eugenics, 258, 289–290

Faber, Knud, 5, 16
Fahr, Theodor, 11–12
Farr, William, 278
Faulkner, James, 28
Ferris, G. N., 114
Flexner, Simon, 159
Flick, Lawrence F., 234, 235–239
Fonda, Jane, 144
Fye, W. Bruce, 53

Gardner, Leroy U., 200
Gilden, Charles, 93
Gitlow, Elsa, 124
Glueck, Sheldon, 220
Goethe, Johann Wolfgang von, 94
Goldman, Emma, 123
Goodfellow, S. J., 8
Gordis, Leon, 21
Gray, John P., 212–213
Green, Thomas Henry, 54–56
Guiteau, Charles, trial of, 212–213

Hale, Matthew, 91
Hall, Murray, 115
Hammer, Adam, 52
Hammond, William A., 120–121, 252, 253
Harden, Victoria, 41–42

Harrison, Tinsley R., 28
Harvey, William, 63
Haygarth, John, 23, 24
heart disease, xix, 20–32, 50–82
Heberden, William, 52
Henry Phipps Institute, 230, 235–237, 239
Herodotus, 305
Herrick, James Bryan, 53, 63–73
Herzog, Alfred W., 110–111
Hesse, Mary, 54, 62
Hewlett, R. Tanner, 55
Hippocrates, xiii, 248
Hirsch, August, 28, 34, 45–46
Hobson, John A., 282, 285, 286
Hodgkin, Thomas, 8
homosexuality, xiv, xv, xviii, xix, 104–133, 258
Howard, William Lee, 116–118
Howell, Joel, 53, 67–72
Hughes, Charles A., 121
Hume, David, 93
Hunter, John, 52
Huxley, Thomas Henry, 289
hysteria, xv, 160, 167–169

Icelandic disease, 161

Jackson, John Hughlings, 251, 255
Jacksonian seizures, 254
Jenner, Edward, 52
Johnson, George, 11
Johnson, J. Taber, 253
Johnson, Samuel, 3, 16
Jones, F. Robertson, 187, 194
Jones, T. Duckett, 21, 22, 27
June, Jennie, see Lind, Earl
juries, 187, 192, 209

Keats, John, 94
Keedy, Edwin, 220
Keys, Thomas E., 53
kidney disease, xx, 3–19
Kiernan, James G., 114–116, 120, 121
Kingsbury, John Adams, 304
Koch, Robert, 21, 34, 41, 229, 233
Kolff, Willem, 14
Kuechenmeister, Friedrich, 37

Laennec, Rene Théophile Hyacinthe, 25
Lanza, Anthony, 187, 193, 194
Lea, Edgar, 73–74
lead poisoning, 186
LeCount, E. R., 60–61
Lehman, Herbert, 197
Leibowitz, J. O., 52–53
Leuckart, Rudolph, 38, 41, 42, 45
Levine, Samuel Albert, 28
Lewis, Thomas, 53, 70–72
Lewis, Timothy, 38
Lind, Earl, 110–113, 119
Lobdell, Reverend Joseph, see Slater, Lucy Ann
Locock, Sir Charles, 252
Looss, Arthur, 42
Lydston, G. Frank, 120
Lyme disease, 173

McFarland, Andrew, 215
Mackenzie, Sir James, 61, 64, 66, 67, 71–72
MacLeod, Sheila, 143
M'Naughten Rule, 210–211
McNee, J. W., 64, 65, 66, 70
Major, Ralph H., 53
malaria, 39–40
Manson, Patrick, 38–44
March, John, 91
Markowitz, Milton, 21
Mayhew, Henry, 278, 281, 286
mental illness, xx, 85–103, 134–154, 206–226
Merry, Robert, 94
Meyer, Adolf, 220–221
mononucleosis, 161, 171
Monroe, George J., 119, 122
Mont Alto Sanatorium, 241–242
Moore, Charles, 97–98
moral insanity, 206–226
Morgagni, Giovanni, 23
Morrow, Prince A., 119
Müller, Johannes, 34, 42–43
Murray, H. Montague, 55–57
Murray, James E., 195
Murray Bill, 195–196
myalgic encephalitis, 155, 157–161, 166, 168, 170–174

Nathanson, M. H., 66–67
National Institute of Allergic and Infectious Disease, 162
National Institutes of Health, 163, 170, 171
nephritis, 11–12, 16
neurasthenia, xv, 155, 157
neurology, 104, 116–125
New Poor Law, see Poor Law Amendment Act
Newsholme, Arthur, 275–292, 297, 304
Newton, Robert S., 119
Nightingale, Florence, 244

occupational disease, 185–205
Oliver, Jean, 13
O'Neill, Cherry Boone, see Boone, Cherry
Ordronaux, John, 212, 257
Osler, William, 11, 60–61, 64, 66, 71, 221, 254

Pardee, Harold E. B., 68–72
Parkinson, John, 51, 65, 72, 74
Parry, Caleb, 52
Pasteur, Louis, 34
Pearson, Karl, 285
Perkins, Frances, 195, 199–200
Peterson, Frederick, 258–259
Phipps, Henry, 235. See also Henry Phipps Institute
Pitcairn, David, 23–24
plague, xviii, xix, xx, 41
Polakov, Walter N., 196
polio, 156, 157–161, 164–166
Poor Law Amendment Act of 1834, 277–278, 290–291
Pope, Alexander, 94
pregnancy, xv, 250
Prichard, James C., 209–210
psychiatry, 104–154, 206–226
Public Health Service, 160, 173, 187, 196
Pye-Smith, Philip Henry, 26–27

Quinton, Wayne, 14

Ray, Isaac, 211–212, 213, 214, 220, 221

rheumatic fever, 20–32
Richards, A. Newton, 13
Roosevelt, Franklin Delano, 199, 304, 307
Rosen, George, 303
Ross, Ronald, 40–41
Rowntree, B. Seebohm, 281–282, 287
Royal Free disease, 161, 165, 169
Rudolphi, Carl, 34–35

St. Vitus's dance, 8
Savill, Thomas D., 57–59, 65
Sayer, Henry D., 187, 193–194
Sayers, R. R., 187
scarlet fever, 29
Scribner, Belding, 14
Scully, Father John, 238
Senior, Nassau, 278
Shanahan, William T., 259, 260
Shaw, J. C., 114
Shrady, George F., 116, 119, 120
Sigerist, Henry E., 297–314
silicosis, 185–205
Slater, Lucy Ann, 116
smallpox, 53
Smith, Fred M., 68, 72
Smith, Homer W., 13
Snellen, H. A., 53
Solis-Cohen, Jacob, 234
Spitzka, Edward C., 121
Stauderman, Jacob, 257
Steenstrup, Japetus, 36–37
Stiles, Charles, 41
Strašun, Ilya Davidovič, 303
suicide, 85–103
Swift, Jonathan, 93–94
syphilis, xxi, 250, 251, 288–289, 301

Taylor, Alfred Swaine, 212, 214
Temkin, Oswei, xxii, 251, 312–313
Texas cattle fever, 40
Trudeau, Edward Livingston, 244
tuberculosis, xv, xix, xx, 41, 168, 186, 194, 229–247, 249, 280, 290, 301

typhoid fever, xv, xx, xxi, 229
typhus, 283
Tyson, Robert, 10

Ulrich, Karl Heinrich, 114, 125

vaccination, xx
van Beneden, Pierre-Joseph, 37
van Swieten, Gerhard, 22, 24
Virchow, Rudolf, 11, 306
Volhard, Franz, 11–12
von Koranyi, Alexander, 12
von Siebold, Carl, 37

Walsh-Healey Act, 196
Wannamaker, Lewis W., 21
Ward, Henry B., 44–45
Watson, Thomas, 8
Watt, Robert J., 187
Wearn, Joseph T., 63, 65, 66
Webb, Beatrice, 304
Webb, Sidney, 304
Weigert, Carl, 52
Welch, William Henry, 302
Wells, William Charles, 22–25, 28
Wharton, Francis, 212, 214, 218–220, 221
White, Paul Dudley, 28
White Haven Sanatorium, 235–241
Whitten, M. B., 72–73
Willius, Frederick A., 53
Winslow, Charles-Edward A., 308–309
Winslow, Randolph, 121–122
Wolfflin, Heinrich, 301
Woodhead, G. Sims, 56
workmen's compensation, 186–205
Wright, George W., 200–201

yellow fever, 35, 42, 229
yuppie flu, see chronic fatigue syndrome

Zimmer, Verne E., 195, 199

Printed in the USA
CPSIA information can be obtained
at www.ICGtesting.com
LVHW05200417102 3
761404LV00003B/24

9 780813 517575